THE PACE OF CHANGE

Studies in Early-Medieval Chronology

THE PACE OF CHANGE

Studies in Early-Medieval Chronology

Edited by

JOHN HINES

KAREN HØILUND NIELSEN

and

FRANK SIEGMUND

Cardiff Studies in Archaeology

Oxbow Books
1999

Published by
Oxbow Books, Park End Place, Oxford OX1 1HN

© The individual authors, 1999

ISBN 1 900188 78 3

This book is available direct from
Oxbow Books, Park End Place, Oxford OX1 1HN
(Phone: 01865-241249; Fax: 01865-794449)

and

The David Brown Book Company
PO Box 511, Oakville, CT 06779, USA
(Phone: 860-945-9329; Fax: 860-945-9468)

or from the Oxbow Books website
www.oxbowbooks.com

*The cover illustrations are of a Visigothic copy of a coin of Justinian,
worn as a pendant, and buried in Gilton, Kent, grave 41. Gold. (Liverpool Museum)*

Printed in Great Britain
at the Short Run Press
Exeter

Contents

John Hines
Introduction: studies in early-medieval chronology vii

SECTION I. THE CONTINENT

1. *Elke Nieveler and Frank Siegmund*
   The Merovingian chronology of the Lower Rhine area: results and problems 3
   Deutsche Zusammenfassung 21

2. *Claudia Theune*
   On the chronology of Merovingian-period grave goods in Alamannia 23
   Deutsche Zusammenfassung 32

Synopsis of discussion (*John Hines*) 34

SECTION II. ANGLO-SAXON ENGLAND

3. *Birte Brugmann*
   The role of Continental artefact-types in sixth-century Kentish chronology 37
   Deutsche Zusammenfassung 51

4. *John Hines*
   The sixth-century transition in Anglian England: an analysis of female graves from Cambridgeshire 65
   Deutsche Zusammenfassung 78

5. *Christopher Scull and Alex Bayliss*
   Dating burials of the seventh and eighth centuries: a case study from Ipswich, Suffolk 80
   Deutsche Zusammenfassung 88

Synopsis of discussion (*John Hines and Karen Høilund Nielsen*) 89

SECTION III. SCANDINAVIA

6. *Siv Kristoffersen*
   Migration Period chronology in Norway 93
   Deutsche Zusammenfassung 110

7. *Bente Magnus*
   The assemblage from Hade in Gästrikland and its relevance for the chronology of
   the late Migration Period in eastern Sweden 115
   Deutsche Zusammenfassung 124

8. *Morten Axboe*
   The chronology of the Scandinavian gold bracteates 126
   Deutsche Zusammenfassung 141

9. *Anne Nørgård Jørgensen*
   A peaceful discussion of a martial topic: the chronology of Scandinavian weapon graves 148
   Deutsche Zusammenfassung 157

10. *Karen Høilund Nielsen*
    Female grave goods of southern and eastern Scandinavia
    from the Late Germanic Iron Age or Vendel Period 160
    Deutsche Zusammenfassung 193

Synopsis of discussion (*John Hines*) 195

Introduction: studies in early-medieval chronology

John Hines

Chronology is so basic an aspect of archaeology that no special justification for the production of a collection of studies devoted to this topic should be necessary. A primary element in the identification of any form of archaeological evidence is its assignation to a context. On the one hand this means confirming the age of the object, so that it can be accepted as representing a certain part of the past, be that Prehistory, the Roman Period, the Middle Ages or whatever; on the other hand this context itself implies a range of other artefacts, features and even events with which the item is synchronically related. With fine chronology one attempts to map delicate sequences of change in the material record, recognizing thereby that while synchronic associations may be frozen in a more or less complex archaeological snapshot, it is to be assumed that historical circumstances and material cultural practices were never static. Archaeological fine chronology is purely and simply the factual basis for the most accurate attainable view of the material reflection of processes of historical development.

Despite this, the techniques and problems of dating have not been a subject of much interest in general archaeology for a long time, irrespective of the period in question. The ambitious interpretative approaches of New (and newer) Archaeology, from the materialist to the symbolic with all of the hybrids and offshots in between, have on the one hand tended to accept rather simplistic and old-fashioned schemes as a good enough basis for their speculative claims, and on the other hand have regarded pure chronological research as a narrow scientific specialism, a service industry playing only a bit part in archaeology – the radiocarbon laboratory may provide interesting dates, but only the 'archaeologist' can create meaning from them. At worst chronological research is treated as an exhausted subject: a useless obsession with the classification and ordering of dead data.

While chronology has suffered from institutional neglect, however, individual and local initiatives have kept the development of the subject alive. The publication of the present book, the first substantial international collection of studies of early-medieval chronology since Kossack and Reichstein's edited volume of *Archäologische Beiträge zur Chronologie der Völkerwanderungszeit* of 1977, is in itself evidence that attitudes are changing over a broad front. But there are still campaigns to be waged to disseminate and gain recognition of what can be done and is being achieved both in and with up-to-date chronological research.

The papers printed here have been developed from presentations precirculated in written form and then extensively discussed at an invitation symposium funded by the Danish Humanities Research Council which was held at Fjordparken, Aalborg, from 24–27 March 1996. The membership of the symposium was intended as far as possible to include those who had most recently been active in archaeological chronological research within three major, inter-related, areas, southern and western Germany, England, and Scandinavia, in respect of the immediately post-Roman or early-medieval period (broadly the 5th to 8th centuries A.D.).

The archaeological chronology for this period is necessarily a structure in which a series of local schemes have to be joined together into progressively more comprehensive and general 'master' chronologies. Prominent in the minds of the initiators and organizers of the Aalborg symposium was the idea that it could profitably serve as a stage in a process, bringing together and focussing disparate research activities that had separately been going on in the various areas for a number of years, and at the same time providing a springboard for more ambitious chronological projects in the future. Selective although the geographical range of the symposium inevitably had to be, it was still large enough for there to have been considerable barriers of communication, in respect of which even a straightforward series of surveys of the current status of chronological studies in the different areas was of great value. Happily, however, there is also a dynamism in the field of study across this whole range which means that current research initiatives, and the special experience of problems that they provide, could also be reported on in every case.

These studies focus on a number of the Germanic or Germanicized populations who achieved cultural and political pre-eminence in northern and western Europe during and after the fall of the Roman Empire in the West. By necessity, the starting point for detailed chronological research must be the copious deliberate depositions of collections of artefacts produced in these contexts – primarily in graves, but to some degree also in other kinds of hoard. Through these, we can attempt to discern sequences of change not only in the forms and the range of artefacts being produced but also in the ritual practices which produced the deposits. Of course, this sheds light on only a part of the cultural life of the periods and places under consideration, but since these rich ritual depositions were *ipso facto* a highly significant feature of the culture of the people who produced them, we need not be unduly embarrassed and over-apologetic on that account. The challenge of correlating these sequences with their most obvious counterpart, a settlement-site chronology, is fully recognized. Here, indeed, we might point to Scull and Bayliss's discussion in this volume of the cemeteries in and around early Ipswich as one example of how the problem might be addressed.

The three geographical areas on which this book concentrates were chosen because they were known to be particularly well suited to comparative study, with broadly similar sets of archaeological data produced by rich funerary or other ritual deposition, and contacts between them attested to by the exchange of artefacts or influence. Despite that, the problems facing chronological research and the results achieved so far vary markedly from area to area. As is illustrated here, in various areas of England and Scandinavia the scope for producing a robust and detailed relative-chronological sequence is considerably weaker and in some cases quite controversial in comparison with the situation in the Rhineland and Alamannia. It is in western and southern Germany too that there is much greater scope for putting absolute-chronological calendrical dates to the relative-chronological sequences on the basis of either the inclusion of an identifiable coin with a *terminus post quem* in the assemblage or of dendrochronology. It is inevitable, then, that the German chronologies (and indeed a closely related sequence from northern France which is being produced and refined: see Delestre and Périn eds. 1998) should appear to offer the yardstick to which other areas' schemes should seek alignment. There, however, a series of further problems soon becomes apparent, the net implications of which are that the enviable successes of Continental chronology offer no easy solution to the problems of Anglo-Saxon or Scandinavian dating. At the heart of these problems is not so much the extensive residual differences in material culture between the areas but rather the uncertainty of the character of the inter-regional relationships implicit in exchange and influence. The mechanisms for the diffusion and adoption of artefact-types across geographical boundaries can be various (e.g. commodity trade, gift exchange, prestige emulation, group migration, exogamy, travelling craftsmen etc.). As is also shown in several places in this book, these render it impossible simply to assume that the 'same' type of object is of the same date in any two areas in which it occurs.

It is not suggested that there is at present any simple and conclusive way of overcoming such problems, but it is obvious that the improvement of chronological understanding, so as to both refine and increase the reliability of our basic map of comparable and associated developments in different areas, can only be a useful step towards a better understanding of the nature of these relationships. It is likewise essential that dating techniques be fully shared and critically compared, for some measure of the integrity of a super-regional chronology will be the degree to which the same methods can be used place by place. The examination of this issue has undoubtedly not been completed, but it is already evident that, for instance, the mere availability of a large number of well-furnished graves in England does not allow seriation to be undertaken with the same results as in contemporary Germany, although both in England and in Scandinavia there is unquestionably more exploratory work that needs to be done, perhaps especially on a regional basis.

In one respect particularly these studies can be claimed to represent a substantial methodological advance on, say, those reported on in Kossack and Reichstein's *Archäologische Beiträge zur Chronologie der Völkerwanderungszeit*, in the impact of the availability of computing facilities and software packages designed specifically for archaeological seriation and correspondence analysis – together with an eagerness to use these tools evident in every area represented here. But, as the work in England shows most clearly, this facility is not a magic wand that can be waved over the data to put them into a reliable chronological sequence. The problems in England are a function of peculiar cultural circumstances which, once again, not only do we perceive all the more clearly, coincidentally as it were, when we attempt chronological analysis, but also themselves seem to have been subject to diachronic change, so that fine chronology remains a vital element in their elucidation and investigation.

At a more general level, meanwhile, the intimate relationship between the latest computerized quantitative and statistical methods and one of archaeology's earliest and most basic techniques, typology, is re-emphasized. Quantitative analyses can work only on typologically classified data. Typological systems, meanwhile, are invariably fluid, as it is in practical terms always possible to classify artefacts in more than one way. The relationship between typology and seriation (however the latter is done) is an interactive one, as it should be virtually standard practice to consider modifying a typology to see what effect that has on, and so whether one might improve, a sequence produced. This does not, however, reduce typology to a vaguely subjective and relativistic level: rather it makes it all the more vital that the parameters within which typological experiments are conducted should be both explicit and rational, and that the decisions taken in the course of the work should be fully explained. These are lessons which perhaps all of us have yet fully to digest. In the process of doing that, and as we seek as archaeologists to make best use of the massive harvest of isolated finds that the metal-detector is delivering to us, it is foreseeable that purely typological and stylistic studies, including their chronological dimension, will themselves regain some of their long and largely lost academic credit.

Specially designed statistical software is not the only computer-enhanced and up-to-date technological method that can be and is being brought to bear on early-medieval chronology. Dendrochronology can yield spectacularly precise results in terms of calendar years. It is representative of the limitations of the buried material primarily considered in this book that only in Germany can we directly associate dendro-dates from graves with the relative-chronological sequences discussed here. In Denmark, meanwhile, a fascinating series of dendro-dates from various engineering works, most them military, and from the earliest layers of the town of Ribe, still awaits full integration with the artefact and burial sequences of the broader cultural region (Axboe 1995). In England, there is some hope that, by way of stratigraphic and ceramic sequences, dendro-dates from urban sites, particularly in London, will come to play a part too (Cowie and Whitehead 1989). Although not made use of in any of the following studies, it was also noted in discussion that thermoluminescence dating has become considerably precise and should therefore be of value in the analysis of cremated material.

In England, meanwhile, pioneering work is now underway to try to clarify at least part of the chronological sequence through a multitechnical approach in which the high-precision radiocarbon dating of skeletal remains will play a central role, alongside other familiar methods such as the typology and seriation of grave good and burial stratigraphy, to allow Bayesian statistical methods to further refine the probable date-range indicated by the carbon-14 levels. Altogether, there is a clear need for archaeologists in the foreseeable future to be able to conceive of and work with chronologies that embrace a number of different types of scale or sequence simultaneously. These are, however, primarily of just two types, both of which are familiar: relative-chronological orders, and the calendrical scale of years in which more or less narrowly focussed estimates of date can be made by various means. In the past, the apparent coarseness of laboratory-produced results of the latter kind has led as far as outright rejection of their utility for Early Medieval archaeology. Advances in the methods available for relative chronology have paradoxically made us all the more aware of how dependent we are on these alternative absolute-chronological dating methods in those contexts where the sophisticated relative-chronological methods still will not work. Quite apart from its hoped-for chronological results, this experiment in Anglo-Saxon archaeology should be methodologically instructive, not least in terms of integrating a number of different dating methods.

At the end of each of the three geographical sections of this book, brief synopses are given of what were in fact the lengthy discussions that followed the papers at the original symposium. In a succinct way, these give quite a good idea of what the participants regarded as the primary problems of their common undertaking and the principal questions for future research, as well as in other cases virtually disposing of hypothetical problems which proved not to be such serious issues in practical terms as far as any of the contributors were concerned. One such matter is the possibility of different chronologies for the production and deposition of artefact-types, which was properly, and quickly, acknowledged, but judged to be no stumbling block for there seemed to be no difficulty in the overwhelming majority of cases in conceiving of our relative and absolute chronologies as some sort of mid-way point between the two (cf. Steuer 1977; below, p. 34). Terminology, and the definitions of major periods, sub-periods and artefact-types, proved conversely to be substantial and abiding problems. Reform in this field is not something to be rushed at, and must be preceded by proper understanding of the traditions and conditions of the different regions, which it is hoped this book will help to provide. Certain specific historical questions – as we might call them – emerge as primary issues too: the continuing obscurity of the 5th-century transition from the Roman Period to the Early Middle Ages, and the controversy over the date and character of profound and extensive changes in all of the areas considered here in the course of the 6th century.

The study of chronology, therefore, requires specialization in the sense that it is a complex subject that needs experience and dedication, not just individually but increasingly on a team basis, in order for it to be understood and developed properly. It is not a specialized field in the reductive and ancillary sense described at the beginning of this introduction – namely something that can either be taken for granted or evaded, or, worst of all, something that is dismissed as a pointless and misguided obsession. Chronological analysis is at the heart of the epistemology of archaeology, not only because it is fundamental to the identification of the material evidence without which archaeology is nothing, but also because time is an ineffable dimension of all relations embodied in the archaeological record. In the case of the Early Medieval Period this means primarily that in seeking to interpret the patterns of the past we have to take account of contextual factors that interfere with the chronological sequence and thus disguise it. Put the other way around, we may identify factors bending or distorting the chronological sequence and therefore read the former from the latter. We cannot in any event separate chronology from context.

None of this removes the desirability of working towards the finest dating possible within a framework of scientific chronological objectivity. The establishment of a detailed and accurate chronology of the archaeological material which represents the end of the Roman Period and the beginning of the Middle Ages in Europe is indeed fundamental to sophisticated historical research. Archaeology has a large role to play in increasing our knowledge and understanding of what was happening then, not least in exploring the inter-relationship between material life and the social and political changes revealed by the historical sources for the period in which the ancient world gave way to the medieval, and the nations of modern Europe began to emerge.

BIBLIOGRAPHY

Axboe, M. 1995. 'Danish kings and dendrochronology: archaeological insights into the early history of the Danish state'. In G. Ausenda (ed.), *After Empire: Towards an Ethnology of Europe's Barbarians*, 217–51. Woodbridge.

Cowie, R. and Whitehead, R. 1989. 'Lundenwic: the archaeological evidence for middle Saxon London'. *Antiquity* 63, 706–18.

Delestre, X. and Périn, P. (eds.) 1998. *La Datation des Structures et des Objets du Haut Moyen Âge: Méthodes et Résultats*. Mémoires publiés par l'Association française d'archéologie mérovingienne VII.

Kossack, G. and Reichstein, J. (eds.) 1977. *Archäologische Beiträge zur Chronologie der Völkerwanderungszeit*. Bonn.

Steuer, H. 1977. 'Bemerkungen zur Chronologie der Merowingerzeit'. *Stud zur Sachsenforschung* 1, 379–403.

I
THE CONTINENT

1. The Merovingian chronology of the Lower Rhine Area: results and problems

Elke Nieveler and Frank Siegmund

THE STATE OF RESEARCH

To this day, the basic chronological system for the Rhineland Frankish find material is the *Stufe* (stage) system of Kurt Böhner. Its starting point was his dissertation written in Munich in 1940 on the subject of the weapon graves around Trier. He enlarged it but, owing to the War, it was not published before 1958 (Böhner 1958). His study was based on the find material around Trier, especially the four large cemeteries of Ehrang, Eisenach, Hohenfels and Rittersdorf, encompassing some 500 graves (Fig. 1.1 nos. 17–20), and other smaller find spots. In contrast to the southern German research tradition centring on costume trappings, which also constitute a large proportion of the finds in Frankish cemeteries, Böhner included the pottery and weaponry in his system and attempted to use them for chronological purposes.

Böhner typologized the entire mass of finds and investigated the combinations of his artefact-types in closed grave contexts. For regularly recurring find assemblages he developed a scheme dividing the material into five stages (*Stufe I – V*) within which the majority of the graves and artefact-types belong to his stages III (6th century) and IV (7th century). He brought these stages into context by considering some 60 graves which contained coins and thus worked towards an absolute chronology (Böhner 1958, 26ff.; Böhner 1978, 11 Abb. 3). The implicit pushing back to the end of the 6th century of a certain horizon that, by virtue of historical considerations, was originally placed in the early 7th century (Werner 1935), soon gained general acceptance. Böhner's system came into use far beyond Trier as a generally serviceable chronological model for Frankish finds.

The gradual study and publication of cemeteries in the western part of Germany resulted in a dramatic growth in the number of finds. Suitable larger cemeteries on the Middle and Lower Rhine provided an opportunity to build on the research in southern Germany (Werner 1953; Christlein 1966). Their combinational system could be studied in more detail, in order to develop more nuanced local chronologies. First, Herman Ament produced an

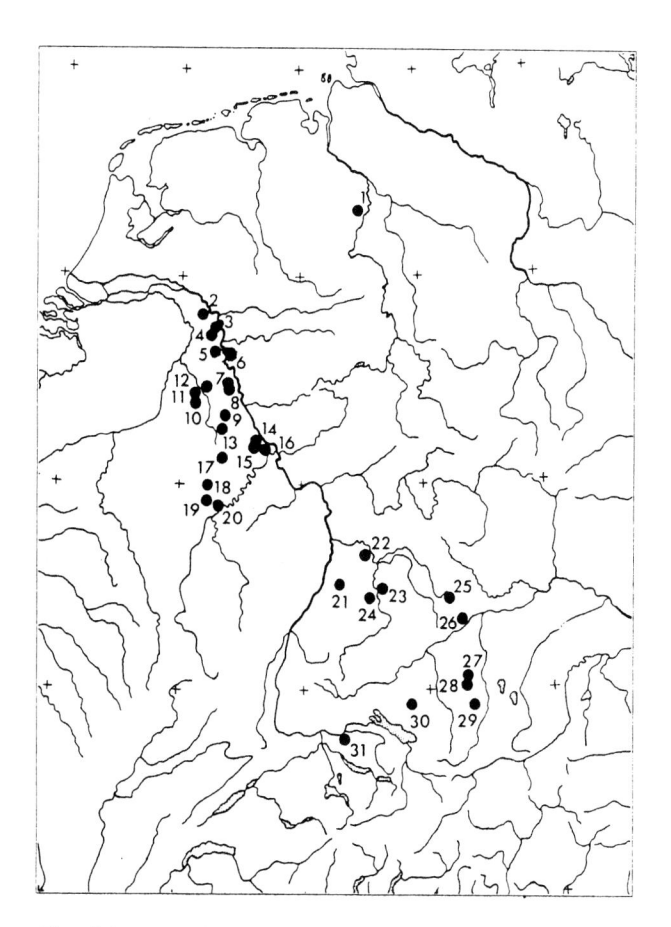

Fig. 1.1 Map of cemeteries mentioned in the text or with important chronological studies. 1 Liebenau, 2 Xanten, 3 Walsum, 4 Eick, 5 Gellep, 6 Stockum, 7 Müngersdorf, 8 Junkersdorf, 9 Lommersum, 10 Lamersdorf, 11 Jülich, 12 Rödingen, 13 Iversheim, 14 Miesenheim, 15 Pommerhof, 16 Rübenach, 17 Hohenfels, 18 Rittersdorf, 19 Eisenach, 20 Ehrang, 21 Berghausen, 22 Bargen, 23 Pleidelsheim, 24 Hemmingen, 25 Lauchheim, 26 Schretzheim, 27 Mindelheim, 28 Dirlewang, 29 Marktoberdorf, 30 Weingarten, 31 Bülach.

Karte der im Text erwähnten Gräberfelder bzw. solcher mit wichtigen chronologischen Studien.

impressive topo-chronological ('chorological') analysis of the cemetery at Rübenach with its 840 grave groups (Neuffer-Müller and Ament 1973; Fig. 1.1 no. 16). This resulted in a division of Böhner's stage III and in the recognition of an important horizon around 600 (Rübenach phase B2). In a study centred in the region around Mayen (Ament 1976a; Fig. 1.1 nos. 14–15) he applied the developed South German scheme of belt-styles of the 7th century to Frankish finds. Thereafter he checked and broadened his model by means of topo-chronological ('chorological') investigations on further Rhinish Merovingian cemeteries and thus, with the help of the belt fashions, established a division of Böhner's stage IV into earlier and later parts (Ament 1976b). He also considered the absolute chronology. He emphasized that the end of the Böhner's stage IV, at that time fixed by means of few graves containing coins 'around 700', could hardly be considered precise, and should lie around 670/680.

On basis of these preliminary works Ament suggested a new general scheme of six phases (AM I–III, JM I–III). In this Ament split each of the phases Böhner III and IV into two sub-phases (Ament 1977). Unfortunately, the 'stage around 600', retrieved from the Rübenach cemetery and an important link to the South German chronologies, was lost once more (Giesler 1983, 508ff. Abb. 19). The contents of Ament's phases AM I–III stayed open, whereas the contents of the phases JM I–III were based on the buckle and brooch sets with other artefact groups only more generally considered. Furthermore, only with difficulty is it possible, in any degree of detail, to connect a critical discussion of Ament's chronological scheme for Rübenach (Giesler 1983; Wieczorek 1987), and further local chronological schemes (Krefeld-Gellep: Siegmund 1982; Köln: Päffgen 1992), with this general scheme. Acceptance of Ament's new phases is thus rather limited.

The picture in the 1980's was thus very varied: the scheme of Kurt Böhner, still valid in his opinion, was frequently still used by scholars for large collections of material from distant geographical regions (Böhner 1978; Pirling 1966; 1974; 1979). Attempts at detailed chronological schemes were (entirely in Böhner's spirit) opposed to theoretical considerations according to which detailed chronology was almost impossible and thus not an attractive aim of research (Steuer 1977; 1990). On the other hand, Ament's handy suggestion, especially concerning the 7th century, was adopted by others (e.g. Janssen 1993). Meanwhile various very detailed chronological schemes for specific cemeteries were developed (Neuffer-Müller and Ament 1973; Ament 1976a; Siegmund 1982; Päffgen 1992). Correlation between these, however, and their extension to other assemblages, appeared very questionable (e.g. Giesler 1983). The reason is probably that local chronological schemes often tend to be based on typological elements very close to the actual material, i.e. too detailed a typological scheme, and this particularity is reflected in the chronological scheme based upon these types. As a result it is almost impossible to encounter the same range of types outside the immediate neighbourhood. This especially affects the undoubtedly locally manufactured pottery and weaponry, which are very common phenomena and typical of the Frankish milieu.

The very early and the late Merovingian period were represented by only few finds in the Trier region and thus also only weakly covered in Böhner's chronological scheme (in stages I–II and V). For the early Merovingian period the dissertation of Horst Wolfgang Böhme, supervised by Joachim Werner, then appeared in Munich in 1969. It is an extraordinarily large-scale study of the 4th- to 5th-century finds from the area between Loire and Elbe (Böhme 1974). Based especially on belt sets, brooches and late-antique glassware, he succeeded in making a chronological scheme for the period *circa* A.D. 330–450 consisting of three phases, I–III, based primarily on find associations. In several smaller studies since then Böhme has extended and consolidated the contents of his chronological scheme, leaning now to a somewhat later absolute chronology (e.g. most recently Böhme 1985; 1986; 1989; 1994).

Using Böhme's chronological scheme on the various regional materials, however, it appears that many of his relevant types occur only rarely. This is especially clear in the Krefeld-Gellep cemetery, which has produced the most substantial collection of 5th-century material. Only a few assemblages here are understandable in Böhme's terms. Renate Pirling's work also constitutes a truly important contribution to a chronological scheme for the Rhine Frankish material, especially in integrating glass and ceramics (summary: Pirling 1979, 159ff. with Abb. 15–16).

The contents and dating of Böhner's phase V ('8th century') were made much clearer through the Munich dissertation of 1961 by Frauke Stein concerned with graves of the 'noblemen' of the 8th century ('Adelsgräber') (Stein 1967). To compensate for the very small quantity of finds from the final phase of the row-grave period, the basic material for the chronological scheme, as in Böhme's work, came from an extremely large area. Stein suggested a division into two regional areas – a northern one ('Nordkreis') and a southern one ('Südkreis') – and a chronological scheme covering each of these in three phases: A–C, dated by Stein to A.D. 680–800. Criticism of this study has predominantly concerned the social interpretation of the graves still containing grave goods as Adelsgräber, while later corrections of her typological and chronological schemes have also been suggested (summary: Ament 1976b, 320ff.). As the supra-regional material was dominated by weapon burials the division into regions and phases was based on these. Only a few Rhineland Frankish grave groups from Stein's southern region are represented in her tables (Stein 1967, Abb. 3). The Rhineland tradition of pottery deposited in the grave, which might have led to further chronological results, cannot be compared on a cross-regional basis, so that Stein could not use this evidence to support her conclusions.

THE BASIS OF THE RHINELAND CHRONOLOGY

For the analysis of the Merovingian Period of the Lower Rhine Area (Reg.-Bez. Düsseldorf) Frank Siegmund's dissertation (1989; 1998) was an attempt to test and transfer to the Lower Rhine area the overlapping and competing chronological schemes produced by the still growing body of research. The attempt appeared problematic. From all the published Rhineland cemeteries, a new typology covering these finds was therefore established. In addition to smaller find spots the analysis includes the cemeteries of Düsseldorf-Stockum (Siegmund 1998: 99 grave groups), Eick (Hinz 1969: 157 grave groups), Krefeld-Gellep West and East (Pirling 1966; 1974; 1979: 192 and 546 grave groups respectively), Köln-Junkersdorf (LaBaume 1967: 544 grave groups), Köln-Müngersdorf (Fremersdorf 1955: 151 grave groups), Orsoy (Böhner 1949; Siegmund 1998: 9 grave groups), Rill (Steeger 1948; Siegmund 1998: 81 grave groups), Walsum (Stampfuss 1939: 44 grave groups), and Xanten-St. Viktor (Siegmund 1998: 150 grave groups): in total about 2,340 grave groups (Fig. 1.1 nos. 2–8). The total material was analysed on the basis of combinations in three separate analyses, each producing a seriation: one of the necklaces, one of the female grave groups, and one of the male grave groups. In addition, topo-chronological ('chorological') analyses were made of all the cemeteries with a chronologically determined topography (Gellep, Junkersdorf, Müngersdorf, Stockum and Walsum). On basis of these arguments the material from the early 5th to the middle of the 8th centuries was arranged in a Lower Rhine Chronological scheme ('NRh-' = Niederrhein) in eleven chronological phases.

Shortly afterwards the 'Franken AG' (*Arbeitsgruppe*) was formed by Heike Aouni, Ulrike Müssemeier, Elke Nieveler and Ruth Plum. As a part of their doctorate preparations supervised by Volker Bierbrauer, Bonn, they intended to analyse predominantly settlement-historical questions in further large areas between the Middle Rhine area, which had been analysed by Ament and his students, and the Lower Rhine area. As the sources in each of these areas consisted predominantly of small find spots and only a few cemeteries which could be analysed in topo-chronological terms it was not possible to establish a chronology for each area, and that in any case would not have been useful in respect of the main question. As a result, the chronology proposed by Siegmund was taken over and, after intensive tests and some modifications, fitted to the material of the areas analysed. Their study added some smaller cemeteries: e.g. Iversheim (Neuffer-Müller 1972: 243 grave groups), Jülich (unpublished: 223 grave groups), Lommersum (Neuffer-Müller 1960: 83 grave groups), Lamersdorf (Piepers 1963: 87 grave groups), and in particular the recently published material from Rödingen (Janssen 1993: 656 grave groups) and Köln-St. Severin (Päffgen 1992) were included (Fig. 1.1 nos. 9–13). With Iversheim, Jülich and Lamersdorf detailed topo-chronological analyses were possible. The seriation analysis also included the total amount of material analysed by Siegmund, which increased the number of grave groups analysed to about 535 assemblages and 187 artefact-types for the male graves and about 400 assemblages and 150 artefact-types for the female graves. The nearly perfect parabola achieved by correspondence analysis emphasizes the high quality of the seriation (Figs. 1.2–1.3; Scollar *et al.* 1992). The repeated test on a larger material basis by the Franken AG generally confirmed the result and the scheme achieved by Siegmund. Discrepancies occur in the typology of the biconical pots and in the phase-division of the second half of the 6th century. This chronological scheme for the 'Kölner Bucht' ('KB') suggests ten phases for the period *circa* 400 to *circa* 740.

Parallel to this, another dissertation with Volker Bierbrauer as supervisor and concerning a new analysis of the late finds from Frauke Stein's 'Nordkreis' was written by Jörg Kleemann (Kleemann 1992). In this extensive study, Stein's three phases were tested on a broad material basis and superseded by a new and more detailed model. This overlaps with Siegmund's area and chronology, which it corroborates.

A synthesis (named 'Rh-': Rheinland) of Siegmund's Lower-Rhine chronology and the Franken AG Kölner-Bucht chronology will be presented in the following pages. It is based on the total material from the above mentioned regions and also includes all the published material from this area (in total about 4,000 grave groups). The bases of the chronological scheme are the above-mentioned seriations of necklaces, female and male grave groups and topo-chronological analyses of the cemeteries Düsseldorf-Stockum, Krefeld-Gellep, Köln-Junkersdorf, Köln-Müngersdorf and Duisburg-Walsum (Siegmund 1989; 1998), and Iversheim, Jülich and Lamersdorf (Franken AG, in press). The broad material basis and the consequently parallel use of topo-chronological ('chorological') and combinational argumentation strengthen the stability of the relative sequence of the chronological scheme presented. The inclusion of as many of grave groups containing coins as possible (about 100) and some dendrochronological dates in the chronological scheme leads to concrete visualisations of the absolute chronology (Siegmund 1998, 200ff.); nevertheless future moves earlier in coin-dating on the basis of new discoveries cannot be excluded.

This Rhineland chronology seems also to be valid beyond the area analysed, as it is possible without contradictions to parallel it with other recent schemes, especially from southern Germany (Fig. 1.4). Its validity ends when and where the constituent types of the chronological scheme (dress assessories, weaponry, pottery and some of the glassware) do not occur in sufficient quantity. Figure 1.4 shows a synoptical table relating our proposal with other chronological systems (Siegmund 1998, 208ff.). This synthesis is based on the archaeological contents of the various phases, stages etc., not on the absolute dates. The absolute dates in this table are the estimates given by the

```
Correspondence Analysis of Männergräber Rheinland
Unit scores
X-Axis: Correlation: 0.9879 (  1.7%)
Y-Axis: Correlation: 0.9795 (  1.7%)
```

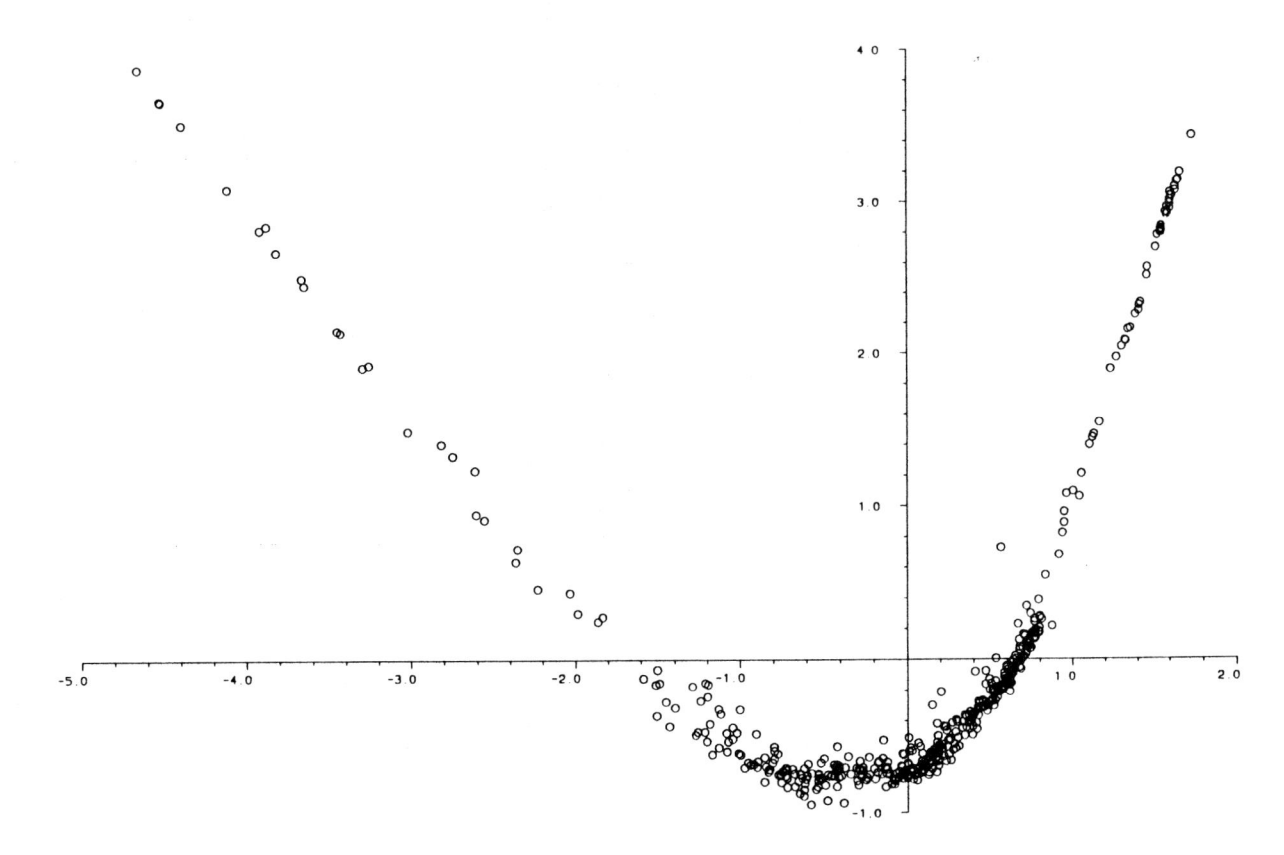

Fig. 1.2 Correspondence analysis of male graves in the Lower Rhine area. Scatterplot of the first two Eigenvalues.
Korrespondenzanalyse der Männergräber vom Niederrhein. Streuungsdiagramm der beiden ersten Eigenvektoren.

different authors and are reproduced without comment here.

THE CONTENTS OF THE RHINELAND CHRONOLOGY

The Rhineland chronology is illustrated here in ten figures (Figs. 1.5–1.14). They portray typical examples of the important artefact-types. Every picture has two lines underneath it: the upper is its codename, the lower its dating. Types defined by Siegmund are given in roman script, those defined by the Franken AG in italics. The lower line shows the phase a type belongs to, or the span of its dating. Underlining indicates the phase in which the type occurs most frequently within this range.

As the male belt sets are best suited to supra-regional synchronization, the development of this large group has also been used for the definition and delimitation of the single phases. In the phases Rh 1–2 late-antique belt buckles are still in use. From the beginning of the phase Rh 2 the earliest local buckles are added. More tangible is the development from the appearance of the buckles with a club-shaped tongue ('Kolbendornschnalle') in phase Rh 3, and replaced in phase Rh 4 by the shield-on-tongue buckles. The buckles with a mushroom tongue and without a plate are only weakly represented in the Rhine Frankish area, but are, however, found in phase Rh 5. This type is followed in Rh 6 by buckles with a hinged triangular plate. In this phase the earliest buckles with semicircular plate also appear. In the following phase, Rh 7, they are a typical element of two-part sets with semicircular counter-plates, predominantly made of iron and in rare cases decorated with mushroom cell inlay. The belt fashion of the 7th century follows a well-known course of development: plain iron sets (Rh 8A) appear well in advance of the belts with monochrome geometrical inlay (Rh 8B). They are followed by the bichrome zoomorphic Style-II inlaid belts of phase Rh 9, of which those with regular Style II are presumably earlier and those with dissolved Style II later. Multipartite belt sets are rare in the Rhineland. The earliest versions appear in the phase Rh 8, but the type is most common in phase Rh 9. No honeycomb-inlay belt sets or belt sets decorated with small single garnet roundels have yet been

```
Correspondence Analysis of Frauengräber Rheinland
Unit scores
X-Axis  Correlation  0.9829  (  2.0%)
Y-Axis  Correlation  0.9711  (  1.9%)
```

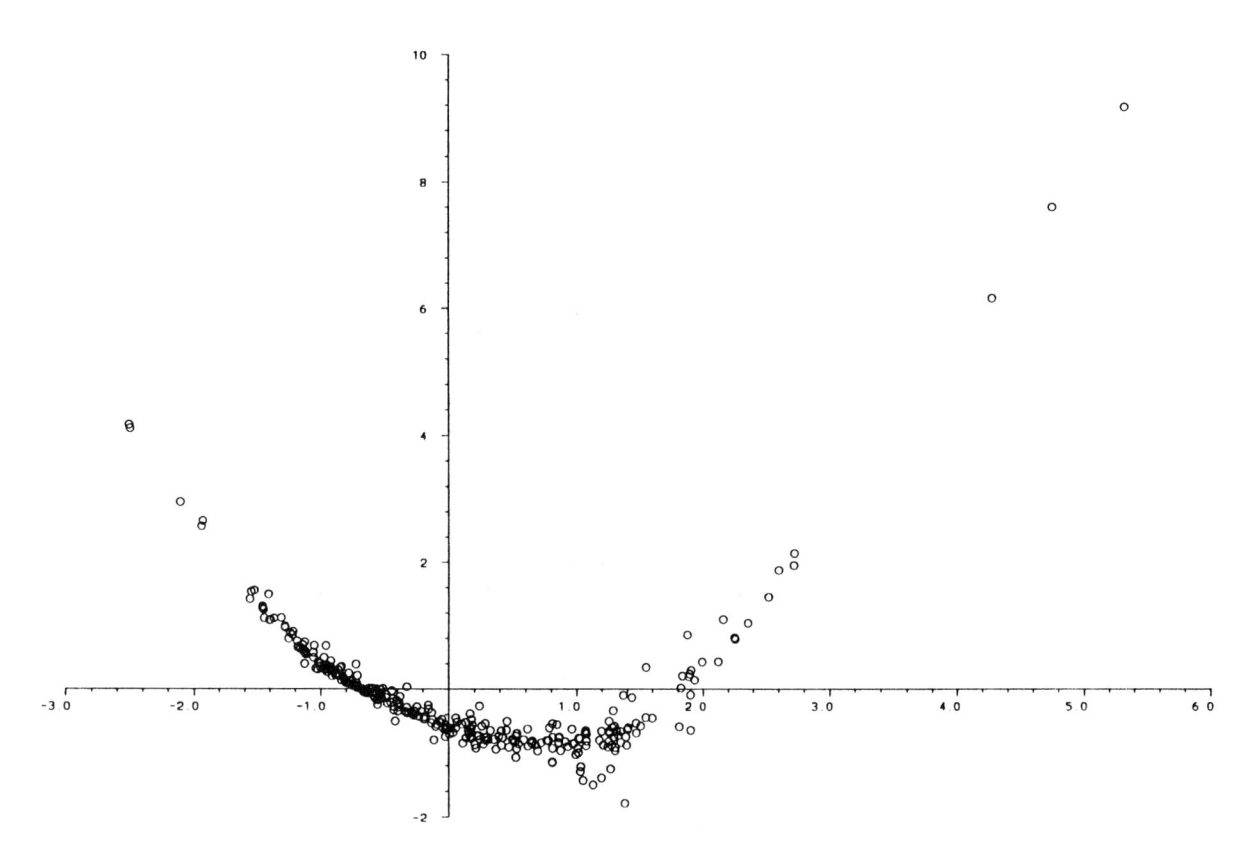

Fig. 1.3 Correspondence analysis of female graves in the Lower Rhine area. Scatterplot of the first two Eigenvalues.
Korrespondenzanalyse der Frauengräber vom Niederrhein. Streuungsdiagramm der beiden ersten Eigenvektoren.

found in the Rhineland, which means that the belt sets mentioned above are followed in phase Rh 10 by more belts with only a few plates and mounts from the end of the row-grave period.

Whilst in the early phases small brooches (as well as small bow-brooches such as type Fib 12.1–5) were mainly worn in pairs in the area of the upper body, from phase Rh 3 onwards the classical '4-brooch' costume (2 small brooches in the upper body, 2 bow-brooches on the pelvis or upper thigh) was dominant. The total of 4 brooches was normally not extended, so that it must be assumed that for a full costume no more than 4 brooches were deemed necessary. It is noticeable that in many find-positions small brooches are regularly present, whereas the number, situation and combination of bow-brooches varies. The already well-known relocation of the bow brooches from the pelvis to the upper thigh in the course of the first half of the 6th century is also clearly visible in the table. Other features of this dress were small garnet brooches (Fib-1.1–3), bird brooches (Fib-7.1–4) and the occasional S-brooch, rare in the Rhineland. The outfit was occasionally enhanced by polyhedrical earrings (mainly in the rich graves, and then in exceptionally fine work, Ohr-2.3–4A) and glass beads as girdle-hangers (GGh-1.1–4). At the end of the 6th century (phases Rh 6 and 7) a decisive change in women's dress seems to have taken place. The now extremely large disc brooches (Fib-1.5–2.3) were worn individually in the neck or breast area. These were accompanied by polyhedrical earrings and hangers (GGh-3–5). The latter are characteristic from phase Rh 8 onwards and often occur in elaborate chain-combinations (GGh-6). Occasionally from phase Rh 8, but mainly in phase Rh 9, they appear together with the large, highly jewelled disc brooches (Fib-2.4). This rich dress is accompanied by shoe buckles (often inlaid) (Sna-2.4–5). As early as phase Rh 9, with the appearance of the first simple bronze brooches (cross brooches), the end of this elaborate fashion is perceptible. Smaller brooch-types appear, such as pressed foil/applied disc brooches (Fib-3), equal-armed brooches (Fib-10) and rectangular brooches (Fib-11), and once again are worn in pairs and in combinations with large earrings with polyhedrons and coiled wire (Ohr-7). The fact that these small brooches with a small chain were also worn in the middle of the breast region seems to

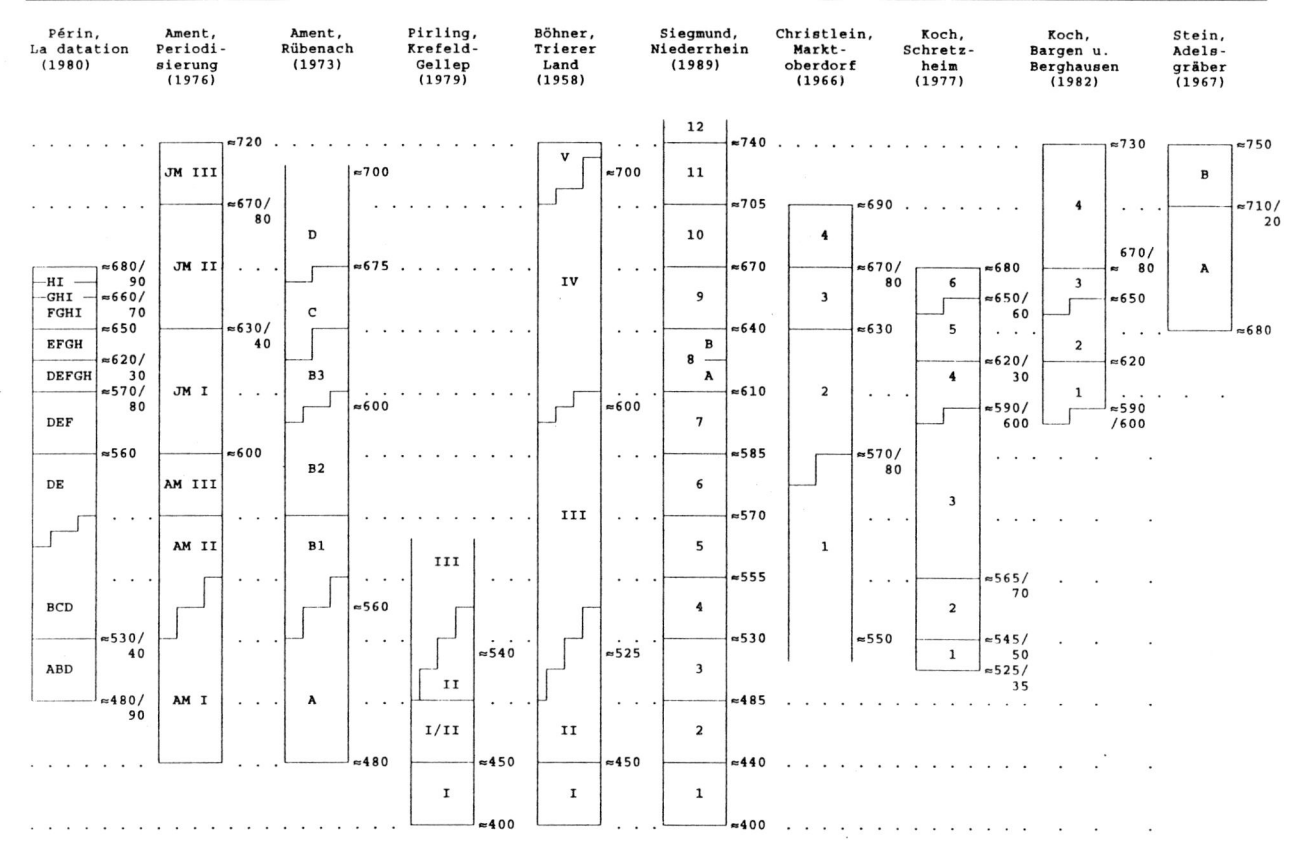

Fig. 1.4 Synoptical table of selected chronological schemes.
Synopse ausgewählter Chronologiesysteme.

indicate the same function as the earlier large disc brooches worn individually (coat- or cloak-fasteners?).

In the male graves, weaponry does not appear regularly as grave goods before phase Rh 4. Weapons in earlier graves are rare and indicate burials of special social status. In graves of phase Rh 1 only occasional axes of the type FBA-2.3 appear, alongside arrows. In phase Rh 2 swords (the *spatha*) with scabbard mouth-pieces of copper alloy and early weapon axes (the *francisca*) type FBA-1.1 appear in Gellep grave 43 and Oberlörick grave 13. Not until phase Rh 3 are the earliest spears found, here the type Lan-1.5 with a narrow blade and parallel edges. They are combined with angons, weapon axes with heavily flanged upper edge FBA 1.2, and shield bosses with flat silver gilt rivets type Sbu-2.

Narrow saxes of type Sax-1 appear occasionally in phase Rh 3 and are typical in phases Rh 4–7. In phase Rh 7 the earliest broad saxes (Sax-2.1) appear. From phase Rh 9 heavy broad saxes with a very broad edge (Sax-2.2) are added, which are then superseded in phase Rh 10 by long saxes. The early 'short saxes' found in the Alamannic area do not occur in the Rhineland.

Angons are found in phases Rh 2–7, but predominantly in graves of phases Rh 4–5. As is usual across the entire Frankish realm, spears with a slit socket (Lan-1) are dominant; those with unslit sockets (Lan-2) do not appear until phase Rh 6 and are dominant from phase Rh 8 onwards. Spears of lengths between 30 and 47 cm with a small blade and long neck, Lan-1.1b, are typical of phase Rh 4. They are replaced in phase Rh 5 by a longer version of this type, Lan-1.2. Typical of phase Rh 6 are short but heavy spears with a long blade (Lan-1.3a) and slit socket. Their counterparts with an unslit socket, Lan-2.3, are typical of phase Rh 7. Parallel short spears with a long blade appear in the form of type Lan-1.4 with a slit socket and type Lan-2.1 with a continuous unslit socket (type 'Dorfmerkingen'). Spears with an unslit socket and a simple, relatively long blade are characteristic of phases Rh 8–9, although the short spears, Lan-2.4, are somewhat earlier (Rh 8) and the longer spears, Lan-2.5, are somewhat later (Rh 8B-9). After phase 9 weaponry is found in the Rhineland only rather sporadically as grave goods, in conservative cemeteries such as Stockum and Walsum. Thus, longer spears with a square socket, Lan-4.1, are used in phase 10. Likewise in phase Rh 10 spears with an octagonal faceted socket occur, first in the short version, Lan-8.1, < 33 cm long, and later, in phase Rh 11, the longer version, Lan-8.2.

In the weaponry of the 6th century various weapon axes occurred. The francisca with a normal upper edge, FBA-1.3, is common in phase Rh 4 but is not seen from phase Rh 6 onwards. Various types of 'bearded' axe (FBA-3.1, FBA-3.2, FBA-4.1, FBA-4.2) occur in phases Rh 5–7. In Phase Rh 6 the simple axes, FBA-2.1, appear, and they are especially common in phase Rh 7. Then – after

The Merovingian chronology of the Lower Rhine Area: results and problems

Rheinland Phase 1 (ca. 400 - 440)	Rheinland Phase 2 (ca. 440 - 480/90)
Gür- 1.1 1-2	Gür- 1.2 2 Gür- 1.3 2-3 Gür- 2.1 2-3 FBA- 1.1 2
FBA- 2.3 1-2	Fib- 12.1 2 Fib- 12.2 2 Fib- 12.3 2 Fib- 12.4 2 Fib- 12.5 2
Fib- 4.1 1	Fib- 13 2-3 Fib- 4.2a 1-2 Nad- 2.1 2 Spa- 5 2
Gla- 1.2 1-2	Gla- 7.1 2 Gla- 4.1 2-3 Gla- 11 1-2 Ger- 4.1 2
Gla- 9 1-4 Gla- 3.1 1	Ger- 4.2 2-3
Sha- 2.12 1-2	
Kru- 1.2 2	Kan- 1.1/2 2-4 Sha- 2.11 2-3 Ger- 2.5 2-3

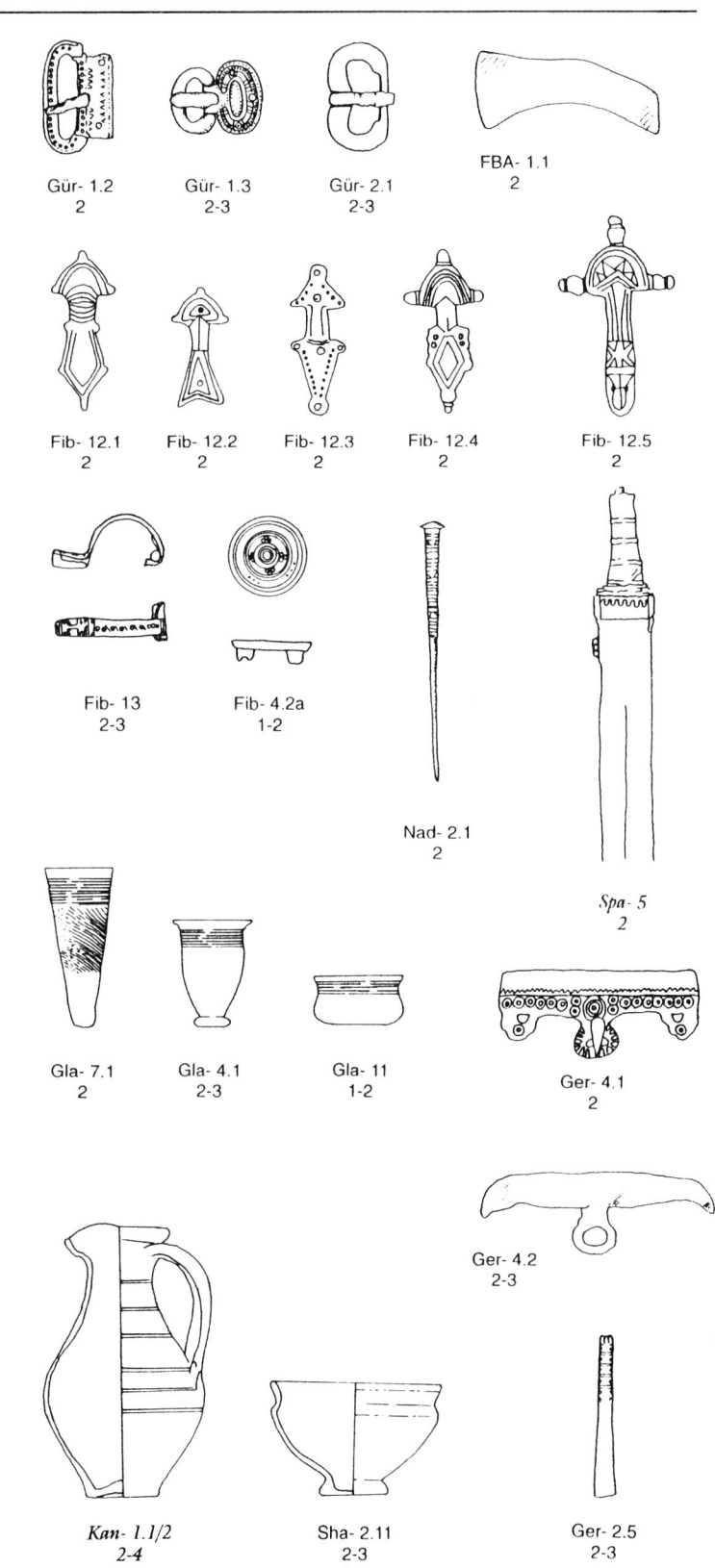

Fig. 1.5 Rheinland/Rheinland Phases 1–2.

Rheinland Phase 3   (480/90 - 530)

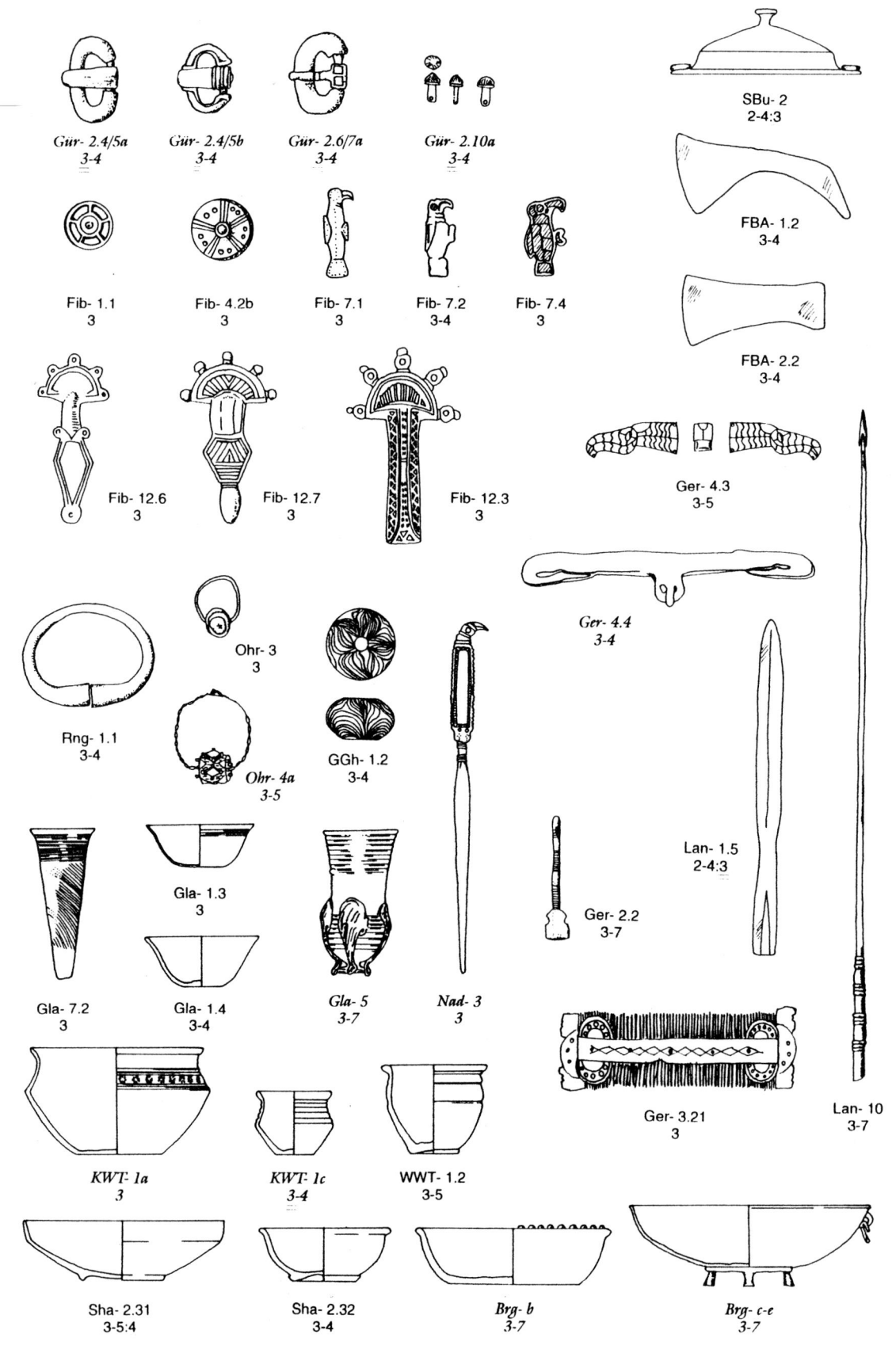

Fig. 1.6 Rhineland/Rheinland Phase 3.

Rheinland Phase 4   (520/30 - 550/60)

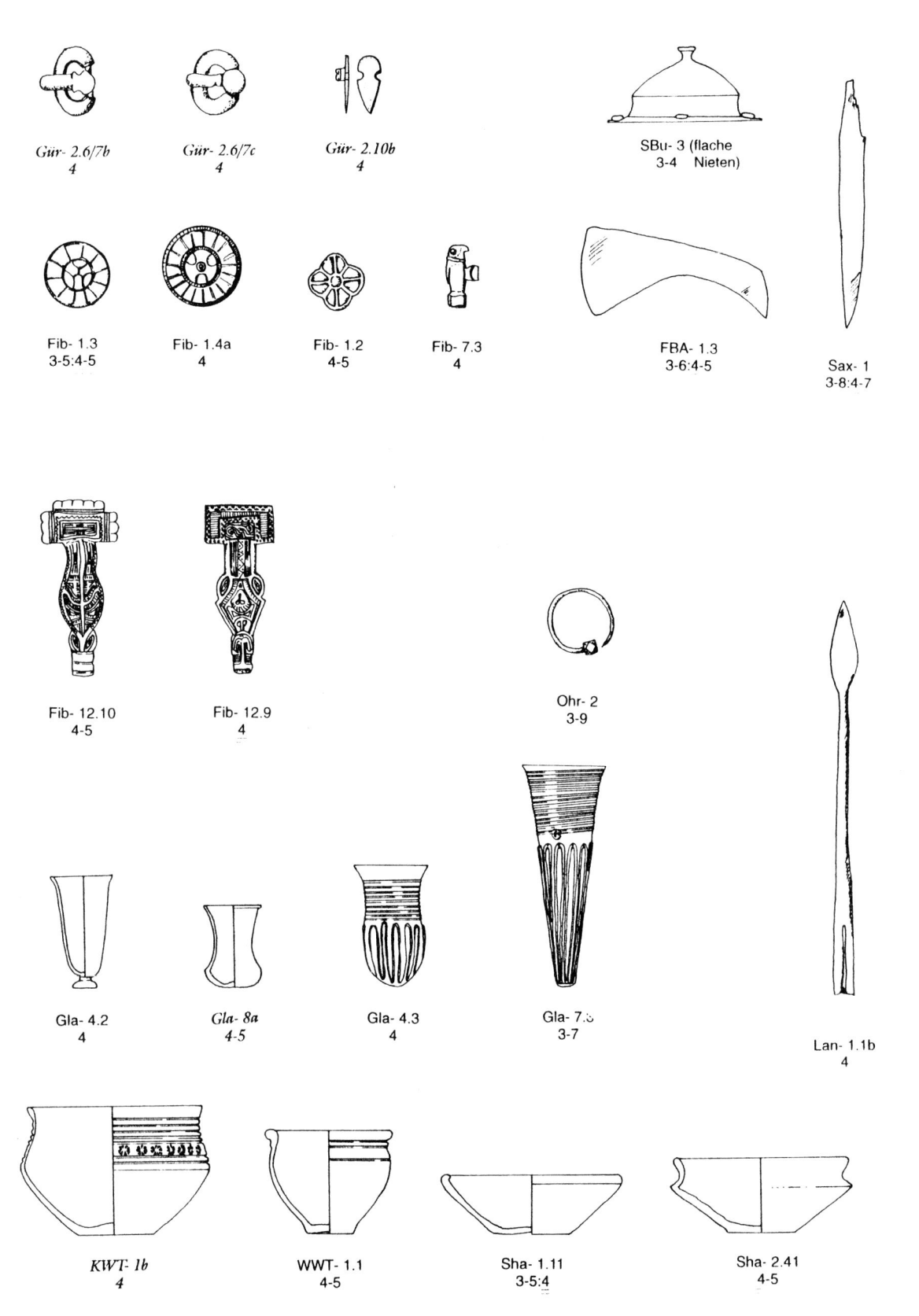

Gür- 2.6/7b
4

Gür- 2.6/7c
4

Gür- 2.10b
4

SBu- 3 (flache
3-4   Nieten)

Fib- 1.3
3-5:4-5

Fib- 1.4a
4

Fib- 1.2
4-5

Fib- 7.3
4

FBA- 1.3
3-6:4-5

Sax- 1
3-8:4-7

Fib- 12.10
4-5

Fib- 12.9
4

Ohr- 2
3-9

Gla- 4.2
4

Gla- 8a
4-5

Gla- 4.3
4

Gla- 7.8
3-7

Lan- 1.1b
4

KWT- 1b
4

WWT- 1.1
4-5

Sha- 1.11
3-5:4

Sha- 2.41
4-5

Fig. 1.7 Rhineland/Rheinland Phase 4.

the appearance of the broad seax – weapon axes are no longer found.

Shield bosses are rare in graves to begin with, their frequency not increasing until phase Rh 4. The version with a disc-headed apex, low wall and flat cone, Sbu-2, is found in phases Rh 2–4. A version with a taller wall and taller cone (Sbu-3) is found in phases Rh 4–7. The bosses Sbu-2 and -3 have flat silver gilt rivets (Rh 2–3), later flat copper-alloy covered rivets (Rh-4). Rivets with a hemispherical head appear occasionally from phase Rh 4 onwards but are not dominant until phases Rh 5–6. Bosses with steep walls, a conical cone and rod apex, Sbu-4, are found in phases Rh 6–7. Typical of phase Rh 8 are shield bosses without an apex feature, with a high wall and a flat cone, Sbu-5, probably a Frankish type. They are replaced in phase 9 by bosses with a high wall and hemispherical cone, Sbu-6. In the late weapon graves the shield bosses a have very low wall and tall cone (Stein 1967: Type Walsum); those lower than 10.5 cm (Sbu-7) are typical of phase Rh 10, the taller ones (Sb-8) typical of phase Rh 11.

The most typical and common form of pottery as grave goods is the so-called '*Knickwandtopf*' (biconical pot). As Böhner (1958, 45) recognized, the hollow-necked pots are the earliest forms and should be basically distinguished from the biconical form. Following in the tradition of the late-Roman 'Terra Nigra' are the small pots with many small ribs, KWT-1c, which appear from phase Rh 2 onwards. From phase 3 the 'classical' large, biconical pot decorated with single stamps (KWT-1a and -1b) occurs.

Within the group of biconical pots with a straight upper wall occurring from phase 4 onwards, it is the decoration and the general development of shape from wide to tall and slim pots which must be the basis of the typological classification. Whilst some stamp patterns were mainly of local origin, there seems to have been a general development from single stamp decoration (KWT-2a, phase Rh 4) through line/wave patterns (KWT-3a and -3b, phase Rh 5) to the earliest rolled stamps (from phase Rh 6 onwards). The chronological development of the rolled stamps from the occasional appearance of single-column square stamps (KWT-5a, Phase Rh 6), following the single stamps and appearing before the pots with multiple columns of rolled stamps and rolled stamps with composite and/or broken patterns (KWT-5b, -5d, -5f; from Rh 7 onwards), is clearly shown on the distribution maps of the various patterns especially in the cemeteries of Junkersdorf and Gellep. Characteristic of the pottery of the 7th century are tall, slim forms with composite rolled stamps and/or double ribs on the upper wall (KWT-5c, 5e; from Rh 7 onwards, predominantly in Rh 8). Occasional undecorated pots of slim shape (KWT-2.43) mark the end of the biconical pots as grave goods in the Rhineland in phase Rh 10. On a large scale this habit had already come to an end in phase Rh 9.

Next to biconical pots only bowls become relatively common as grave goods in phases Rh 4–7; other pottery-types are rather rare. Bowls with a rough texture and inturned rim, Sha-1.11, belong to phases Rh 4 and 5, when smooth textured, red-painted biconical bowls, Sha-2.31 (Pirling type 129–130) are also used. The bowls of the following phases Rh 6–9 are hard to classify and seem to have been in use for a long time, although the biconical bowls of smooth texture (predominantly type Sha-2.21) are in general earlier (mostly Rh 5–8) and those with a rough texture generally later (mostly Rh 8–9).

Bottles are very rarely used as grave goods in the 6th century, so that analyses of their typological and chronological development are rather limited. When they finally appear as grave goods in phase Rh 7, and increasingly in phase Rh 8, it is in the form of a bottle with a rough texture and a relatively broad body and cylindrical neck (Fla-1.1). Similar bottles, however, with a short neck and profiled inner rim (Fla-1.2) belong to phase Rh 10. Large bottles with a relatively open neck are typical of the late Merovingian Period; bottles with a long body (Fla-2.1), often with ornamental ribs, belong to phase Rh 10, and those with short body (Fla-2.2) mostly to phase Rh 11.

At this time an important change appears in the fabric of all pots with a rough texture: more often than before the clay appears in light colours – it is much more tempered, the tempering finer, and the fabric is soft. These characteristics, linked to the introduction of the 'Badorf ware', occur from phase Rh 10 onwards (cf. Bridger and Siegmund 1987, Abb. 5–6).

While in most cemeteries ceramic grave goods decline and disappear in the course of phase Rh 9, the tradition continues in some of the Lower Rhine cemeteries. This group of pottery especially involves types that were previously either missing or very rare. Large pots with two or three handles and without a spout, so-called '*Mehrhenkelkrüge*', appear in phase Rh 9. The early versions have a rounded body (Kru-2.1), the later ones are slimmer and often carry ribbed decoration (Kru-2.21: Rh 10). These are followed by pots with high shoulders (Kru-2.22: Rh 11). A further characteristic type is the jug/pitcher with a small spout and one handle: forms with a flat bottom, a faceted and profiled rim and ribbed decoration (Kan-2.1) appear in phase Rh 10, forms with '*Linsenboden*' (a lentoid bottom), round rim and often also rolled stamp decoration (Kan-2.2) in phase Rh 11.

Only in phases Rh 3–5 and 9–11 does round-bodied pottery occur in the graves. The early and late types can be divided on basis of their rims. Within the phases Rh 3–5 there is no point to further sub-classification (WWT-1.1/3) as the types are not chronologically sensitive. In the group of late (and often large) round-bodied pots the broad types (WWT-2.21) appear in phase Rh 10, the slim types (WWT-2.22) in phase Rh 11.

The glass beaker with a beaten rim (Gla-3.1), the ribbed glass bowl (Gla-1.2) and the cone beaker (Gla-7.1), appearing in the graves of the 5th century, are basically derived from glass beakers of the Roman tradition. Cone beakers of the 6th century are larger and higher than the

Rheinland Phase 5   (550/60 - 570)

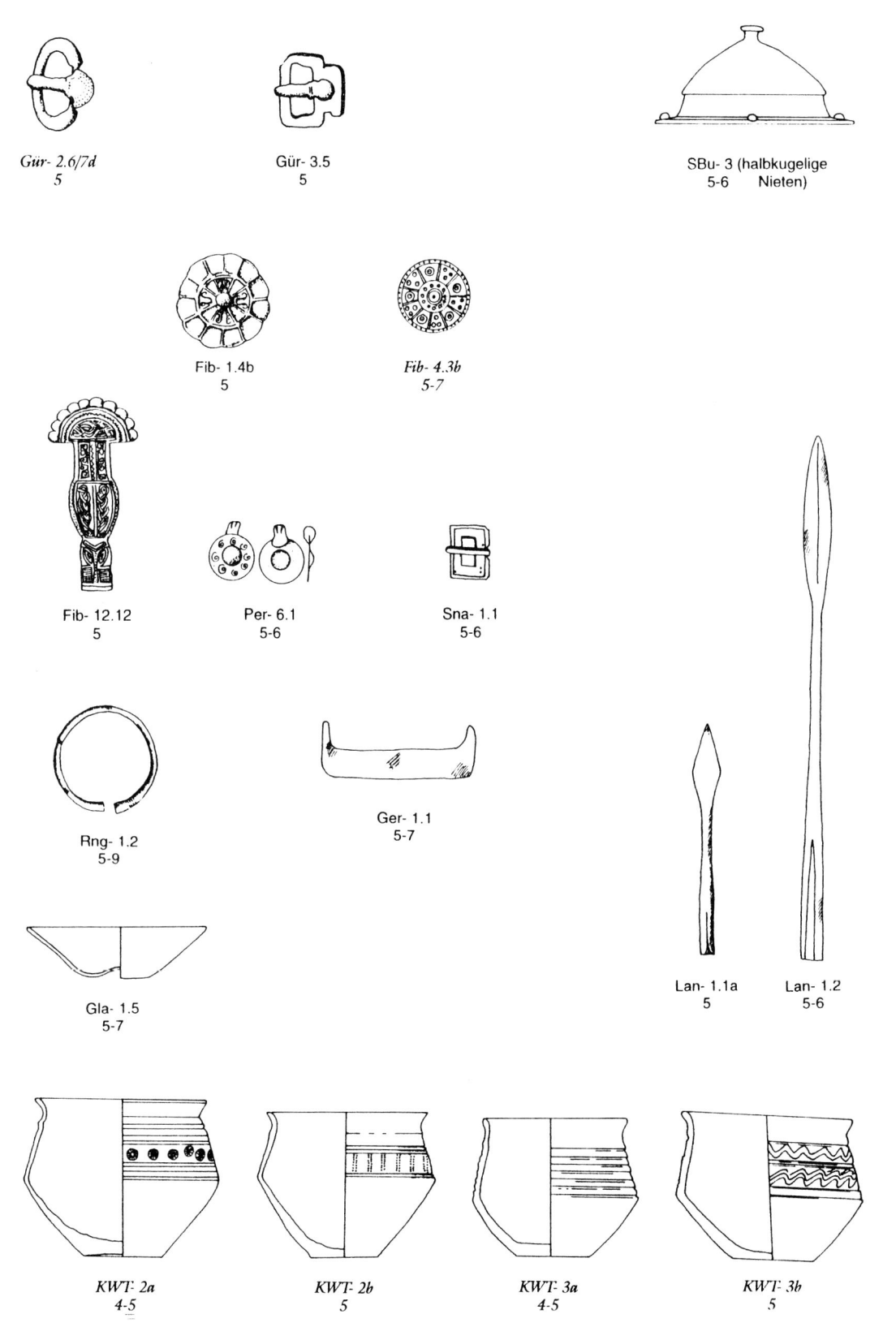

Gür- 2.6/7d
5

Gür- 3.5
5

SBu- 3 (halbkugelige
5-6   Nieten)

Fib- 1.4b
5

Fib- 4.3b
5-7

Fib- 12.12
5

Per- 6.1
5-6

Sna- 1.1
5-6

Rng- 1.2
5-9

Ger- 1.1
5-7

Lan- 1.1a
5

Lan- 1.2
5-6

Gla- 1.5
5-7

KWT- 2a
4-5

KWT- 2b
5

KWT- 3a
4-5

KWT- 3b
5

Fig. 1.8 Rhineland/Rheinland Phase 5.

Rheinland Phase 6 (570 - 580/90)

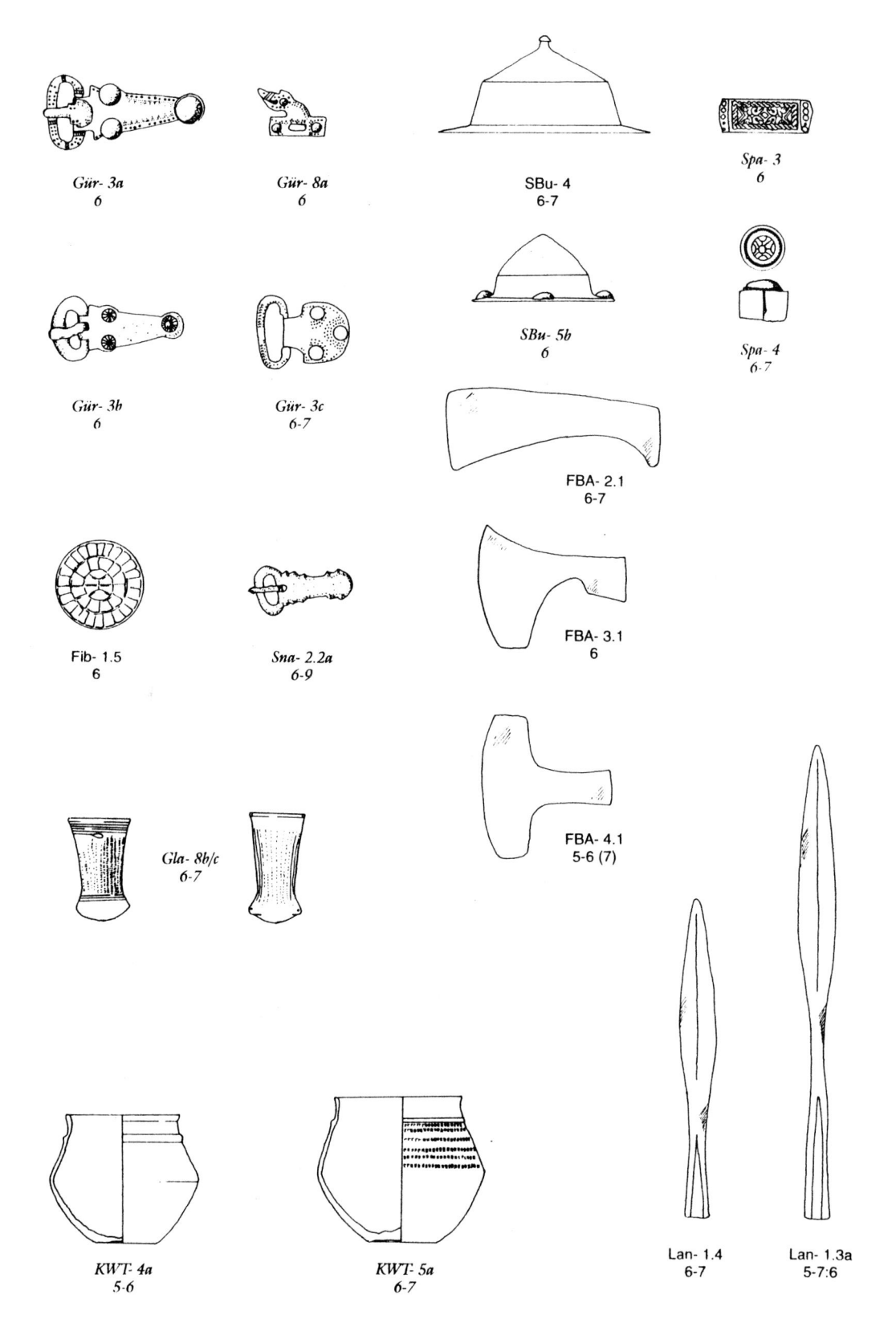

Gür- 3a
6

Gür- 8a
6

SBu- 4
6-7

Spa- 3
6

Gür- 3b
6

Gür- 3c
6-7

SBu- 5b
6

Spa- 4
6-7

FBA- 2.1
6-7

Fib- 1.5
6

Sna- 2.2a
6-9

FBA- 3.1
6

Gla- 8b/c
6-7

FBA- 4.1
5-6 (7)

KWT- 4a
5-6

KWT- 5a
6-7

Lan- 1.4
6-7

Lan- 1.3a
5-7:6

Fig. 1.9 Rhineland/Rheinland Phase 6.

The Merovingian chronology of the Lower Rhine Area: results and problems

Rheinland Phase 7 (580/90 - 610)

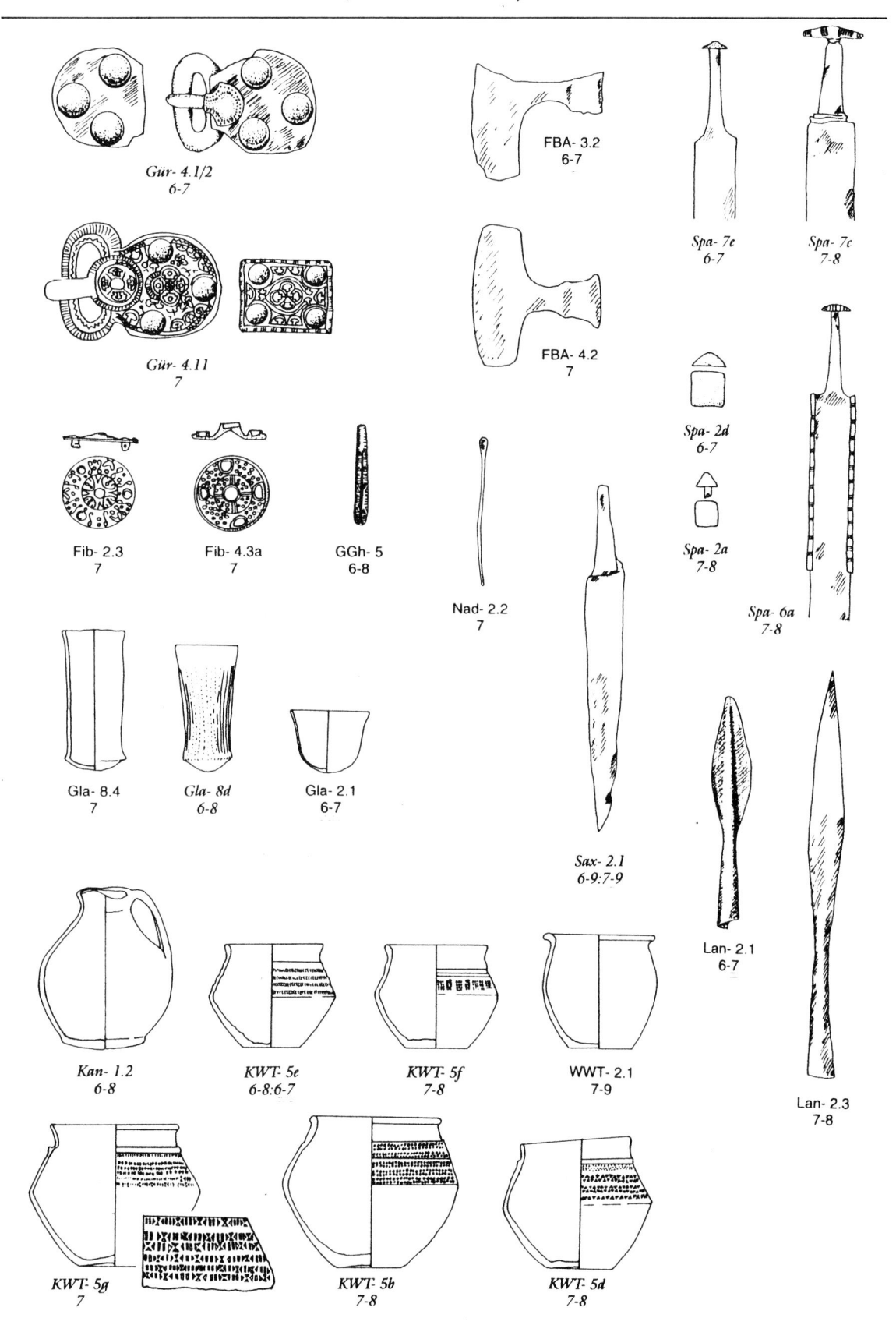

*Fig. 1.10 Rhineland/*Rheinland Phase 7.

Roman forms and are often decorated with glass trails (Gla-7.2 and -7.3). Roman glass bottles (Gla-9) were commonly used throughout the 6th century. Typical of the graves of phase Rh 3 are the slightly rounded glass bowls with opaque white stripes on the rim (Gla-1.3 and -1.4), and strictly conical pieces (Gla-1.5) are from the second half of the 6th century. The first small, undecorated bell beakers (so-called '*Sturzbecher*') appear as early as phase Rh 4, whilst the majority of developed examples appear from phases Rh 6 to 8. The range of development ends here with high, slim pieces (Gla-8.4). Palm cups, appearing in the late 6th century (phase Rh 6), and some with plain, thickened rims, others with wide, down-folded rims, represent a typical form of the 7th century (phase Rh 9). Glass as grave goods ends in the Rhineland with the pointed palm cup (Gla-2.3) of phases 10 and 11.

To give a closer date to the final phase of the Merovingian Period, phase Rh 11, a phase Rh 12 was described, providing a *terminus post quem* for everything later. On the basis of the pottery fabric an end-dating for the light-fired pottery of 'Badorf' type is defined, as pottery from phase Rh 12 is all hard-fired. At Rill the burials in dug-out wooden coffins with hand-made 'Frisian' spherical pots belong to this post-Merovingian phase Rh 12. In other places too it is possible to see a clear change in burial-form. The still broad, late-Merovingian graves of phase Rh 11 at Rommerskirchen-St. Peter are followed in phase Rh 12 by narrow, slightly trapezoid graves and graves with a 'head-niche'. The same development is seen at Xanten-St. Viktor: instead of the previously usual coffin-less or wooden coffin graves, stone cists appear more frequently from phase Rh 8, and especially those composed strictly of six side-stones; from phase 10 onwards four side-stones are usual, i.e. only one long stone slab at each side. Both grave-types have the usual Merovingian measurements. They are followed in phase Rh 12 by typically narrow, trapezoid sarcophagi and 'head-niche' graves. At both Rommers-kirchen and Xanten the arms and hands of the dead from phase Rh 12 lie alongside the body; 'praying' hands lying in the pelvis area is a later phenomenon.

PROBLEMS AND PERSPECTIVES

The material basis of the model presented ought to have been of sufficient size. From the tables of this particular study it appears that the limits of clarity have been reached. In a way the relative chronology is steady, and only in the case of very rare types are modifications as to the perceived period of use possible.

Material remains of the 5th century (Rh 1–3) and of the final phases of the Merovingian Period (Rh 10–11) are, however, rare. In particular, complete cemeteries or larger parts of cemeteries suitable for topo-chronological analysis are missing. In this respect an expansion of the material basis through purposeful excavations might lead to a modification of our model. It is, for example, impossible in the Rhineland to establish the division into a Childeric-period phase and a Chlodewig-period phase (*circa* Rh 3) as can be done in South Germany primarily on basis of well-equipped burials (Müller 1976; most recently Quast 1993) because we have too few grave groups.

Another problem is the absolute fixing of the relative chronology. Here, indeed, all graves containing coins even outside the Rhineland were included as far as possible. The basis, however, remained slender in the well-known problematic periods of the 5th and 7th-8th centuries. This is especially clear in relation to the still open discussion about the absolute dating of the border between the stages II and III, Rh 3 and Rh 4 respectively (Martin 1989). Only a few coin-dated graves from the questionable periods, or dendrochronologically dated complexes, could affect our ideas of the absolute chronology considerably. New good data are expected from the already excavated cemetery of Lauchheim (Stork 1997; fig. 1 no. 25) where some more dendro-dated graves will be obtained.

Because of bad state of preservation there are few skeletal remains in the graves of the Lower Rhine area. Extensive skeletal analyses could enhance understanding of the composition of the grave goods. On the basis of analyses from other areas there seem to have been distinct rules for the acquisition and acquisition-time of the inventories of the graves. Knowledge about such rules and traditions combined with skeletal age determinations would improve the scope for interpreting the grave assemblages.

A statistical attempt, allowing a combined analysis of chorological and combinational information, only reveals known information, and nothing new (Herzog and Siegmund 1991). Still, using statistical methods, it may be possible to improve the results produced. The correlation and phase-division of the tables of the female and the male graves were here done in conventional way and in close contact with the topo-chronological phases of the cemeteries, which is probably a legitimate and practicable method. The Prokrustes-Rotation suggested by Andreas Zimmerman seems, however, much more elegant and should be tested for its practicability (Zimmermann 1994; 1995). The correspondence analysis used gave centres of gravity and Eigenvalues for graves and types respectively; wanting, however, is information about the quality of this assessment and a measure for the dispersion around the average value, i.e. a parameter like the standard deviation. Here the actual development within statistics should be followed with a watchful eye, as useful improvements of this kind are to be expected (Vach 1994).

Acknowledgements

Drawings: Gabriele Körber, Landesamt für archäologische Denkmalpflege/Landesmuseum für Vorgeschichte Halle.

Translation: Paul Yule, Bonn; Alan Brown, Titz; Karen Høilund Nielsen, Århus.

Rheinland Phase 8 (610 - 640)

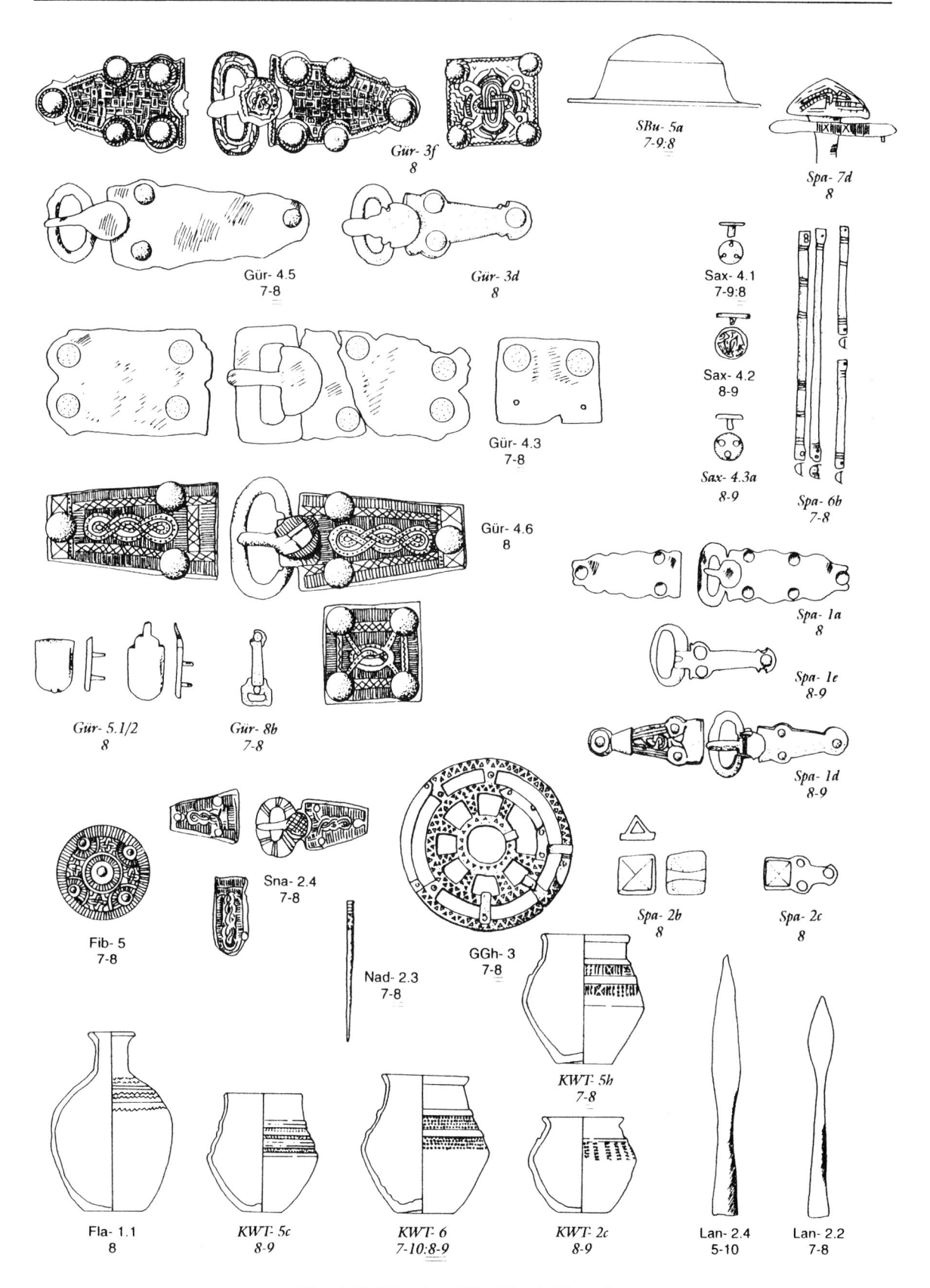

Fig. 1.11 Rhineland/Rheinland Phase 8.

Rheinland Phase 9 (640 - 670)

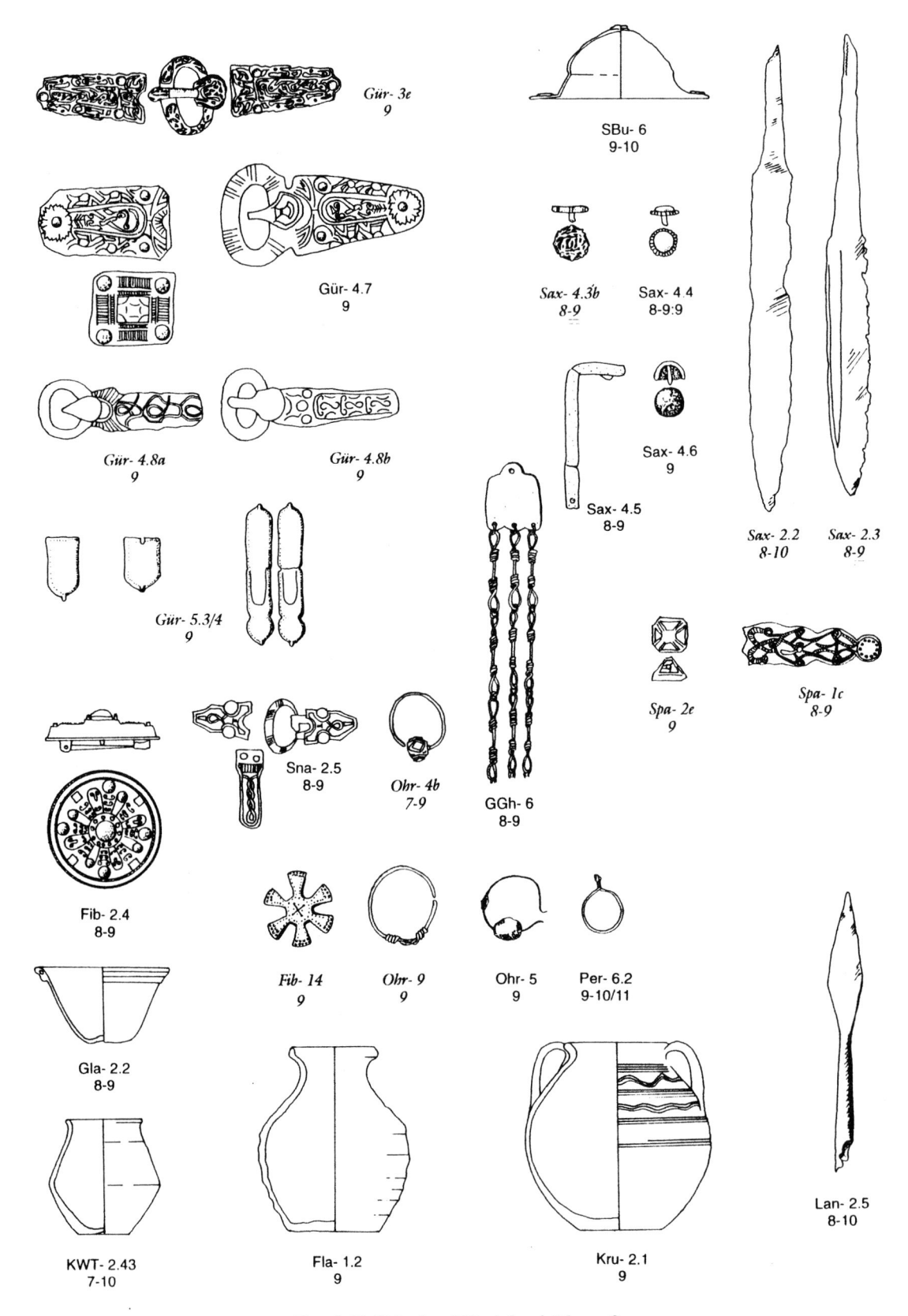

Fig. 1.12 Rhineland/Rheinland Phase 9.

Rheinland Phase 10 (670 - 710)

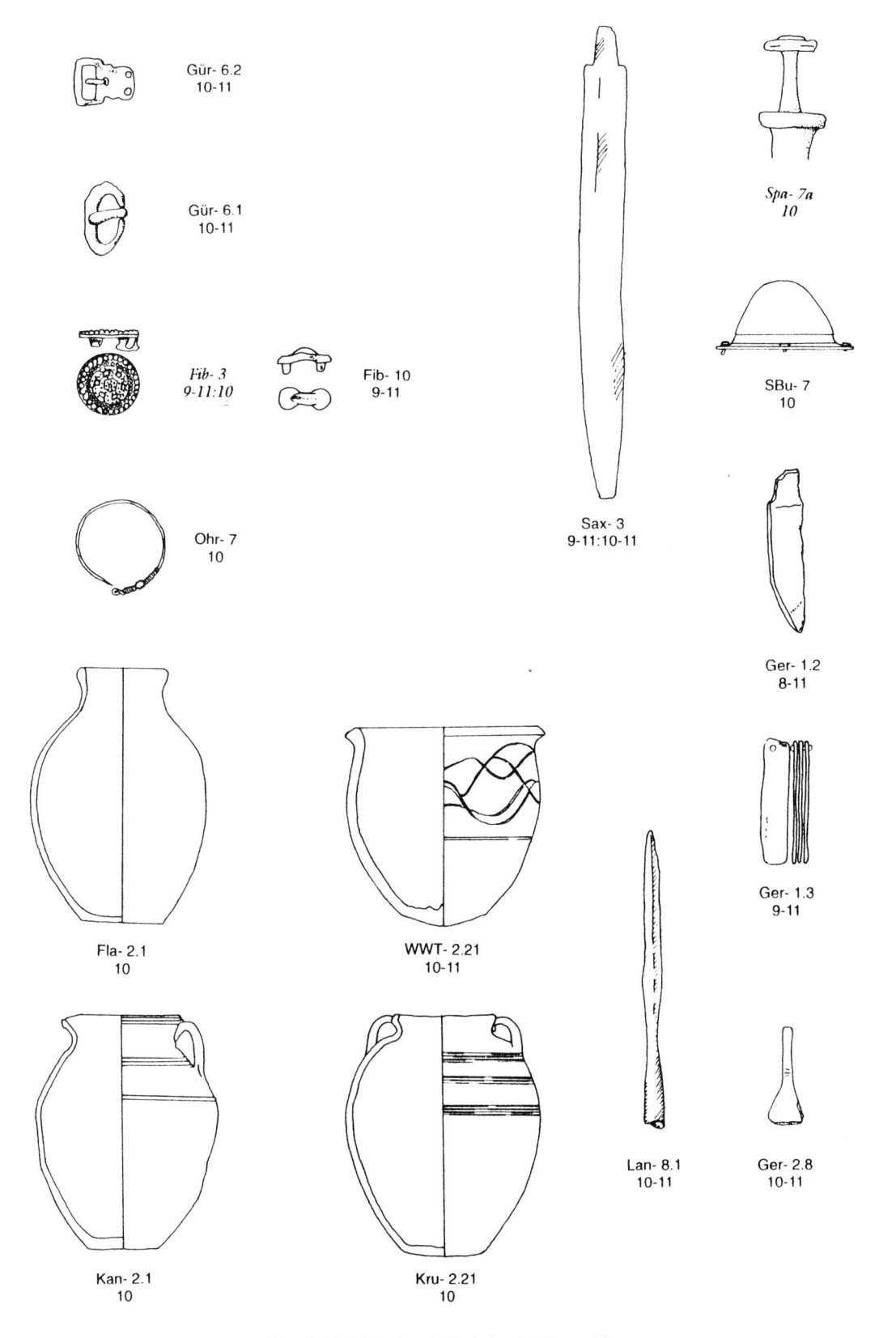

Gür- 6.2
10-11

Gür- 6.1
10-11

Fib- 3
9-11:10

Fib- 10
9-11

Ohr- 7
10

Sax- 3
9-11:10-11

Spa- 7a
10

SBu- 7
10

Ger- 1.2
8-11

Fla- 2.1
10

WWT- 2.21
10-11

Ger- 1.3
9-11

Kan- 2.1
10

Kru- 2.21
10

Lan- 8.1
10-11

Ger- 2.8
10-11

Fig. 1.13 Rhineland/Rheinland Phase 10.

Rheinland Phase 11   (710 - 740)

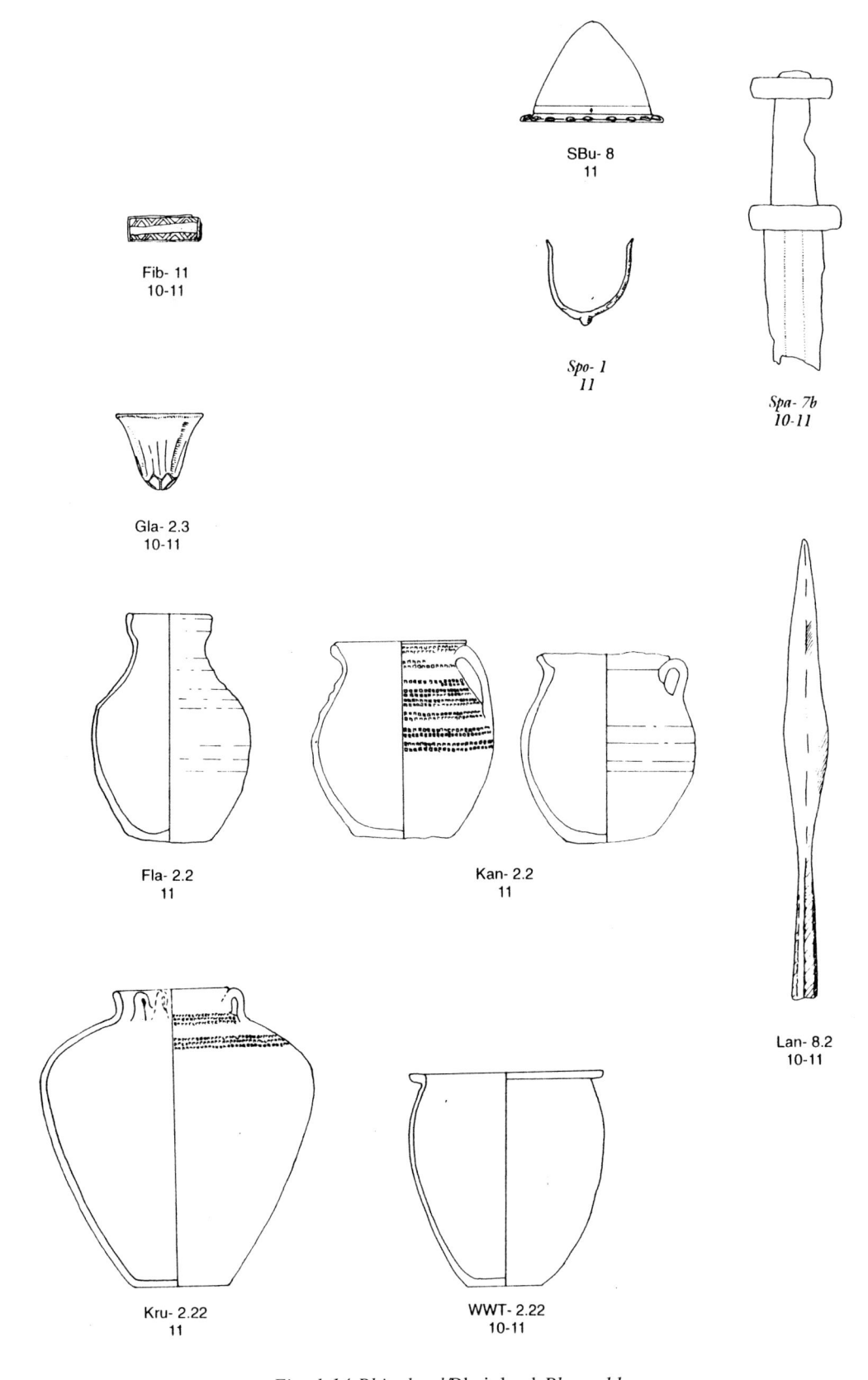

SBu- 8
11

Fib- 11
10-11

Spo- 1
11

Spa- 7b
10-11

Gla- 2.3
10-11

Fla- 2.2
11

Kan- 2.2
11

Lan- 8.2
10-11

Kru- 2.22
11

WWT- 2.22
10-11

*Fig. 1.14 Rhineland/*Rheinland *Phase 11.*

DEUTSCHE ZUSAMMENFASSUNG

Der Beitrag stellt das Chronologiesystem für die nördlichen Rheinlande vor (Abb. 1.5–1.14); es umfaßt 11 Phasen für die Zeit vom Beginn des 5. Jahrhunderts bis in die Mitte des 8. Jahrhunderts. Das System gründet auf der Dissertation von F. Siegmund über den Niederrhein (Siegmund 1989, 1998; Abb. 1 Nr. 2–8) und der Chronologiestudie der 'Franken AG' (H. Aouni, U. Müssemeier, E. Nieveler, R. Plum) zur südlichen Kölner Bucht (Franken AG, im Druck; Abb. 1 Nr. 9–13). Vorgestellt wird eine Synthese beider Studien. Grundlage des Systems sind chorologische Unter-suchungen aller geeigneten Gräberfelder (Düsseldorf-Stockum, Krefeld-Gellep, Köln-Junkersdorf, Köln-Müngersdorf, Walsum und Iversheim, Jülich, Lamersdorf, Rödingen) sowie zwei Korrespondenzanalysen für alle geeigneten Frauen- und Männergräber (ca. 390 Inventare mit 145 Typen bzw. ca. 530 Inventare mit 185 Typen). Die vorgeschlagene relative Ordnung ist stabil, was auch durch die nahezu ideale Form der Parabeln beider Korrespondenzanalysen unterstrichen wird (Abb. 1.2–1.3). Die Verknüpfung der Männer- und Frauenchronologie erfolgt über die chorologischen Analysen und die beiden Geschlechtern gemeinsamen Typen. Die Phasengliederung orientiert sich vor allem an den wechselnden Gürtelmoden sowie neu auftretenden Keramikformen oder -verzierungen.

Diese Rheinland-Chronologie (kurz 'Rh 1–11') läßt sich weitgehend konfliktfrei mit anderen modernen Systemen verknüpfen (Abb. 1.4). Ihre regionale Gültigkeit endet dort, wo die phasendefinierenden Typen (Trachtbestandteile, Waffen, Keramik) in nicht mehr ausreichender Zahl vertreten sind. Für die vorgeschlagene absolute Chronologie wurden auch über das konkrete Arbeitsgebiet hinaus alle verfügbaren und in diese Systematik einordnenbaren münzführenden Gräber herangezogen.

BIBLIOGRAPHY

Ament, H. 1976a. *Die fränkischen Grabfunde aus Mayen und der Pellenz*. Germanische Denkmäler der Völkerwanderungszeit B 9. Berlin.

Ament, H. 1976b. 'Chronologische Untersuchungen an fränkischen Gräberfeldern der jüngeren Merowingerzeit im Rheinland'. *Berichte Römisch-Germanische Komm* 57, 285–336.

Ament, H. 1977. 'Zur archäologischen Periodisierung der Merowingerzeit'. *Germania* 55, 131–40.

Böhme, H. W. 1974. *Germanische Grabfunde des 4. bis 5. Jahrhunderts zwischen unterer Elbe und Loire. Studien zur Chronologie und Bevölkerungsgeschichte*. Münchner Beiträge zur Vor- und Frühgeschichte 19. München.

Böhme, H. W. 1985. 'Les découvertes du Bas-Empire à Vireux-Molhain. Considérations générales'. In J.-P. Lémant, *Le cimetière et la fortification du Bas-Empire de Vireux-Molhain, Dép. Ardennes*, 76–88. Römisch-Germanische Zentralmus Monographien 7. Mainz.

Böhme, H. W. 1986. 'Studien zu Gallien in der Spätantike'. *Jahrb Römisch-Germanisch Zentralmus* 33, 844–8.

Böhme, H. W. 1989. 'Gallien in der Spätantike. Forschungen zum Ende der Römerherrschaft in den westlichen Provinzen'. *Jahrb Römisch-Germanisch Zentralmus* 34, 770–3.

Böhme, H. W. 1994. 'Der Frankenkönig Childerich zwischen Attila und Aëtius. Zu den Goldgriffspathen der Merowingerzeit'. In C. Dobiat (ed.), *Festschrift für Otto-Herman Frey zum 65. Geburtstag*, 69–110. Marburger Studien zur Vor- und Frühgeschichte 16. Marburg.

Böhner, K. 1949. 'Die fränkischen Gräber von Orsoy, Kreis Mörs'. *Bonner Jahrb* 149, 146–96.

Böhner, K. 1958. *Die fränkischen Altertümer des Trierer Landes*. Germanische Denkmäler der Völkerwanderungszeit B 1. Berlin.

Böhner, K. 1978. 'La chronologie des antiquités funéraires d'époque mérovingienne en Austrasie'. In M. Fleury and P. Périn (eds.), *Problèmes de chronologie relative et absolue concernant les cimetières mérovingiens d'entre Loire et Rhin. Actes du IIe colloque archéologique de la IVe Section de l'Ecole pratique des Hautes Etudes, Paris 1973*, 7–12. Paris.

Bridger, C. J. and F. Siegmund. 1987. 'Funde des 8. Jahrhunderts aus Xanten'. *Bonner Jahrb* 187, 543–62.

Christlein, R. 1966. *Das alamannische Reihengräberfeld von Marktoberdorf im Allgäu*. Materialhefte zur bayerischen Vorgeschichte 21. Kallmünz.

Franken AG, in press. *Chronologie der merowingerzeitlichen Grabfunde vom linken Niederrhein und der nördlichen Eifel*. By H. Aouni, U. Müssemeier, E. Nieveler and R. Plum. Bonner Beiträge zur Vorgeschichte. Bonn.

Fremersdorf, F. 1955. *Das fränkische Reihengräberfeld von Köln-Müngersdorf*. Germanische Denkmäler der Völkerwanderungszeit 6. Berlin.

Giesler, J. 1983. 'Frühmittelalterliche Funde aus Niederkassel, Rhein-Sieg-Kreis'. *Bonner Jahrb* 183, 475–590.

Herzog, I. and Siegmund, F. 1991. 'Clusteranalyse räumlicher Daten mit Hilfe der Gemeinsamen-Nachbarschafts-Gruppierung (GN-Gruppierung)'. *Prähistorische Zeitschrift* 66, 219–34.

Hinz, H. 1969. *Das fränkische Gräberfeld von Eick, Gem. Rheinkamp, Kr. Moers*. Germanische Denkmäler der Völkerwanderungszeit B 4. Berlin.

Janssen, W. 1993. *Das fränkische Reihengräberfeld von Rödingen, Kr. Düren*. Germanische Denkmäler der Völkerwanderungszeit B 16. Stuttgart.

Kleemann, J. 1992. *Grabfunde des 8. und 9. Jahrhunderts im nördlichen Randgebiet des Karolingerreiches*. Ph.D. thesis, University of Bonn.

LaBaume, P. 1967. *Das fränkische Gräberfeld von Junkersdorf bei Köln*. Germanische Denkmäler der Völkerwanderungszeit B 3. Berlin.

Martin, M. 1989. 'Bemerkungen zur chronologischen Gliederung der frühen Merowingerzeit'. *Germania* 67, 121–41.

Müller, H. F. 1976. *Das alamannische Gräberfeld von Hemmingen (Kreis Ludwigsburg)*. Forschungen und Berichte zur Vor- und Frühgeschichte in Baden-Württemberg 7. Stuttgart.

Neuffer-Müller, C. 1960. 'Das fränkische Gräberfeld von Lommersum, Kreis Euskirchen'. *Bonner Jahrb* 160, 204–64.

Neuffer-Müller, Chr. 1972. *Das fränkische Gräberfeld von Iversheim*. Germanische Denkmäler der Völkerwanderungszeit B 6. Berlin.

Neuffer-Müller, C. and Ament, H. 1973. *Das fränkische Gräberfeld von Rübenach, Stadt Koblenz*. Germanische Denkmäler der Völkerwanderungszeit B 7. Berlin.

Päffgen, B. 1992. *Die Ausgrabungen in St. Severin zu Köln*. Kölner Forschungen 5,1–3. Mainz.

Périn, P. 1980. *La datation des tombes mérovingiennes. Historique – Méthodes – Applications*. Geneva.

Piepers, W. 1963. 'Ein fränkisches Gräberfeld bei Lamersdorf, Kreis Düren'. *Bonner Jahrb* 163, 424–68.

Pirling, R. 1966. *Das römisch-fränkische Gräberfeld von Krefeld-Gellep*. Germanische Denkmäler der Völkerwanderungszeit B 2. Berlin.

Pirling, R. 1974. *Das römisch-fränkische Gräberfeld von Krefeld-*

Gellep 1960–1963. Germanische Denkmäler der Völkerwanderungszeit B 8. Berlin.

Pirling, R. 1979. *Das römisch-fränkische Gräberfeld von Krefeld-Gellep 1964–1965*. Germanische Denkmäler der Völkerwanderungszeit B 10. Berlin.

Quast, D. 1993. *Die merowingerzeitlichen Grabfunde aus Gültlingen (Stadt Wildberg, Kreis Calw)*. Forschungen und Berichte zur Vor- und Frühgeschichte in Baden-Württemberg 52. Stuttgart.

Scollar, I., Herzog, I., Rehmet J. and Greenacre, M. J. 1992. *The Bonn Archaeological Statistics Package Vers. 4.5*. Remagen.

Siegmund, F. 1982. 'Zum Belegungsablauf auf dem fränkischen Gräberfeld von Krefeld-Gellep'. *Jahrb Römisch-Germanische Zentralmus* 29, 249–65.

Siegmund, F. 1989. *Fränkische Funde vom deutschen Niederrhein und der nördlichen Kölner Bucht*. Ph.D. thesis, Universität Köln.

Siegmund, F. 1998. *Merowingerzeit am Niederrhein. Die frühmittelalterlichen Funde aus dem Regierungsbezirk Düsseldorf und dem Kreis Heinsberg*. Rheinische Ausgrabungen 34. Köln.

Stampfuß, R. 1939. *Der spätfränkische Sippenfriedhof von Walsum*. Quellenschriften zur westdeutschen Vor- und Frühgeschichte 1. Leipzig.

Steeger, A. 1948. 'Der fränkische Friedhof in Rill bei Xanten'. *Bonner Jahrb* 148, 249–98.

Stein, F. 1967. *Adelsgräber des achten Jahrhunderts in Deutschland*. Germanische Denkmäler der Völkerwanderungszeit A 9. Berlin.

Steuer, H. 1977. 'Bemerkungen zur Chronologie der Merowingerzeit'. *Stud zur Sachsenforschung* 1, 379–402.

Steuer, H. 1990. Review of: H. Beumann and W. Schröder (eds.), Die transalpinen Verbindungen der Bayern, Alemannen und Franken bis zum 10. Jahrhundert. *Fundberichte Baden-Württemberg* 15, 494–503.

Stork, I. 1997. 'Friedhof und Dorf, Herrenhof und Adelsgrab'. In *Die Alamannen*, 290–310. Stuttgart.

Vach, W. 1994. *Erweiterung der mathematisch-statistischen Methodik für Seriationsprobleme in archäologischen Anwendungen*. Unpublished dissertation, Freiburg.

Werner, J. 1935. *Münzdatierte austrasische Grabfunde*. Germanische Denkmäler der Völkerwanderungszeit A 3. Berlin.

Werner, J. 1953. *Das alamannische Gräberfeld von Bülach*. Monographien zur Ur- und Frühgeschichte der Schweiz 9. Basel.

Wieczorek, A. 1987. 'Die frühmerowingischen Phasen des Gräberfeldes von Rübenach. Mit einem Vorschlag zur chronologischen Gliederung des Belegungsareales A'. *Berichte Römisch-Germanische Komm* 68, 353–492.

Zimmermann, A. 1994. 'Prokrustes-Rotation als Technik zur Synchronisierung lokaler Chronologien'. In J. Lüning and P. Stehli (eds.), *Die Bandkeramik im Merzbachtal auf der Aldenhovener Platte*, 194–205. Rheinische Ausgrabungen 36. Köln.

Zimmermann, A. 1995. 'Some aspects of the application of correspondence analysis in archaeology'. In M. Kuna and N. Venclová (eds.), *Whither Archaeology? Papers in honour of Evzen Neustupny*, 255–263. Prague.

2. On the chronology of Merovingian-period grave goods in Alamannia

Claudia Theune

HISTORY OF RESEARCH

Merovingian-period cemeteries are a unique source for many aspects of archaeological research, not least into chronological questions. Early in the present century, E. Brenner (1912) and W. Veeck (1924/26; 1931) divided the Merovingian-period finds from some cemeteries in south-western Germany into coarse 6th- and 7th-century categories. In the 1930's J. Werner (1935) produced a finer division of the grave goods into four groups on the basis of the graves containing coins. The absolute dates of these groups were not, however, derived solely from the *termini post quos* of the coins found in the graves but also by reference to historical events (Werner 1935, 23ff.). Werner was of the opinion that the material from Langobard-occupied Italy could only have crossed the Alps to northern Italy after the end of the war between the Franks and the Langobards in 591 even if the coins were clearly older pieces.

These absolute dates were corrected by K. Böhner in his study of the Frankish remains of the Trier (Trèves) region (Böhner 1958), which both was and continues to be a study of much wider relevance. While for Werner the absolute dates of the individual phases clearly lay after the *termini post quos* of the coins, for Böhner they lay within the range of the currency of the coins.

A massive increase in the number of grave finds from the 1960's onwards injected a great deal of impetus into early medieval archaeology. For the further development of research into the chronology of the Merovingian-period deposits in south-western Germany the studies of Werner and his pupils in Munich were of special significance, producing a number of results using the method of horizontal stratigraphy. Werner himself used this method in analysing the finds from the cemeteries of Bülach (1953) and Mindelheim (1955), where the chronological sequencing of belt fittings also played a special role. Werner divided the belt fittings of the male graves into a number of chronologically diagnostic types on the basis of their form and decoration and their distribution in the plan of the cemetery. The absolute datings matched those of the study of coin-dated Austrasian graves (Werner 1953, 70). The following belt-types were distinguished:

- Shield-on-tongue buckles (copper-alloy): second half of the 6th century and first half of the 7th (Werner 1953, 22ff.).
- Wide tripartite iron fittings, some with inlaid plaited ribbon patterns (Typ Bülach): second half of the 7th century, early (Werner 1953, 25ff.); second half of the 7th century (Werner 1955, 11ff.).
- Narrow tripartite faceted fittings, some decorated with Style II: late 7th century (Werner 1955, 35ff.).
- Five-sectioned fittings with degenerate Style-II ornament (Typ Bern-Solothurn), found mostly west of the Rhine in the Bodensee region and in Switzerland: second half of the 7th century, late (Werner 1953, 34f.).
- Multipartite fittings found mostly in the remainder of Alamannia: third quarter of the 7th century (Werner 1953, 13).

This classification was partly confirmed and partly modified by several of Werner's pupils, such as G. Fingerlin (1971), U. Koch (1968; 1977), and R. Christlein, who published the cemetery of Marktoberdorf (1966). The chronological schemes produced in connexion with these studies were based primarily on the grave goods from male graves of the 7th century, and female dress accessories were only marginally involved. Only Christlein (1966) made particular efforts to discover special chronological reference points for the finds from female graves. Through her work on the cemetery of Schretzheim, it proved possible for Koch (1977) to refine the familiar schemes even more and to extend them much further into the 6th century than hitherto (Fig. 2.1). She distinguished six phases (*Stufen*) in all, beginning in the middle third of the 6th century and ending in the third quarter of the 7th. While each chronological phase in Böhner's scheme covered approximately 70 years, the finds were now assigned to phases of about 30 years or the span of one generation. A similar and concurrent development can also

24 *Claudia Theune*

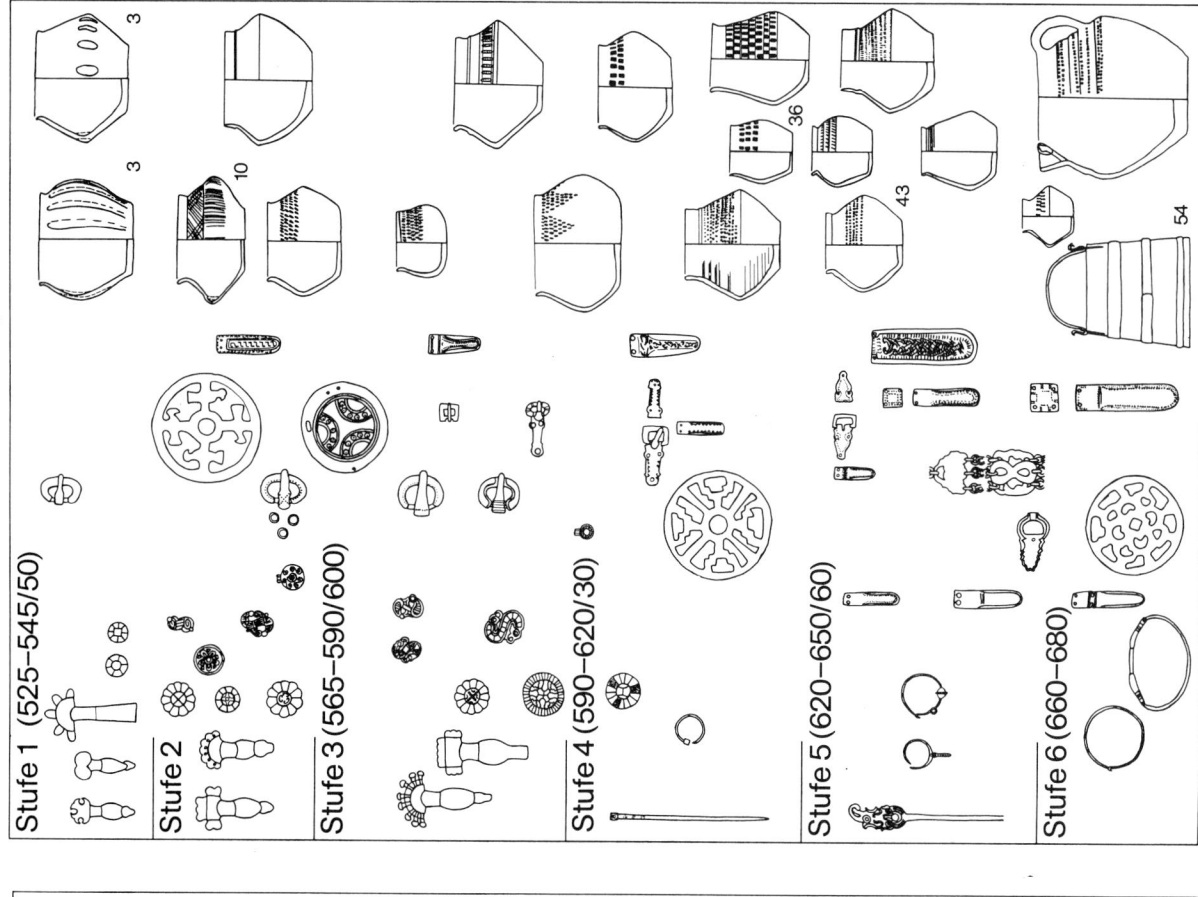

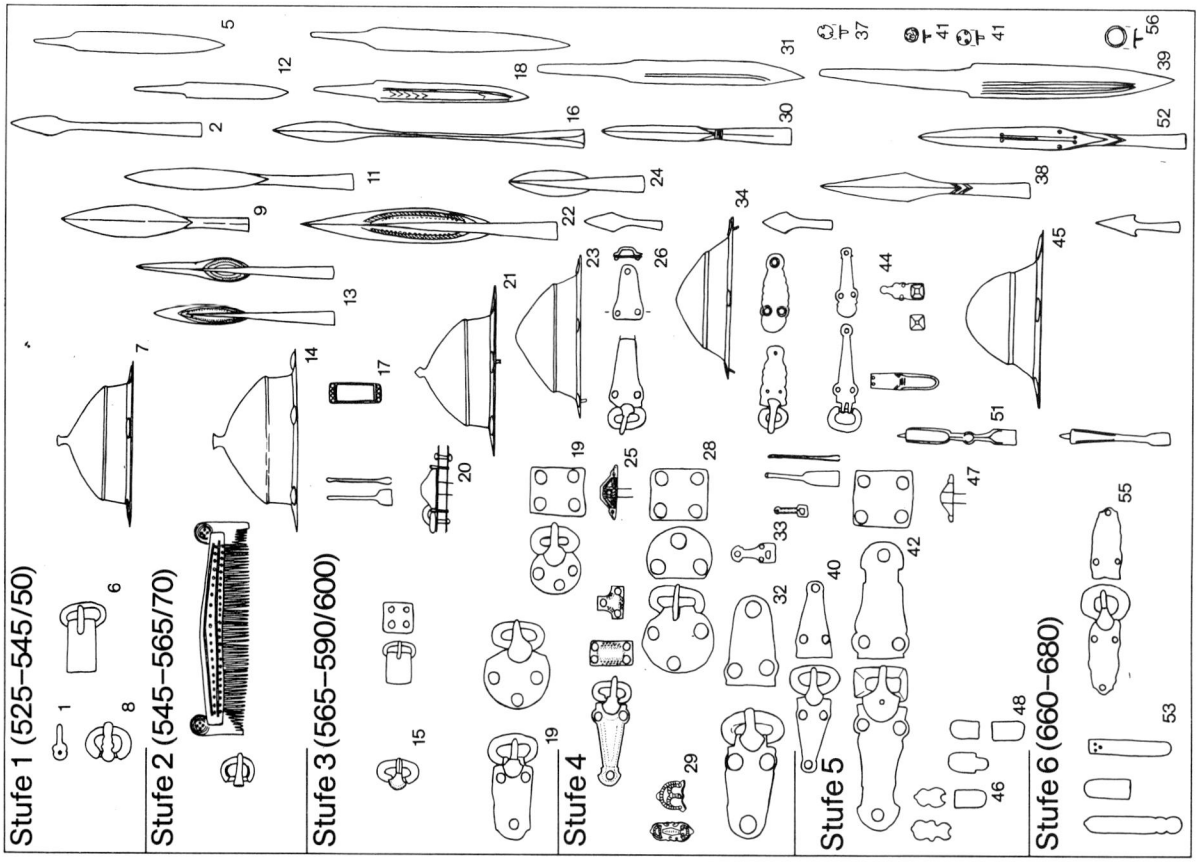

Fig. 2.1 Conspectus of the chronologically diagnostic grave goods from Schretzheim (after Koch 1977, Abb. 8).
Übersicht der chronologisch wichtigen Beigaben aus Schretzheim.

be seen with the chronology of the Rhineland, in the work, for instance, of H. Ament (Neuffer-Müller and Ament 1973; Ament 1976; 1977). Problematic aspects of the dating of graves, concerning extended periods of production, the period of use of artefacts, or the age of death, were, and were to remain, a neglected subject (Steuer 1977).

As a rule, the individual chronological schemes were worked out for particular cemeteries, without, to begin with, any claim to cross-regional significance. In some cases, however, reference was made to local similarity to other cemeteries, and chronological schemes already worked out for such sites adopted (Christlein 1966, 19; 1971, 10). In particular, Koch's chronological system for Schretzheim has been adopted by several scholars working on cemeteries in south-western Germany since the end of the 1970's. Attempts to synchronize the various separate chronological schemes are evident in the large number of ever broader chronological tables (e.g. Ament 1976, 319 Abb. 14; 1977, 135, Abb. 1–2; Ament in *RGA* 1981, s.v. Chronologie, 665 Abb. 165; Neuffer-Müller 1982, 19; Knaut 1993, 189). While the relative chronological groups, phases, strata, horizons or stages (for the terminology, see Steuer 1977, 279ff.) from individual monographs may be linked to various absolute dates – albeit sometimes with barely significant differences – in the synthetic tables these segments are joined together into an absolute-chronological system of apparently general validity. The question is, whether equivalent or similar find groups really have been put side by side in this way. The varying datings of the phase boundaries indicate the uncertainty inherent in the basis for the dating, even when the absolute figures vary by only 5, 10 or 20 years. In principle, all of the absolute datings derive from the same, repeatedly cited group of about 150 graves which either through precise dendrochronological datings (less than 10% of the total), or by means of a coin, provide us with a more or less accurate fixed point for the dating, a *terminus post quem*. The questions of where the coins were issued and their quality, and their deposition in the grave as primary, secondary or even tertiary use, are some of the problems which need to be considered in attempting to establish absolute dates on the basis of coins as grave goods. How very few of such absolute dates stand on really secure ground is well illustrated by the current debate over the early dating of the boundary between Böhner's Stufen II and III (Martin 1989).

A further problem lies the differing periods of use of the cemeteries. There are some that were used only in the second half of the 5th century and then abandoned around 500 or in the early 6th century, while at others burial begins in the course of the 6th century, in the early 7th century, or not before the middle of the 7th century. In this late period some cemeteries seem to no longer to have grave goods while at others furnished graves continue up to the end of the 7th century if not even into the 8th.

THE CEMETERY OF WEINGARTEN

The cemetery of Weingarten, Ldk. Ravensburg, Baden-Württemberg, with its 800 or so burials, was in continual use throughout the Merowingian Period and thus is particularly well suited to various forms of analysis, including chronological study (Roth and Theune 1988; 1995 (with refs.); Sasse and Theune 1996). Unfortunately no complete cemetery plan is available, so that horizontal-stratigraphic (chorological) analysis is impossible. The finds have therefore been analysed by a number of seriations, some which are already published.

The women's graves

As a first step, the dress accessories from the female graves were seriated (Roth and Theune 1988). The types included were brooches, pins, ear rings, shoe and garter fittings, decorative hangers, buckles, bracelets and pendants. From amongst the beads, only millefiore and amethyst beads were included as individual types. Each of the horizontal lines represents a single closed grave find, and the groups of various numbers of lines that cluster together because of the similarity of separate grave assemblages were interpreted as chronologically sequential, though overlapping, phases (*Modephasen*: mode phases) (Roth and Theune 1988 Tab. 1, 4 and 6; Theune 1995). The vertical columns represent the date-ranges of the individual artefact-types (Roth and Theune 1988 Tab. 2, 3 and 5). The 9 mode phases (A-I) represent successive costume-types, determined by the brooch and pin groups in particular.

Next, the 27 female graves from south-western Germany and Switzerland containing coins were added in. With this development, some of the artefact-types represented at Weingarten could be further subclassified or included within the analysis for the first time. This seriation resulted in a division into 11 (12) mode phases. The chronological interpretation of these mode phases is confirmed by the sequence of the coin dates. In order to test the cross-regional relevance of these results, the finds from six further cemeteries (Hemmingen: Müller 1976; Schretzheim: Koch 1977; Marktoberdorf: Christlein 1966; Fridingen a.d. Donau: von Schnurbein 1987 – see also Quast 1996; Güttingen: Fingerlin 1971; and Kirchheim/Ries: Neuffer-Müller 1982) were added. Still included were the graves with coins. The existing division into 12 mode phases was again confirmed, and these could now be grouped into five major steps (SW I–V), each comprising two or three mode phases. The coin datings of the various mode phases once again shows that the result of the seriation is to be accepted as a chronological sequence:

Mode phase I A   411
Mode phase I B   493
Mode phase I C1   425–527
Mode phase I C2   (425) 518–527

Mode phase II D   527
Mode phase II E   (493) 555
Mode phase III F   (493) 541–578
Mode phase III G, I (584)
Mode phase V K   (584) 692

The problems of coin dating are nevertheless clearly visible here. Reference is often made to the relatively unreliable dates of coin pendants. This is apparently reflected by the coin pendants involving issues of Theoderic, which occur in graves from phase B to phase F. However both mounted and unmounted coins of Justinian occur side by side in the seriation (Phase D Dischingen, no. 7, Straubing, no. 21; Phase F Herbrechtingen, no. 11, Klepsau, no. 13, Unterthürheim, no. 25: Roth and Theune 1988, Tab. 5.6). Also problematic are long stretches of the 7th century during which no coins appear in the graves. At this date, however, the dendrochronologically dated graves of the late 7th century and the period around 700 from Lauchheim may clarify matters in the future (Stork 1993; 1997).

The interesting question was not only whether the chronological subdivision produced would still prove to be valid but also whether the chronological schemes relating to individual cemeteries would prove to be mutually comparable. This was a matter first and foremost of the finds characteristic of the individual relative-chronological phases or stages, and less one of the absolute-chronological dates assigned to them. It transpired that the range of forms in the various phases and steps could often be correlated, and different authors and also to some degree new find groups added in, in order to define the character of a relative-chronological phase (Tab. 2.1).

The beads were examined separately (Theune-Vogt 1990; 1991; Sasse and Theune 1996). While in the past beads have usually been only summarily described and only the very long date-ranges of a few conspicuously unusual types of bead presented, in her analysis of the cemeteries of Bargen and Berghausen, Koch has, for the first time, demonstrated the importance of bead sets (1982, 59ff.). This was taken as the basis for the examination of the approximately 8,000 beads from Weingarten. Rather like the seriation of the closed grave groups, only the bead sets from closed finds were classified and seriated. Some types of bead have a very long date-range, but some bead-combination groups, which are always defined on the basis of the combination of particular types of bead, can be much more narrowly dated. The bead sets from Weingarten could be subdivided into seven bead-combination groups (Fig. 2.2 – see foldout page). This means that even poorly furnished graves can be well dated. Here again, the results from Weingarten were compared with the results of bead analyses from Schretzheim (Koch 1977), Bargen and Berghausen (Koch 1982) and Eichstetten (Sasse and Theune 1996). Although Eichstetten, where burial began later, lacks the early bead-combination group A, its successive combination groups 1–5 correlate extremely well to Weingarten groups B-G. Weingarten's combination groups D, E and F correspond to combination groups A, B and C at Bargen and Berghausen.

As several graves from Weingarten had been included in both the seriation of the dress accessories and the seriation of beads, it was reasonable to integrate the results of the bead seriation, namely bead-combination groups A-G, into the seriation of female dress accessories from Weingarten. Yet again the sequence of mode phases was confirmed. The correlation of the seriations is, at 0.992 (following Spearman), very high.

In general, we can confirm that brooches were basic factors in rapid changes of fashion and thus have only short date-ranges. They are often characteristic of no more than one or two phases, while pins, shoe fittings, garter fittings, ear rings, bead sets and/or belt buckles remained in use during several mode phases. The definition of a phase, or the assignation of a grave to a phase, is thus dependent on the one hand on the relevant leading types (*Leittypen*) and on the other on the combination of particular types. The individual phases are now described (Fig. 2.3 – see foldout page).

South-western Germany	Weingarten	Hemmingen	Schretzheim	Marktoberdorf	Fridingen	Kirchheim/Reis	Güttingen
Phase SW I A	Phase A	Childerichzeitlich					
Phase SW I B							
Phase SW I C	Phase B	Chlodwigzeitlich	Stufe 1				
Phase SW II D	Phase C						
Phase SW II E	Phase D		Stufe 2	Stage 1	Stage 1	Stage 1	
Phase SW III F	Phase E		Stufe 3	Stage 2	Stage 2	Stage 2	Phase 1
Phase SW III G			Stufe 4				Phase 2
Phase SW IV H	Phase F		Stufe 5	Stage 3	Stage 3	Stage 3	Phase 3
Phase SW IV I	Phase G		Stufe 6				
Phase SW V J	Phase H			Stage 4	Stage 4	Stage 4	(final) Phase 4
Phase SW V K	Phase I					Stage 5	

Tab. 2.1 Relative-chronological scheme for south-western Germany.
Relativchronologisches Schema für Südwestdeutschland.

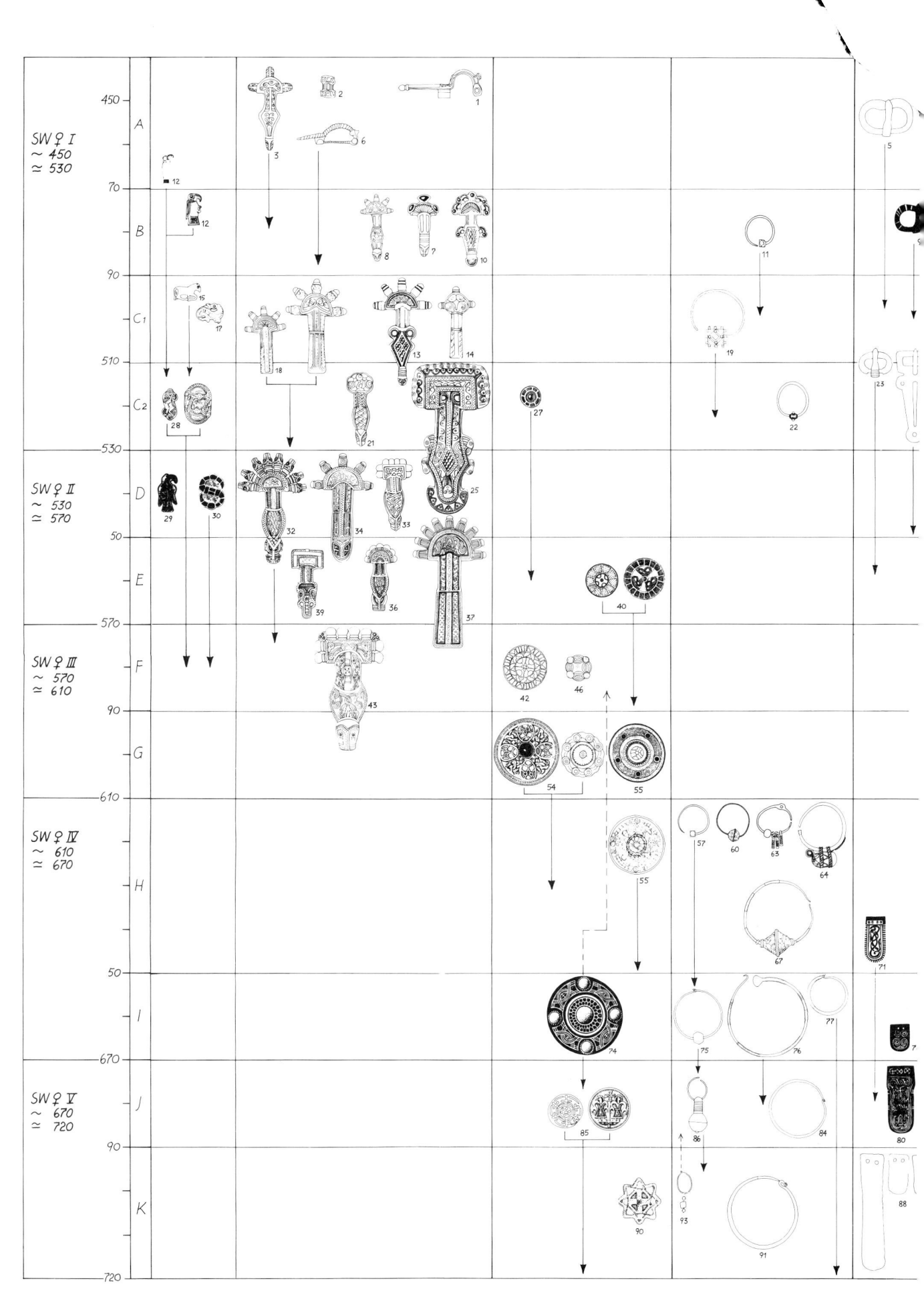

Fig. 2.3 Dress jewellery from the female graves of south-western Germany (after Roth and Theue 1988, Ta
Trachtschmuck aus den Frauengräbern Südwestdeutschlands.

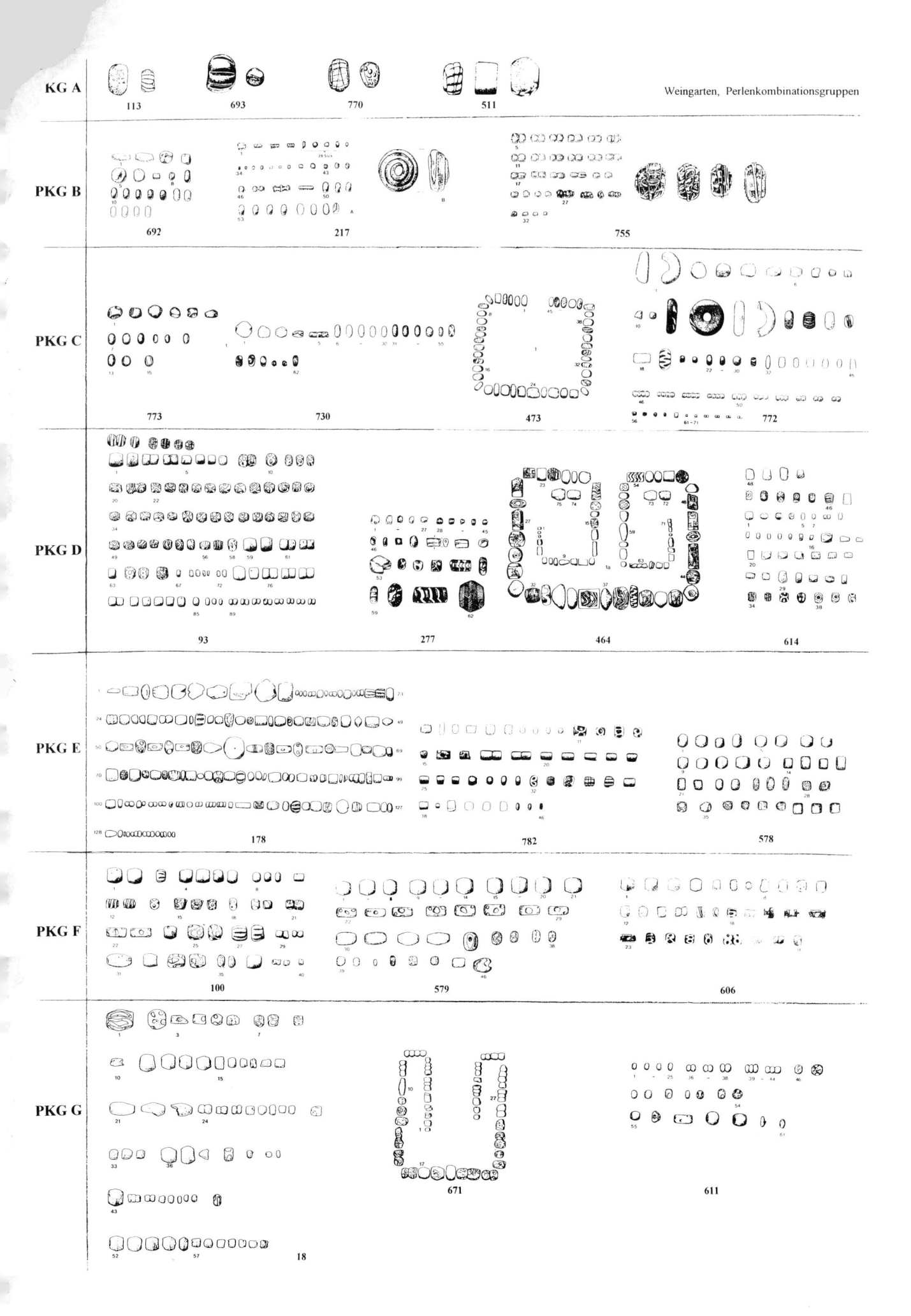

Weingarten, Perlenkombinationsgruppen

Fig. 2.2 Typical bead sets of bead-combination groups A–G from Weingarten.
Typische Perlenketten der Perlenkombinationen A–G aus Weingarten.

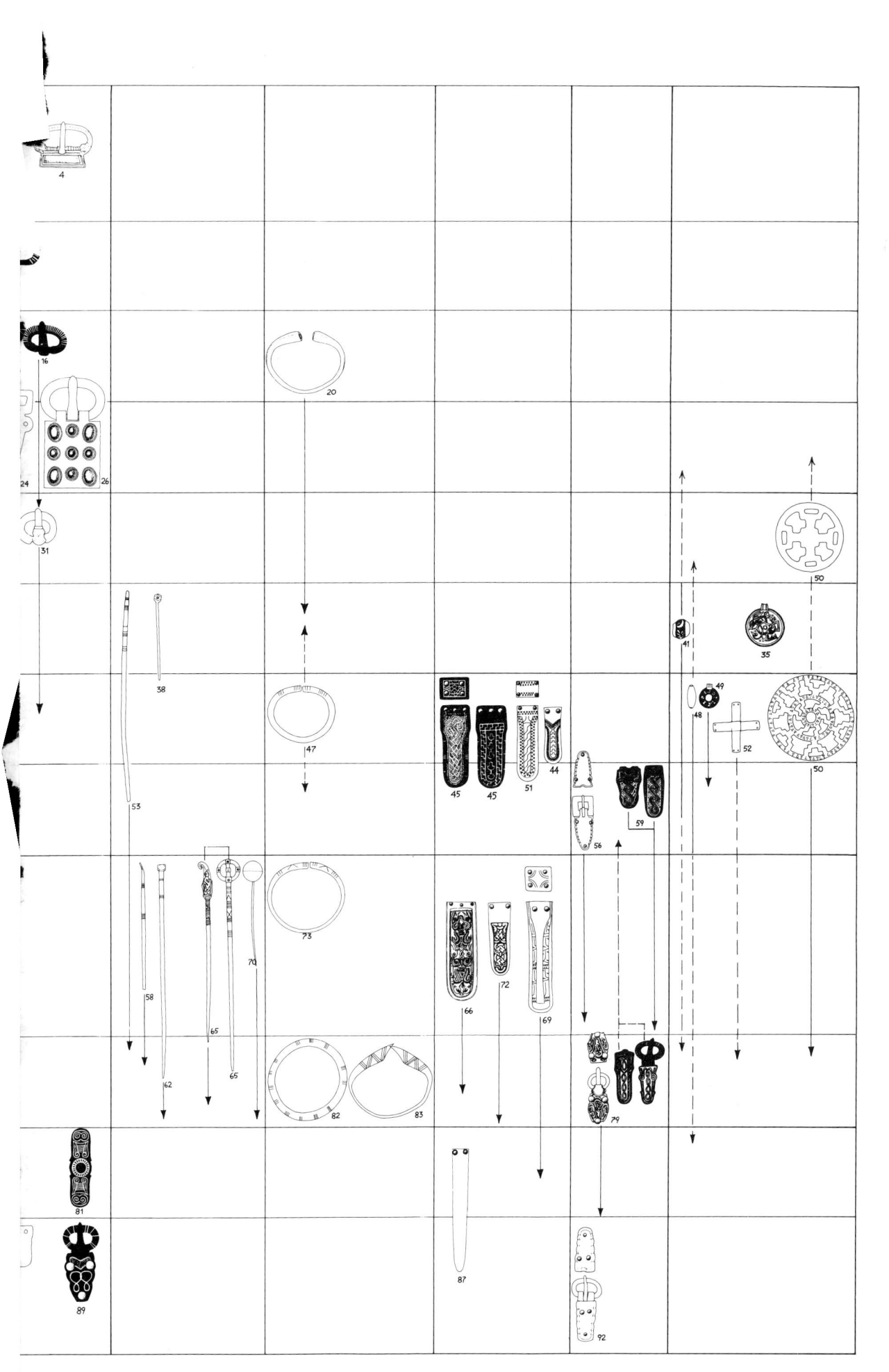

tb. 7, slightly modified).

Mode phase SW I A, B and C
Phase SW I is still influenced by Roman form traditions in its earliest phase, while later the early four-brooch costume, in combination with the earliest necklaces, is typical. Characteristic of phase A are brooches of the late-Roman tradition such as knob-on-bow brooches, two-part crossbow brooches or miniature brooches. Alongside these there are already bow brooches with semicircular headplates, three knobs and a rhomboidal foot. Late-Roman animal-head buckles and kidney-shaped iron buckles are the buckle-types found. There are no bead necklaces, only a few large beads which formed part of the women's hanger ornaments (bead-combination group A). Typical are meerschaum beads or large translucent dark blue beads with small colourful spots. In phase B there are virtually no late-Roman forms still to be found, but a number of new forms of bow brooch and the first bird brooches. These bird brooches were worn either alone or combined with early bow brooches. Some of the kidney-shaped iron buckles have inlaid strips. Beads of bead-combination group A are still frequent.

The Childeric period at Hemmingen comprises a range of finds corresponding to mode phases A and B of the south-west German female-grave chronology. In addition to late-Roman forms which can be aligned with phase A, Müller dated the early bird brooches from Hemmingen and the early bow brooches (graves 11, 14 and 59) to the period around 500 (Müller 1976, 149f.: Childeric period, or ca. 500). One should also note in respect of Hemmingen that hardly any beads were found in women's graves in the area of the neck, indicating that no bead necklaces were worn here.

Radiate-head brooches can be identified as the leading type of mode phase I C, mostly found in combination with bird brooches in the classic four-brooch costume. Somewhat later (C2) other small brooches such as S-brooches, animal brooches or small and simple garnet disc brooches appear too. The strip-inlaid belt buckles are no longer kidney-shaped but oval. Amongst the buckles, forms with club-shaped tongues, which are virtually restricted to phase C, and copper-alloy buckles with fixed triangular plates are also to be noted. An innovation is the bead necklace. Leading types here are opaque yellow and red-brown, rarely white, beads, together with gold-in-glass beads and small black miniature beads. Any other jewellery is rare, although ear rings with polyhedral terminals and inlaid stones may be mentioned.

Finds of the early 6th century at Hemmingen are the latest finds from the cemetery, e.g. graves 20 and 52, which belong to phase C1 of the south-west German female chronology. Burial in the Schretzheim cemetery began in this period. As the leading types of her Stufe 1 Koch counted radiate-head brooches, bird brooches, animal brooches, small garnet disc brooches and buckles with a narrow club-shaped tongue. The beads of Stufe 1, which also occur in Stufe 2, again include gold-in-glass beads (Koch 1977, 16). Schretzheim Stufe 1, however, extends as far as phase SW II D of south-western Germany.

Mode phases SW II D and E
Phase SW II is characterized by the later four-brooch costume with developed bow brooches and various small brooches. Shield-on-tongue buckles stand out amongst the belt fittings. Radiate-head brooches are already out of use by phase D, superseded by other types such as brooches with a rectangular headplate, oval footplate and animal-head terminal. There are also no longer any bird brooches as small brooches, although S-brooches, small and simple garnet disc brooches and eagle brooches remain (the later four-brooch costume). Typical buckles are the shield-on-tongue types. Typical bead sets are combination groups B and C, with large numbers of round, opaque monochrome forms and gold-in-glass beads. In addition, in bead-combination group C, a few polychrome beads occur for the first time, including reticella beads, millefiore beads (mainly prismatic) or beads with broad intertwined wavy bands.

In mode phase SW II E there are, in addition to developed bow brooches (see SW II D), also now developed garnet disc brooches and S-brooches. These are larger than the earlier specimens and often carry applied filigree ornament. Shield-on-tongue buckles are still common (one with a mushroom tongue). Alongside beads of combination group C the first necklaces of bead-combination group D now appear. This combination group is characterized by a large number of polychrome beads with intertwined wavy bands enclosing spots or eyes. There are also reticella, (round) millefiore and amethyst beads. Segmented yellow beads are encountered for the first time. New, if infrequent, features are other dress accessories such as small decorative discs and small pins with polyhedral heads.

Phase SW II E can be aligned with Schretzheim Stufe 2. As leading types of this phase Koch identified developed bow brooches, S-brooches and paired disc brooches and no other types of bow brooch. Amongst the beads, gold-in-glass beads are still listed along with polychrome types with broad intertwined wavy bands (Koch group 33). From Stufe 2 onwards reticella and millefiore beads are found, continuing into Stufe 3. Shield-on-tongue buckles are noted as typical belt fasteners (Koch 1977, 18f.). The finds which Christlein identified as typical of stage 1 at Marktoberdorf correspond to the types of phase SW II E. These include beads of group A (gold-in-glass beads, millefiore beads, opaque smooth spherical beads and small amber beads: Christlein 1966, 71ff.) and a variety of small brooches (Christlein 1966, 83). The same is the case for stage 1 at Fridingen a.d. Donau (von Schnurbein 1987, 84f.) and stage 1 at Kirchheim/Ries (Neuffer-Müller 1982, 102).

Mode phase III F and G
In mode phase III a clear change in costume is encountered. It is represented by a change to a single disc brooch worn

at the neck or on the breast to fasten a mantel-like garment. Bow brooches and the various small brooches are extremely rare by phase F. An exception occurs in the form of the large filigree-decorated garnet disc brooches, which significantly almost always occur singly, and disc brooches with narrow cellwork. Alongside these there are other, simplified dress accessories such as shoe buckles and garter fittings and ornamented discs in the graves. The characteristic ornament is narrow cloisonné cellwork on the brooches and mushroom-cell patterns on the hosiery. The typical bead sets are those of combination groups D and E. Similar to bead-combination group D, the sets of combination group E are characterized by a large number of polychrome beads. Especially frequent now are beads with narrow intertwined wavy bands. New monochrome forms are short cylindrical beads, sub-melon beads, and the first biconical beads.

Phase F can be aligned with Stufe 3 at Schretzheim. The narrow-cell cloisonné disc brooches noted, most of which occur singly, count as leading types, while various polychrome beads and decorated discs are also typical, along with the earliest female graves with shoe buckles and garter fittings (Koch 1977, 22f.). These artefact-types also occur in the first phase of burial at Güttingen. In addition to the definitive grave goods of grave 38, which is placed in phase F of the seriation, Fingerlin counted artefact-types and brooches of the 6th century (such as small brooches) and shield-on-tongue brooches amongst the typical finds of this burial phase (1971, 151ff.).

In mode phase G earlier brooch-types such as bow and small brooches no longer occur. Typical of phase G are gold foil disc brooches and pressed foil disc brooches together with simple, undecorated, copper-alloy shoe buckles. The other dress accessories now occur as common elements in a large number of women's graves. Mode phase G corresponds to Stufe 4 at Schretzheim and stage 2 at Marktoberdorf, Fridingen a.d. Donau and Kirchheim/Ries as well as phase 2 at Güttingen. It must, however, be noted that many types included here, such as ear rings and various forms of pin, first become numerous and regular in phase H, so that these stages 2 (Marktoberdorf, Fridingen a.d. Donau and Kirchheim/Reis), Stufe 4 (Schretzheim) and phase 2 (Güttingen) extend into phase H.

The beads with intertwined wavy bands and enclosed dots (Koch group 2) which have now been present since combination group D are dated to Stufe 4 at Schretzheim. This is also the case with other dress accessories such as ear rings, pins with polyhedral heads, and also developed decorative discs, garter mounts and shoe buckles (Koch 1977, 25ff.). Stage 2 at Kirchheim/Ries runs parallel. Neuffer-Müller lists polychrome beads, ear rings (here already with polyhedral terminals) and garter mounts (1982, 103). Amongst other things, Christlein notes as typical finds of stage 2 at Marktoberdorf the beads of group B (amethysts, polychrome forms with intertwined wavy bands and enclosed dots, and 'die' beads) (Christlein 1966, 71ff.). While the latter are first characteristic of mode phase SW IV, the other types occur already in phases F and G. In respect of Fridingen a.d. Donau, von Schnurbein lists primarily ear rings, decorative discs and simple shoe buckles as typical forms of stage 2 (1987, 84f.). For the cemetery of Güttingen, Fingerlin cites ear rings, undecorated shoe buckles and pressed foil disc brooches for phase 2: finds which are also characteristic of our phase G (Fingerlin 1971, 151ff.).

Mode phase IV H and I
In mode phase IV a further change in female dress can be demonstrated. Pins now occur much more frequently in the graves. Women also regularly wore ear rings. Amongst the typical artefact-types found in mode phase H are developed gold disc brooches and large pins with a polyhedral or relief-decorated bird's head. The shoe buckles and garter mounts – in inlaid iron, cast copper alloy or pressed foil – display a new style of decoration, animal Style II. Ear rings worn in pairs are regularly found as part of the fixed range of female jewellery, particularly small basketwork ear rings or the type with polyhedral terminals. Typical are bead sets of combination groups E and F. Alongside undecorated beads, patterns such as narrow intertwined wavy bands with dots are common. Typical monochrome beads are the biconical forms and, for the first time, almond-shaped pieces.

Mode phase H of south-western Germany corresponds to Schretzheim Stufe 5. Identified as leading types or typical finds there are paired ear rings, relief-decorated pins, smooth, prismatic (almond-shaped) green beads, and ashlar-shaped beads with yellow dots (Koch 1977, 29ff.).

The familiar pin-types also occur in mode phase I. The shoe buckles and garter mounts match those of phase H. Only cast copper-alloy shoe buckles with Style-II ornament are new. Typical ear rings are now up to 5 cm in diameter with simple hook-and-eye fastenings. The bead sets are unchanged.

Matching types are characteristic of Schretzheim Stufe 6. Here large 'wire ear rings' and large 'punch-decorated' garter mounts, such as had already been known since phase H of south-western Germany, are also typical (Koch 1977, 32ff.). The latest graves at Schretzheim extend, in relative-chronological terms, into phase SW V J.

Phase SW IV in south-western Germany could be divided into sub-phases H and I. The same division was apparent at Schretzheim (Stufen 5 and 6). This division was not applied to the other cemeteries. At Marktoberdorf the types cited for phases H and I appear in stage 3. Especially characteristic are simple hooked ear rings and beads of group C ('drum' beads and 'die' beads) (Christlein 1966, 70ff.). Ear rings with hooked fastenings and shoe buckles with Style-II decoration are also cited by von Schnurbein as typical finds of stage 3 at Fridingen a.d. Donau (1987, 84f.). Similar dress accessories occur in the graves of stage 3 at Kirchheim/Ries (Neuffer-Müller 1982, 104). The same range of finds as in phase IV H and I (gold disc brooches, pins, large garter mounts and Style II-decorated shoe buckles) are assigned to the women's graves of phase 3 at Güttingen (Fingerlin 1971, 151ff.).

Mode phase SW V J and K

Mode phase V sees a sharp reduction in grave goods placed in women's graves. The pins that had been characteristic of phases IV H and I no longer occur in phase J although there are bracteate brooches or late pressed foil disc brooches. Other new types amongst the ear rings are forms more than 5 cm in diameter, which may have a simple hooked fastening or an S-shaped catch, and drop ear rings. Bead sets of bead-combination group G are general. These sets are represented by small, mostly monochrome beads, always in large numbers. The beads are usually segmented yellow or green beads, almond-shaped beads or ashlar-shaped beads with 12 dots, or small, round, black beads with bright dots. The women's graves of phase K are poorly furnished because of the declining practice of burying grave goods. Brooches are extremely rare, as are hosiery fittings. Exceptions, however, are copper-alloy shoe buckles with a turned rim or beaded rim rivets. Large ear rings are also still current, which may have an S-shaped catch, and drop ear rings. The bead sets are assigned to bead-combination group G.

A similar division of finds was worked out in respect of the cemetery of Kirchheim/Ries. Phase SW V J is to be equated with stage 4. Typical finds are pressed foil disc brooches, costume ear rings and drop ear rings, and large numbers of orange-coloured beads, mostly biconical in form, in the bead sets. In stage 5 the pressed foil disc brooches carry figural ornament, sometimes of Christian character (Neuffer-Müller 1982, 105ff.). These ear rings with S-shaped catches were also cited by Fingerlin as characteristic types of the final phase at Güttingen (1971, 151ff.). The shoe buckles with a turned rim that occur together with such ear rings, for instance from Güttingen grave 71, could also be counted (Fingerlin 1971, Taf. 38,76). Drop ear rings and late beads of group C which are matched in bead-combination groups F and G are also considered typical of stage 4 at Marktoberdorf (Christlein 1966, 70ff.). At Fridingen a.d. Donau, large ear rings with S-shaped catches and late gold disc brooches from grave 278 are counted as the typical forms (von Schnurbein 1987, 84f.).

The men's graves

The finds from the male graves were analysed in a separate seriation. As with the women's graves, a variety of seriations were carried out: one with the male graves from Weingarten, and another with the male graves from Weingarten together with the graves with coins from south-western Germany and Switzerland. Here again the chronological validity of the seriation was confirmed by the order of the *termini post quos* of the coins. The tabulated seriation of the male graves from Weingarten can, like that of the female graves, be divided into 9 phases, and that of the coin graves into 10 (11) phases. Regular leading types in male burials are the various belt fittings. Here too, however, the combination of different finds is decisive in identifying a particular phase (Fig. 2.4).[1]

Phase I A, B and C

Matching the situation in female graves, men's graves of phase I may contain late-Roman forms (as, for instance, at Hemmingen) alongside Migration-period types. The clearly distinguishable sub-phases A and B of the female burials are barely, if at all, separable for the men. Typical of early phase A/B are buckles with inlaid garnets, some of which have a tongue with a box frame, and kidney-shaped buckles. Amongst the weapons the very long and narrow sax is to be noted. Other sporadically occurring weapon-types are franciscas (Böhner Type A) and shield bosses with a low dome and pointed apex, together with small spurs which occur alone. Male equipment also includes purse mounts with cellwork inlays, or with small buckles and markedly in-turned ends.

The same find groups with shield bosses with a low, conical dome and pointed apex, strip-inlaid buckles and franciscas of Böhner Type A belong to the Childeric-period phase at Hemmingen (Müller 1976, 142ff.). Hemmingen grave 2 also yielded a late-Roman animal-head buckle with a trapezoid mount similar to the specimen from the female grave 241 at Weingarten (Roth and Theune 1995, Taf. 76B) which is also dated to Phase A.

SW I C

In addition to the kidney-shaped types, typical buckles are oval, strip-inlaid buckles and the first specimens with a club-shaped tongue. The short sax (ca. 25 cm) has superseded the narrow long sax, and only franciscas of Böhner Type B are found. The other weapon-types and purse mounts are unchanged.

The combination of a francisca of Böhner Type B with a buckle with a club-shaped tonuge and a short sax is also dated by Müller to the post-Childeric phase at Hemmingen, i.e. to the early 6th century (Müller 1976, 142ff.). The earliest graves from Schretzheim are furnished with buckles with club-shaped tongues or shield-tongues. As weapons, there are shield bosses with a pointed apex and either long spearheads with short, angular blades or spearheads with pointed oval blades and short sockets (Koch 1977, 16), forms which also occur in Stufe 2.

Phase SW II D and E

The Migration-period range of material no longer occurs. Typical of phase II are shield-on-tongue buckles and short or narrow saxes. As well as buckles with club-shaped tongues, shield-on-tongue buckles are the characteristic belt fasteners of phase D. Other, if infrequent, forms are small rectangular copper-alloy buckles and mounts of Typ Langenenslingen. Amongst the weaponry, not only late franciscas (Böhner Type B) but also various forms of axe can be noted. The short sax remains familiar, the low

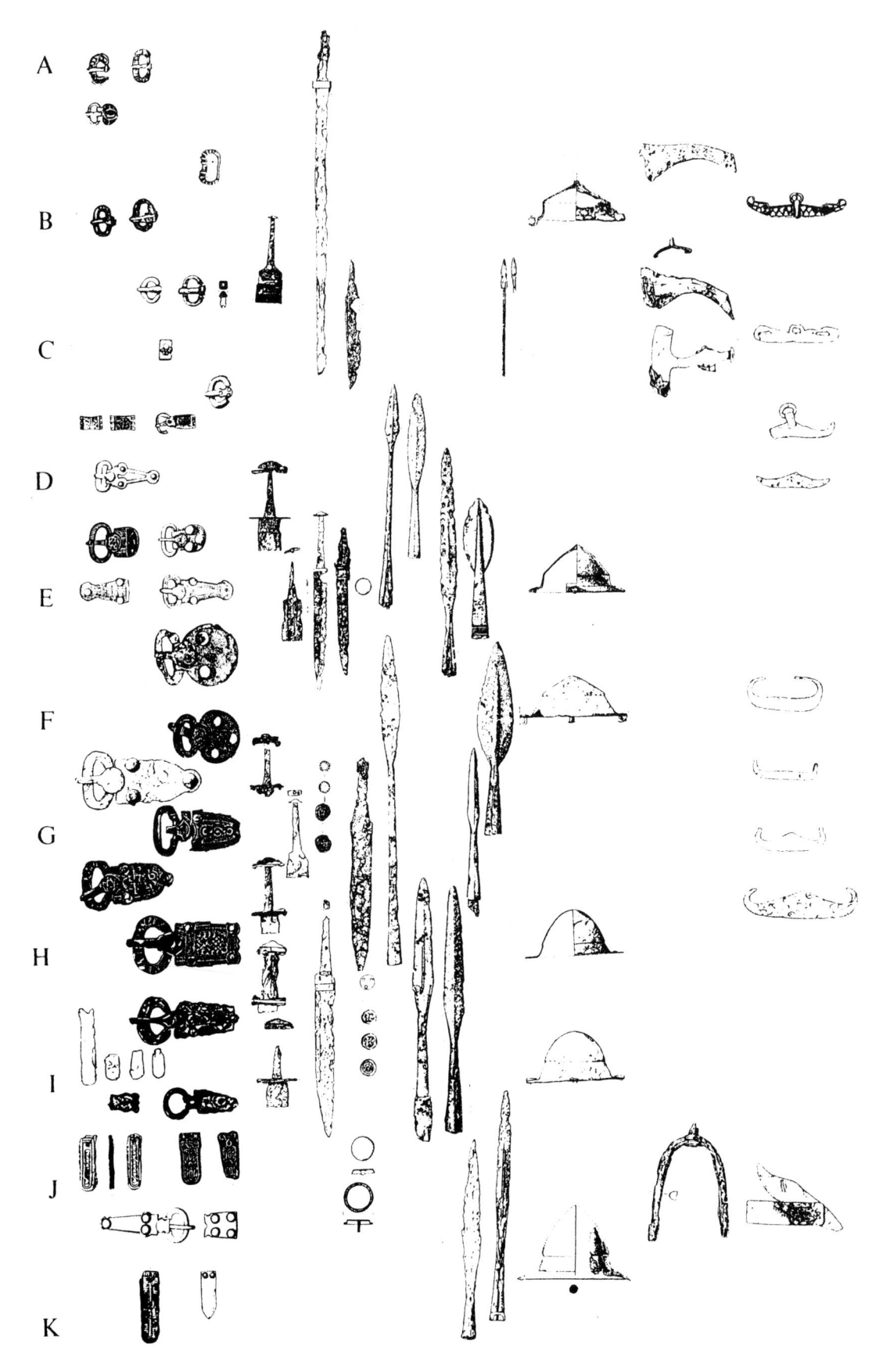

Fig. 2.4 Weaponry and belt fittings from the male graves of Weingarten.
Waffen und Gürtel aus den Männergräbern von Weingarten.

shield boss with pointed apex, and spearheads with a long, cleft socket and narrow, angular blade, or with a narrow oval blade and a short cleft socket. Firesteels are plain triangular forms with turned ends and sometimes with a small buckle. Phase D thus displays a range of artefact-types matching those of Stufe 1 at Schretzheim.

In phase E, the older buckle-types occur only occasionally, as copper-alloy buckles with round or triangular plates, sometimes punch-decorated, are now typical. Amongst the weaponry, the early shield bosses and short saxes have disappeared and their places are taken by shield bosses with a higher conical dome and pointed apex and narrow saxes with a blade 2.5 to 4 cm wide and 20 to 35 cm long and a grip up to 15 cm in length. Their sheaths are sometimes decorated with simple copper-alloy scabbard buttons. The remaining weapon-types of phase D continue in use.

These belt- and weapon-types are matched in Stufe 2 at Schretzheim. Punch-decorated buckles and shield bosses with a pointed apex, small punch-decorated spearheads and short saxes of variant B are frequent forms (Koch 1977, 18ff.). The earliest graves of stage 1 at Marktoberdorf, Fridingen a.d. Donau and Kirchheim/Ries also carry a range of finds comparable to those of phase E. Noteworthy are simple buckles of round cross-section, which include shield-on-tongue buckles, and saxes with short and narrow blades and simple scabbard studs (Christlein 1966, 19f.; Neuffer-Müller 1982, 19). In addition to these, franciscas occur at Fridingen a.d. Donau (von Schnurbein 1987, 84f.).

Phase SW III F and G
At the transition between SW II and SW III a more marked change in armament and the belt-style can be observed. The difference in artefact-types between phases E and F is very clear. Only the new narrow saxes of phase E and the shield bosses with high conical domes are still to be found, and all the other types, both in terms of weaponry and belt fittings, are new. Amongst the leading features of phase F is counted the combination of one- to three-part iron belt sets with semicircular and rectangular mounts and narrow saxes. Other newly introduced weapons are two-edged swords (*spathae*) with small pyramidical iron pommels or with curvilinear trilobed inlaid pommels decorated with mushroom motifs. The firesteels either have a high triangular form with turned ends or simply the turned ends.

Narrow saxes, extremely long spearheads and sword scabbard mounts with niello ornament are the leading types of Stufe 3 at Schretzheim. Typical – but long-lived – are one-part iron strap fittings (Koch 1977, 22), of which Koch dates the one-part types earlier than the multipart specimens. This distinction could not be made at Weingarten. Nevertheless Stufe 3 can be aligned with phase SW III E. Saxes with short blades are also counted amongst the typical finds of the first phase of burial at Güttingen by Fingerlin (1971, 151ff.).

While there are examples of iron belt fittings with semicircular and rectangular mounts in phase G they have become rarer, and examples with inlaid mushroom cells occur. At the same time belt fittings with triangular mounts become frequent. Here too inlaid patterns are encountered such as simple twisted ribbons (Typ Bülach). In addition to the narrow sax the so-called broad sax (*Breitsax*: blade-width up to 5.5 cm, length 25–57 cm, length of grip up to 15 cm) is found. The associated scabbard studs are decorated with punchmarks or have three cavities. Broad saxes with longer grips (blade-width 4–6 cm, length 25–53 cm, length of grip 15–28.5 cm) are also occasionally found.

The same belt fittings and weapons also occur in Stufe 4 at Schretzheim (Koch 1977, 25f.) and in phase 2 at Güttingen (Fingerlin 1971, 151ff.), and stage 2 at Marktoberdorf (Christlein 1966, 20), Fridingen a.d. Donau (von Schnurbein 1987, 84f.) and Kirchheim/Ries (Neuffer-Müller 1982, 102f.).

Phase IV H and I
In addition to further continuous changes in the type inventory, the frequent occurrence of Style-II decoration is particularly characteristic of this phase. The belt sets with triangular mounts already encountered in phase G still occur in the graves. The inlaid decoration, however, has changed. The ribbon interlace is more complicated, and, predictably, Style-II motifs now appear. The broad saxes with shorter and longer grips are unchanged, but alongside scabbard studs with three cavities there are now also examples with Style-II decoration. The swords have smooth triangular pommels, the shield bosses a relatively low, skull-shaped dome. The typical spearheads have an angular blade and short socket, or an angular blade with stroked and ring decoration.

The graves from Güttingen which Fingerlin regards as typical of his phase 3 contain heavy broad saxes and inlaid belt fittings of Typ Bülach (with simple ribbon twists) or Style II (Fingerlin 1971, 151ff) and are thus readily synchronized with phase H. Such forms also occur, however, in phase I. At Schretzheim, heavy broad saxes with scabbards decorated with studs with three cavities or relief ornament, sword fittings of Typ Civezzano, and early multipartite belt fittings with short strap ends, are assigned to Stufe 5.

These belt-types, and those of Typ Bern-Solothurn, do not appear in the seriation of the Weingarten men's graves before the following phase, phase I; meanwhile the tripartite belt sets with triangular mounts become rare here. Animal Style II is found in degenerate form. The broad saxes with longer and shorter grips are still frequent. The shield boss with a relatively low, skull-shaped dome is superseded by bosses with high skull-shaped domes, while at a later stage sugarloaf shield bosses appear. Spearheads have a narrow blade and chevron ornament. Shield bosses and spearheads of this kind occur in the final Stufe at Schretzheim. Typical belts of this period have multipartite belt sets with long strap ends. Schretzheim Stufe 6 thus corresponds to Weingarten phase I, although some graves from Schretzheim probably extend into phase J.

Multipartite belt sets, or Typ Bern-Solothurn, often decorated in late Style II, associated with saxes with long and wide blades, are typical forms of stage 3 at Marktoberdorf (Christlein 1966, 27ff.), Fridingen a.d. Donau (von Schnurbein 1987, 84f.) and Kirchheim/Ries (Neuffer-Müller 1982, 103).

Phase V J and K
As with the female graves, the practice of depositing grave goods comes to an end in phase V. There are now only a few weapons and belt-types. New forms of phase J are spurs and folding knives. Besides these, men wore multipartite belt sets with plated strap ends carrying threadlike Style II or honeycomb decoration. Broad saxes with shorter grips are now rare while broad saxes with longer grips still occur regularly. Their scabbard studs have a turned rim which is sometimes grooved. The spearheads often have an octagonal socket.

Folding knives and spurs, together with sugarloaf shield bosses and octagonally faceted spearheads, are regular finds in the graves which Fingerlin assigns to his phase 4, which is to be aligned with phase V J. At Marktoberdorf too these spearheads, long saxes and spurs, plus extremely long strap ends, are regular in stage 4 (Christlein 1966, 20ff. & 30ff.). To the inventory of stage 4 von Schnurbein also adds high skull-shaped or sugarloaf shield bosses, and sax scabbard studs with turned and notched rims, spurs and folding knives (1987, 84f.). At Marktoberdorf, Fridingen a.d. Donau and Güttingen these are the final phases of burial. Some graves may be datable to phase K.

The range of finds from phase K at Weingarten is, however, small. There is no longer any signicant weaponry. Belt fittings of Typ Bern-Solothurn are no longer found, likewise Typ Civezzano, only plated strap ends with honeycomb decoration or threadlike Style II and beaded rim rivets. Folding knives fill out these last grave assemblages. There is a corresponding phase of burial at Kirchheim/Ries. Typical weapons are noted as the long sax, together with spurs and their associated fittings and simple narrow straps (Neuffer-Müller 1982, 20).

A comparison of the male and female series
With just a few exceptions, the finds from the male and female graves are thoroughly different, as female jewellery does not occur with men and weapons and heavy belt fittings are not found in female graves. As a result, the chronological development of the different forms may also follow separate paths. Only a few buckle-types or pottery are shared by the two sexes. A further point of reference for a connexion between the separate chronological series lies in the decorative motifs which occur in some measure on the shoe buckles of the women and the belt fittings of the men.

A comparison based upon these practical criteria shows that the phases of the male and female chronologies do run in parallel. In phases A and B both the women and the men have kidney-shaped iron buckles, some of which have inlaid strip decoration. Both sexes first have oval, strip-inlaid buckles and buckles with a club-shaped tongue in phase C. The further development of shield-on-tongue buckles also runs in tandem in phase D. While, however, these are still present with women in phase E, new forms occur with men in this phase, such as copper-alloy belt fittings with punched decoration. In the subsequent phases F and G the same decorative motifs can be found with both women and men, such as mushroom-cell patterns and the earliest simple ribbon interlace. In phases H, I and J both sexes are associated with decoration in Germanic Style II which becomes degenerate in phases I and J or occurs in a threadlike form. Beaded rim rivets on belt fittings and turned rims on sax scabbard studs are found in male graves of phase J, while with women beaded rim rivets on strap ends and turned rims on shoe-buckle parts first appear in phase K. In this way one can demonstrate that the sequential development from phase to phase ran very largely in parallel.

DEUTSCHE ZUSAMMENFASSUNG

Die Chronologie der merowingerzeitlichen Funde aus Südwestdeutschland wurde in den letzten Jahrzehnten immer weiter verfeinert, wobei man meist für einzelne Gräberfelder jeweils eigene Chronologiesysteme entwickelte. Ziel war es nun, mit der Methode der Seriation ein überregionales Schema zu erarbeiten. Ausgehend von dem großen Gräberfeld von Weingarten, Baden-Württemberg, wurden zunächst die münzführenden Gräber aus Südwestdeutschland und der Schweiz mit in die Analyse der Männer- bzw. der Frauengräber einbezogen. Für die Frauengräber wurden dann noch in einem weiteren Schritt sechs bedeutende Gräberfelder der Region (Fridingen a.d. Donau, Güttingen, Hemmingen, Kirchheim/Ries, Marktoberdorf und Schretzheim) mit in die Seriation aufgenommen. Die Seriation erbrachte sowohl bei den Männergräbern als auch bei den Frauengräbern eine Untergliederung in fünf Hauptphasen (I – V). Diese Hauptphasen konnten jeweils noch weiter unterteilt werden (insgesamt 11 Phasen). Die Verknüpfung der jeweils getrennt erstellten Chronologien für Männer und Frauen gelang durch einige gemeinsam auftretende Merkmale wie bestimmte Gürtelschnallen und Verzierungsmuster (Tierstildekor, u.a.m.). In einem letzten Schritt wurden sowohl die Einzelchronologien als auch die übergreifende südwestdeutsche Chronlogie miteinander verglichen. In der Regel stimmen die relativchronolgischen Abfolgen überein, lediglich die absoluten Zeitansätzen variieren um einige Jahre.

NOTES

1   A comprehensive analysis of the chronology of the male graves will appear in volume II of H. Roth and C. Theune, *Das frühmittelalterliche Gräberfeld bei Weingarten II*, Forschungen und Berichte zur Vor- und Frühgeschichte in Baden-Württemberg, 44/2.

BIBLIOGRAPHY

Ament, H. 1976. 'Chronologische Untersuchungen an fränkischen Gräberfelder der jüngeren Merowingerzeit im Rheinland'. *Berichte Römisch-Germanische Komm* 57, 285–336.

Ament, H. 1977. 'Zur archäologischen Periodisierung der Merowingerzeit'. *Germania* 55, 133–40.

Ament, H. 1981. 'Merowingerzeit'. *Reallexikon der germanischen Altertumskunde*. 2nd ed. Bd. 4, 665. Berlin.

Böhner, K. 1985. *Die fränkischen Altertümer des Trierer Landes*. Germanische Denkmäler der Völkerwanderungszeit B 1. Berlin.

Brenner, E. 1912. 'Der Stand der Forschung über die Kultur der Merowingerzeit'. *Berichte Römisch-Germanische Komm* 7, 253–350.

Christlein, R. 1966. *Das alamannische Reihengräberfeld von Marktoberdorf*. Materialhefte zur Bayerischen Vorgeschichte A 21. Kallmünz.

Christlein, R. 1971. *Das alamannische Gräberfeld von Dirlewang bei Mindelheim*. Materialhefte zur Bayerischen Vorgeschichte A 25. Kallmünz.

Fingerlin, G. 1971. *Die alamannische Gräberfelder von Güttingen und Merdingen in Südbaden*. Germanische Denkmäler der Völkerwanderungszeit A 12. Berlin.

Knaut, M. 1993. *Die alamannischen Gräberfelder von Neresheim und Kösingen, Ostalbkreis, Württemberg*. Forschungen und Berichte zur Vor- und Frühgeschichte in Baden-Württemberg 48. Stuttgart.

Koch, U. 1968. *Die Grabfunde der Merowingerzeit aus dem Donauland um Regensburg*. Germanische Denkmäler der Völkerwanderungszeit A 10. Berlin.

Koch, U. 1977. *Das Reihengräberfeld bei Schretzheim*. Germanische Denkmäler der Völkerwanderungszeit A 13. Berlin.

Koch, U. 1982. *Die fränkischen Gräberfelder von Bargen und Berghausen in Nordbaden*. Forschungen und Berichte zur Vor- und Frühgeschichte in Baden-Württemberg 12. Stuttgart.

Martin, M. 1989. 'Bemerkungen zur chronologischen Gliederung der frühen Merowingerzeit'. *Germania* 67, 121–41.

Müller, H. F. 1976. *Das alamannische Gräberfeld von Hemmingen (Kr. Ludwigsburg)*. Forschungen und Berichte zur Vor- und Frühgeschichte in Baden-Württemberg 7. Stuttgart.

Neuffer-Müller, C. and Ament, H. 1973. *Das fränkische Gräberfeld von Rübenach*. Germanische Denkmäler der Völkerwanderungszeit B 7. Berlin.

Neuffer-Müller, C. 1982. *Der alamannische Adelsbestattungsplatz und die Reihengräberfriedhöfe von Kirchheim am Ries (Ostalbkreis)*. Forschungen und Berichte zur Vor- und Frühgeschichte in Baden-Württemberg 15. Stuttgart.

Quast, D. 1996. 'Bemerkungen zum merowingerzeitlichen Gräberfeld bei Fridingen an der Donau, Kreis Tuttlingen'. *Fundberichte Baden-Württemberg* 20, 803–36.

Roth, H. and Theune, C. 1988. *SW ♀ I–V: Zur Chronologie merowingerzeitlicher Frauengräber in Südwestdeutschland*. Archäologische Informationen aus Baden-Württemberg 6. Stuttgart.

Roth, H. and Theune, C. 1995. *Das frühmittelalterliche Gräberfeld bei Weingarten I*. Forschungen und Berichte zur Vor- und Frühgeschichte in Baden-Württemberg 44/1. Stuttgart.

Sasse, B. and Theune, C. 1996. 'Perlen als Leittypen der Merowingerzeit'. *Germania* 74, 187–231.

von Schnurbein, A. 1987. *Der alamannische Friedhof bei Fridingen an der Donau (Kreis Tuttlingen)*. Forschungen und Berichte zur Vor- und Frühgeschichte in Baden-Württemberg 21. Stuttgart.

Steuer, H. 1977. 'Bemerkungen zur Chronologie der Merowingerzeit'. *Studien zur Sachsenforschung* 1, 379–403.

Stork, I. 1993. 'Neue Siedlungsstrukturen und Holzbefunde in Lauchheim, Ostalbkreis'. *Archäol Ausgrabungen in Baden-Württemberg* 1993, 227–31.

Stork, I. 1997. 'Friedhof und Dorf, Herrenhof und Adelsgrab'. In K. Fuchs *et al.* (eds), *Die Alamannen*, 290–310. Stuttgart.

Theune-Vogt, C. 1990. *Chronologische Ergebnisse zur den Perlen aus dem alamannischen Gräberfeld von Weingarten, Kr. Ravensburg*. Kleine Schriften Vorgeschichtliches Seminar Marburg 33. Marburg.

Theune-Vogt, C. 1991. 'An analysis of beads found in the Merovingian cemetery of Weingarten'. In H.-H. Bock and P. Ihm (eds), *Classification, Data, Analysis and Knowledge Organisation: Models and Methods with Applications*, 352–61. Berlin.

Theune, C. 1995. 'Möglichkeit und Grenzen der Seriation'. *Ethnograph-Archäol Zeitschrift* 36, 323–41.

Veeck, W. 1924/26. 'Der Reihengräberfriedhof von Holzgerlingen'. *Fundberichte Schwaben* N.F. 3, 154–201.

Veeck, W. 1931. *Die Alamannen in Württemberg*. Germanische Denkmäler der Völkerwanderungszeit A 1. Berlin.

Werner, J. 1935. *Münzdatierte austrasische Grabfunde*. Germanische Denkmäler der Völkerwanderungszeit A 3. Berlin.

Werner, J. 1953. *Das alamannische Gräberfeld von Bülach*. Monographien Ur- und Frühgeschichte der Schweiz 9. Basel.

Werner, J. 1955. *Das alamannische Gräberfeld von Mindelheim*. Materialhelft zur Bayerischen Vorgeschichte A 6. Kallmünz.

Synopsis of discussion

One of the major topics of discussion in relation to this section was how one might understand or interpret the character of the phases within the long and detailed sequences offered for the Rhineland and Alamannic Germany. The question of whether there were any reason to worry about local variation within these two major regions was answered negatively.

It was also asked whether the phases and their date ranges should be regarded primarily as periods of production or deposition of the types assigned to them, or as some form of average of the two. It was recognized that the fact that the phasing relied on combinations of material within grave groups implied the primacy of depositional patterns, but there was little support for the idea that the sequence of the production was likely to have had a radically different pattern. The particularly sceptical position adopted by Steuer was explicitly disputed on the grounds that it is false to assume that the worst possible case is even likely to be true.

It was further noted that the short phases of these sequences did not allow for leading types (*Leittypen*) restricted to a single phase, but did allow for main types (*Haupttypen*) that are particularly characteristic of a certain phase.

The question of other dating evidence was also raised, in particular that of dendrochronology. It was noted that the conditions for wood-survival in the Rhineland are generally poor, and that there was no new evidence from this region. From further south, however, new dates from Lauchheim and other sites were referred to.

The comparability of the various sequences was also discussed. In respect of combining the separate male and female sequences, the significant role played by topo-chronology (horizontal stratigraphy) was emphasized. Between the regions, the question of which types overlapped most was also raised. The distinctly local character of pottery-types was noted. In respect of metalwork, the consistency of the type-definitions between the two schemes was queried. It was judged that these were largely equivalent, although it was also argued that it was important not to disguise local variation even within what might be a supra-regional general type.

JH

II
ENGLAND

3. The role of Continental artefact-types in sixth-century Kentish chronology

Birte Brugmann

Early Anglo-Saxon Kentish archaeology is overwhelmingly dependent upon cemetery evidence. Systematic research on late 5th- and 7th-century graves started as early as the 18th century when local scholars excavated graves which they assumed to date to the Roman or Viking Period. Searching for battlefields, the Reverend Bryan Faussett excavated barrow burials at Kingston,[1] Sibertswold, Barfreston, Adisham, Breach Down and Chartham Downs, and also in the flat cemetery at Gilton, which was being destroyed by sand extraction at the time (Faussett 1856; Fig. 3.1). Faussett was not only interested in barrows: he also used a prod to detect grave cuts in the chalk. However, it was Captain James Douglas, the excavator at Chatham Lines, who was the first to understand that he was not dealing with *Britons Romanized or Romans Britonized*, as Faussett had put it, but with Early Anglo-Saxon archaeology. The records both Douglas and Faussett made are remarkably detailed and systematic for their time, and modern research still relies on their major publications, the *Nenia Britannica* (Douglas 1793) and the *Inventorium Sepulchrale* edited and published in 1856 by Charles Roach Smith (Faussett 1856).

Although there was also interest in flat graves, barrows were easier to find and therefore were excavated in larger numbers in the 18th century. In the 19th century, chance finds of flat graves at Sarre, Bifrons, Stowting and other sites led to systematic excavations of several hundred graves. The *Collectanea Antiqua* and *Archaeologia Cantiana* became major sources of information on Charles Roach Smith´s and T. G. Godfrey-Faussett´s activities in this field of research.[2] From other sites such as Faversham, Ozengell, Howletts and Westbere, which were completely or partly destroyed during building works, rich finds have survived but little or nothing is known about their grave contexts.

Buckland, Lyminge, Bekesbourne, Finglesham, Broadstairs St. Peters, Bradstow (Broadstairs) School,[3] Monkton, Lord of the Manor[4] and Mill Hill I (Deal) were mainly rescue excavations carried out to modern standards. Publications of Lyminge (Warhurst 1955), Bekesbourne (Jenkins 1957; Härke 1992, 247f.; cat. no. 5), Finglesham (Chadwick 1958; Hawkes and Pollard 1981), and Broadstairs St. Peters (Hogarth 1973) are incomplete, and the excavations at Bradstow School and Lord of the Manor (Perkins, n.d.) await detailed publication. Vera Evison's report on and interpretation of the 1950's excavations at Buckland published in 1987 has become a major reference point for the archaeology of Anglo-Saxon Kent (Evison 1987).

The majority of excavated graves – almost all the barrows and the larger part of the flat graves – can be dated to the 7th century. The main sites relevant to late 5th- and 6th-century research are Bifrons, Sarre, Lyminge, Bekesbourne, Finglesham, Mill Hill I and Buckland. When Mill Hill I was excavated in 1986–89 the site produced the largest number of 6th-century graves excavated to modern standards in Kent at that time. The 79 graves at Mill Hill I were then joined by another 244 graves discovered at Buckland in 1994 (Parfitt 1995), most of them dated to the 6th century. A report on Mill Hill I is published (Parfitt and Brugmann 1997) and a second report on Buckland is in preparation.

From the very beginning, the Kentish material has been interpreted in a larger geographical context. As soon as it was understood that the cemetery evidence dated after the Romans and before the Vikings, it was used to fill gaps in the written sources on the 5th and 6th centuries. Though it seemed to be relatively easy to recognize the regions settled in A.D. 449 by the Angles and Saxons according to Bede's Ecclesiastical History, the Jutish invasion posed some problems. The Continental influence on the Kentish material culture appeared much stronger than those elements considered to be Jutish. Edward Thurlow Leeds (1913; 1936, 43ff.) identified the River Medway as a dividing line between Jutish East Kent and Saxon West Kent and developed a chronological system of three phases, a Jutish Phase (*circa* A.D. 450–500), a Frankish Phase (early to late 6th century) and a Kentish Phase (late 6th century onwards). For a long time, Bede's invasion date of A.D. 449 was accepted as a *terminus post quem* for any archaeological evidence on Anglo-Saxon cemeteries in England.

Leeds (1913, 121ff.) tried to match Bede's report of a Jutish invasion in Kent with the fact that he found the best parallels for some of his Kentish material in the Rhineland. At first he believed that 'there is nothing inconceivable in a band of Ripuarian Franks moving down the Old Rhine and after joining themselves with a Jutish contingent, descending the shores of Kent'(ibid., 137) but he later changed his mind (Leeds 1936).

Myres (1971, 27f.) described the discussion on the roots and dating of 5th- to 7th-century Kentish material culture as follows: 'The dating of the rich culture of Kent was further complicated by the recognition that some elements in it echoed Late Antique fashions of the early 5th century and might even be interpreted as a final flowering of Romano-British art. In all this concentration of learned interest and of art-historical expertise upon the more spectacular elements in Kentish jewellery it became for a while almost a heresy to think that the Jutes, who were supposed to have used it, ever had anything to do with Jutland at all. [...] there has been a growing reaction against the former over-emphasis on these splendid jewels as providing clues to the origin of Kentish culture. It has been realised that their characteristic styles owe much to secondary Frankish influences which, whether they are interpreted in terms of trade or settlement, belong rather to the 6th and 7th centuries than to the original Germanic invaders in the 5th'.

There is some archaeological material to confirm Bede's statement that the Jutes settled in Kent but it is not overwhelming. Pottery, cruciform brooches, bracteates and Jutish-Kentish square-headed brooches are the main types of objects which link Kent to Jutland but there is little evidence for settlement activity.[5] During the course of the 6th century, Jutish influence on material culture in Kent was confined to bracteates and square-headed brooches with remnants of Jutish animal-style artwork, which are considered to have been high-status objects treasured over generations (Hawkes and Pollard 1981).

The 6th-century Continental influence on Kent is strong but selective. Glass and copper-alloy vessels were deposited in both men's and women's graves. Women also had brooch-types originating from almost every part of the Frankish domain, and Continental types of buckles. Polychrome glass beads, but not ear-rings, hairpins or metal fittings from gaiters, were a regular part of Kentish women's dress. The choice of objects suggests that Kentish dress was based on Anglo-Saxon fashion which had use for brooches, buckles and beads but not for ear-rings, hairpins and gaiters.

The use of Continental objects tended to be unorthodox. Brooches were worn in unusual numbers and combinations. Decorative belt sets, mostly buckles with a club-shaped tongue sometimes combined with kidney-shaped belt plates (*Kolbendornschnallen mit nierenförmigem Beschlag*) or shield-on-tongue buckles (*Schilddornschnallen*) often combined with one to three shoe-shaped rivets (*schildförmige Hafteln*) were used by women more often than men, though on the Continent they were for men.

The approximately 100 Continental brooches excavated in Kent include eleven pairs cast from the same mould, one group of three brooches cast from the same mould and 78 single brooches (Brugmann 1994, appendix 2). Fifty-one of these brooches can be assigned to a total of 37 graves at twelve sites and were recorded with associated grave goods.

The brooch-types found in Kent occur mainly in northern France, the central and lower Rhine valley and South-western Germany, suggesting strong contacts with the Frankish domain especially in the first half of the 6th century.

There are some indications that most or all of the Continental brooches were exported to Kent but only metal analysis could give certain proof of this. Imitations can give themselves away on stylistical grounds, such as the copy of a Hahnheim-type brooch found at Great Chesterford (Evison 1994, Fig. 28, pl. 10 f). It has a spatulate foot terminal which is never found on Continental type-Hahnheim brooches but frequently occurs on regional brooch-types. The type-Hahnheim brooches from grave 86 at Mill Hill I, in contrast, seem to be cast from the same mould as a pair of brooches from Biblis II Wattenheim (Kr. Bergstraße; Möller 1987, Taf. 121, 1, 2), a pair from Mainz-Finthen I (pers. comm., Christoph Engels) and another brooch in the British Museum labelled 'Continent' (Kühn 1965, 460 Taf. 46:166). Kentish jewellers used the same techniques as their Continental colleagues but on the whole achieved a better quality than the Continental brooches found in Kent. From a technical point of view, Kentish jewellers should have found it easy to reproduce Continental types of object, but we do not know whether they were required to do so.

About two-thirds of the Continental types are bow brooches (*Bügelfibeln*) and one-third small brooches (*Kleinfibeln*). They were in use at the same time as the 70 or so Kentish bow brooches (Leigh 1980) and at least 14 of the 75 Kentish disc brooches of Avent's Classes 1 and 2 (Avent 1975) also found in Kentish graves. Grave groups involving Kentish and Continental brooch-types, bracteates, buckles with club-shaped tongue or shield-on-tongue buckles and datable polychrome glass beads (Tab. 3.1, phase II–IV) were followed by a Kentish fashion identified by a single Kentish disc brooch of Avent's Classes 3ff. or composite disc brooches, pendants other than D-bracteates and scutiform pendants, and amethyst beads (Tab. 3.1, phase V).

In comparison with Merovingian studies, there have been very few attempts to develop a systematic chronological framework for Kentish graves. There are three main reasons for this. Firstly, the evidence comes mostly from old excavations with inadequate records. Secondly, research requires a detailed understanding of the Continental chronological systems, which are themselves already complicated (see below). Thirdly, studies need to take account of the chronologies of the neighbouring Anglian and Saxon regions, but these are not yet sufficiently developed to be of much assistance (see Hines, this volume).

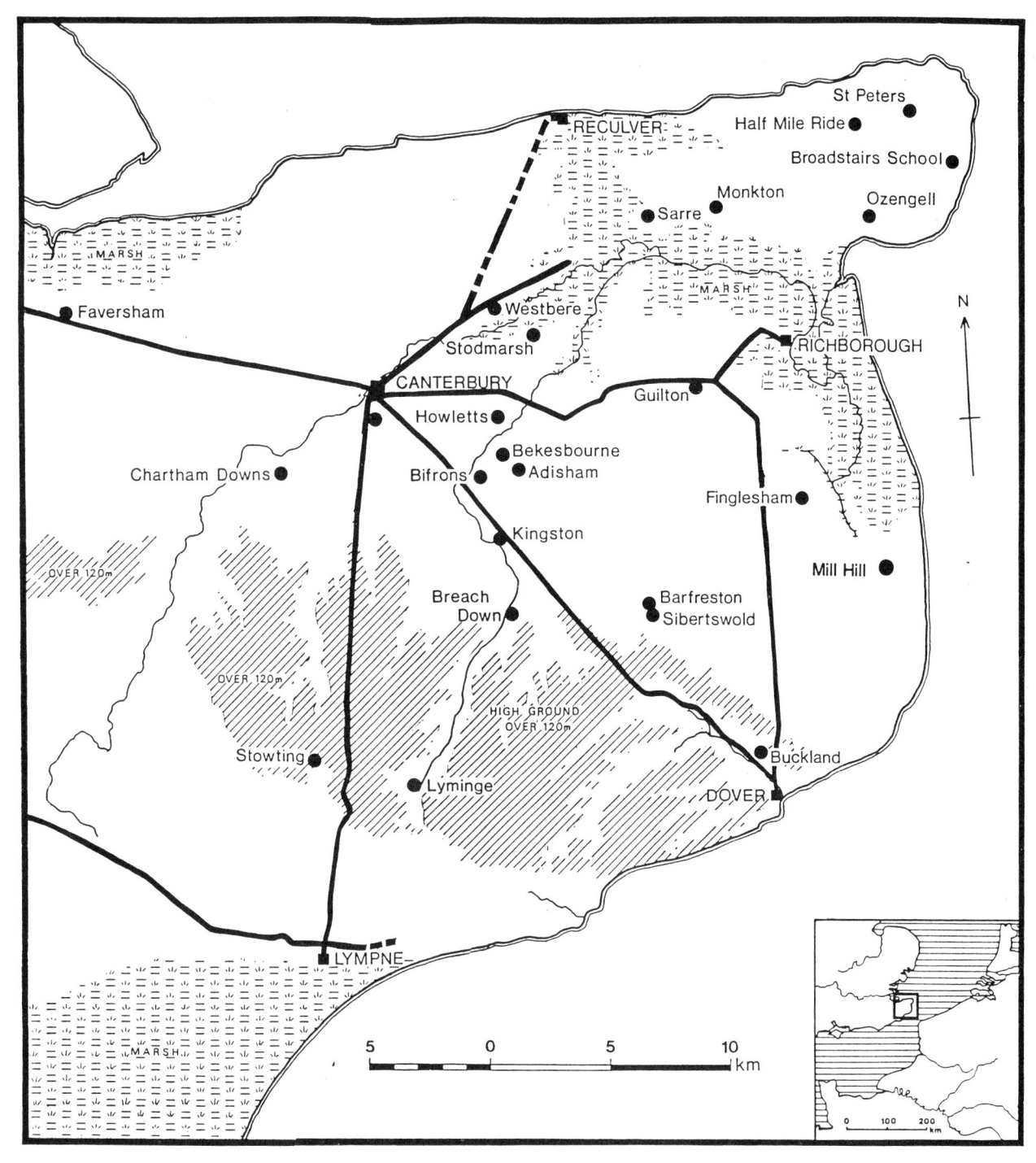

Fig. 3.1 Major cemetery excavations in East Kent (Parfitt and Brugmann 1997 Fig. 1).
Größere Gräberfelduntersuchungen im östlichen Kent.

An additional problem is that Kentish dating has traditionally been dealt with in absolute terms. Since the basis for this dating has often not been made sufficiently clear, or has lost its relevance in the course of further research, the chronology has become a maze that can only be understood if its many strands are meticulously followed back to their various origins.

In 1981 Bakka (1981, 16ff.) divided Leeds's 'Frankish phase' into two: '1, the beginning of the square-headed brooch and Salin's Style I in Kent (Bakka 1958), and 2, the beginning of the Kentish keystone-garnet disc brooch'. He also used the term 'Kentish stage II' for the period starting with the square-headed brooch, and Kentish stage III for that which starts with the Kentish keystone-garnet disc brooch. He then related these two 'stages' to Böhner's Continental chronological framework (Böhner 1958) which had been further developed by Ament (1977). The subdivision of Böhner's Stufe III into early (Stufe IIIa) and late (Stufe IIIb) corresponds to Ament's AM II and AM III. Bakka (1981, 21) concluded that 'Kentish stage III, starting with the beginning of the keystone-garnet disc brooches and finishing with the end of Kentish square-headed brooches, can be firmly dated as contemporary with the Frankish Stufe IIIa, probably starting no later than the beginning of IIIa (about A.D. 525), and probably not outliving it [A.D. 560]' (Tab. 3.3).

In her publication of the Buckland cemetery Evison (1987) defined seven phases covering the period A.D. 475–750 and gave an overview of datable grave goods from the site (Tab. 3.3; Fig. 3.2).

The definition of three 6th-century Kentish phases in the Mill Hill I cemetery evidence (Tab. 3.1; Fig. 3.4; Parfitt and Brugmann 1997) is largely based on the following study.

The majority of the datable finds from Kentish cemeteries come from women's graves. Dates for men's graves are much more difficult to determine because weapons are not so closely datable as jewellery. Continental types of object in these graves are generally limited to buckles and vessels. Vessels were often used over a long period of time and are therefore of limited use in chronological studies. Shield-on-tongue buckles and reticella (sword) beads are more closely datable but were worn by fewer men than women.[6] Therefore only a framework for women's graves is discussed in the following.

Ideally, any chronological framework should be built up independently from those developed for other regions, but this is not feasible with the evidence currently available from 6th-century Kent because there are too few graves available for study. However the recent excavations at Buckland may double the number of these graves and thus profoundly change the situation.

Continental datings which must be involved with any discussion of Kentish material can be difficult to deal with because various frameworks are in use.[7]

Further difficulties are posed by the fact that Continental publications do not always relate absolute to relative dates, nor distinguish explicitly between the time of manufacture and the date of burial. The general tendency over the last few decades has been to place absolute dates earlier than was previously assumed, necessitating constant updating (cf. Martin 1989). Both the relative and absolute dates given to Kentish material are inevitably tied into this difficult Continental background.

In spite of all this, research on Continental chronology has reached a level of precision which can be of great help in the study of the Kentish material. This does not, however, mean that Continental datings for certain types of objects also found in Kentish graves can automatically be used to date the latter with the same precision. The 'heirloom factor' which seems to play a significant role in the use and deposition of Jutish types of objects in Kentish graves (Hawkes and Pollard 1981) may also apply to Continental types of object. Therefore, before Continental datings are used to bolster a Kentish frame-

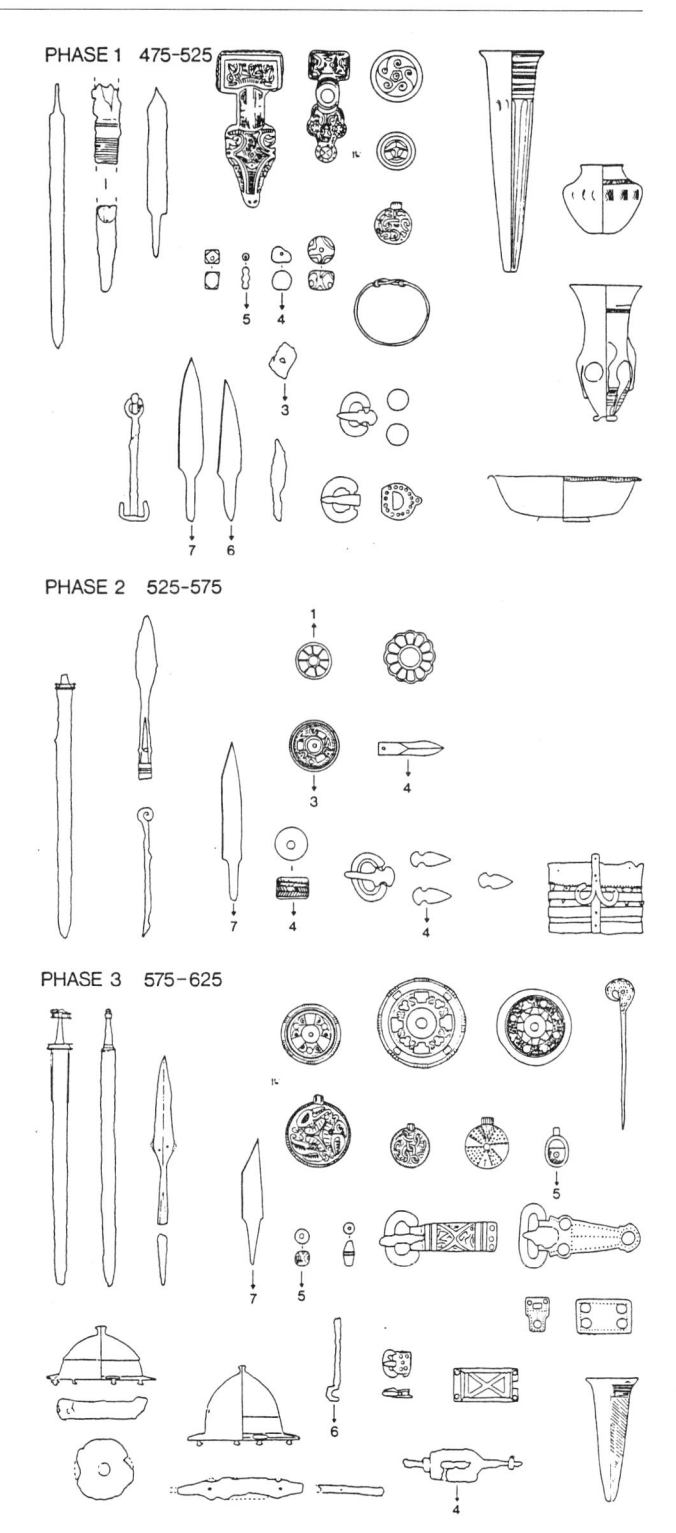

Fig. 3.2 Phases 1–3 at Buckland. After Evison (1987, text figs. 25–6). Objects not to scale.
Phasen 1–3 in Buckland.

work, it is necessary to check whether this 'heirloom factor' is likely to distort the evidence significantly. For this purpose, the dating of relatively closely datable Continental objects found associated in Kentish burials has been assessed.

Bifrons 29 (Fig. 3.16. All graves mentioned in the text are catalogued, with references, in Appendix 1):

Pair of radiate-head brooches (1–2): dated to the second quarter of the 6th century (Hawkes and Pollard 1981, 346).

Large shield-on-tongue buckle with flattish square loop with zigzag decoration and probably an iron loop to its massive tongue (6): shield-on-tongue buckles with square loops do not seem to be found among the early types of shield-on-tongue buckles already in use in Böhner's phase II. Square loops of the second quarter of the 6th century often have squarish or trapezoid cross-sections, e.g. buckle (l) in grave 81 at Mill Hill (Parfitt and Brugmann 1997). The buckle loops in grave 29 at Bifrons and grave 30 at Buckland (see below) are faceted and of proportions similar to the shield-on-tongue buckle in graves D3 at Finglesham dated to the second quarter of the 6th century (see below).

Bifrons 41 (Fig. 3.8):
Bird brooch (3): In south-western Germany bird brooches were most common in SD Phases 3 and 4 and went out of fashion during SD Phase 5 (A.D. 530–50) when S-shaped brooches became more fashionable (Koch, in press).[8] This general date seems also to apply to northern France, where two brooches of the same type were found (Thiery 1939, No. 475; 477).

Shield-on-tongue buckle and shoe-shaped rivet (5): the buckle is small and has an usually thin loop and tip of a tongue with a fully developed shield with a copper-alloy loop. Stylistically it is an early type which may have been in use in Böhner's phase II.

Bifrons 42 (Fig. 3.9):
Disc brooches (2–3): disc brooch (3) with a central inlay and wedge-shaped garnets which had lost its pin and with holes drilled into its edge, probably to be sewn on to a garment. The twelve garnets are cut unusually narrow to fit on to the brooch 2.2 cm in diameter. Brooch (2) has a circular central cell and four partitions intercut by half-circles and is similar in size and decoration to a pair of disc brooches found in Grave 46 at Villey-Saint-Etienne and dated to the second quarter of the 6th century (Wieczorek, Périn, Welck and Menghin 1996, 887ff.). Small disc brooches exclusively with inlays (as opposed to filigree decoration) with simple patterns including no more than one concentric partition are generally dated to Böhner's phase II and early phase III (cf. Koch 1977, 58ff.; 1990, 140f.).

Shield-on-tongue buckle (4): with a width of 2.4 cm the buckle is extremely small for a shield-on-tongue buckle and has a roundish loop section. It is therefore likely to be of an early date, possibly in Böhner's phase II.

Bifrons 74:
Massive faceted ?tinned shield-on-tongue buckle with two shoe-shaped rivets (2): shield-on-tongue buckles with oval loops with a roundish section and shoe-shaped rivets were introduced late in Böhner's phase II (Böhner 1958, 179). The main occurrence of shield-on-tongue buckles is dated by Martin (1989, 132ff.; Fig. 3.9) to the middle third of the 6th century. In Koch's chronological sequence developed for southwestern Germany (SD Phases 1–8; Koch, in press), shield-on-tongue buckles are dated mainly to Phase 5 (A.D. 530–550). Close parallels to the Bifrons buckle from Basel-Bernerring (Martin 1989, Fig. 3.9) with faceted buckle loops and tongues with iron loops were in use in the second quarter of the 6th century (ibid.; 1976, 61f.). The copper-alloy shoe-shaped rivets from grave 74 were probably made for an earlier type of shield-on-tongue buckle such as found in grave 89 at Mill Hill (Parfitt and Brugmann 1997) but the same combination of rivets and buckles with a whitemetal surface was found, for example, in grave D3 at Finglesham (Fig. 3.15, 4) and Grave 94 at Mill Hill (Fig. 3.32, c).

Reticella bead (3): reticella beads occur from SD Phase 5 (A.D. 530–550; Koch, in press) onwards.

Buckland 20 (Fig. 3.11):
Garnet disc brooch with inlaid rim (2): small garnet disc brooches with a circular cell in the centre, surrounded by between four and ten keystone garnets, are generally dated to the late 5th and first half of the 6th century (see, e.g., Martin 1976, 81ff.; Koch 1977, 58 and 1968, 170; Wieczorek 1987, 430f.).

Shield-on-tongue buckle with two rivets (9): is a small type comparable to the buckles from graves 41 and 42 at Bifrons (Figs. 3.8–3.9), with a copper-alloy loop to the flat tongue and disc-shaped rivets. The buckle is dated to Buckland phase 1 (A.D. 475–525) by Evison (1987, text Fig. 25).

Buckland 30 (Fig. 3.29):
Buckle (6): square loop with zigzag decoration and iron tongue, probably replacing a shield-on-tongue such as found in Grave 29 at Bifrons (Fig. 3.16, 6; see above Evison 1987, 98). The buckle was probably a 'heirloom'.

Biconical 'striped' mosaic beads (s–t): a common type of mosaic bead at Schretzheim dated from phase 2 onwards (M65–67; Koch 1977, Colour Pl. 6). At Marktoberdorf biconical striped mosaic beads are dated to *Schicht 1* (*circa* A.D. 550–570/90; Christlein 1966, 72); at Weingarten they were part of a group dated *circa* A.D. 570–610 (type 73; Theune-Vogt 1990, 49).

Seven rust-red short cylindrical/disc-shaped beads with yellow crossing trails and dots (two double) and two rust-red barrel-shaped double beads with white cros-

sing trails and dots (p–r): at Schretzheim this type occurs from phase 3 onwards (Koch 1977, Colour Pl. 2, 20:1–7, especially 20:1 and 4).

Buckland 92 (Fig. 3.18):
Garnet-set disc brooch with inlaid rim (2): see brooch (2) from grave 20 at Buckland (above).
Reticella bead (d): see comment on bead in grave 74 at Bifrons (above).

Finglesham D3 (Fig. 3.15):
Pair of radiate-headed brooches (2) from grave D3: manufactured around A.D. 500 (Hawkes and Pollard 1981, 331f.). Grave 94 at Basel-Kleinhüningen containing the closest parallel is dated to SD Phase 4 (A.D. 510–530; Koch, in press).
Belt set (4): the buckle is described as 'thickly silvered' and probably 'being made of iron heavily plated with silver' (Chadwick 1958, 17). It is not quite as massive as the iron silver-coated buckle from grave 39 at Bifrons but is unlikely to pre-date Böhner's phase III.

Finglesham 203 (Fig. 3.20):
Belt set (16–19): late-5th/early-6th-century type (Hawkes and Pollard 1981, 337).
Square-headed brooch (22): there is no exact parallel known so far, but its shape and scroll decoration suggests that it is related to Series IIB of Nordic type brooches defined by Nissen-Fett (1941) and dated to Schretzheim Phase 3 (A.D. 565–590/600) by Koch (1977, 56f.).
Rosette brooch (23): a type first manufactured in the mid-6th century (Hawkes and Pollard 1981, 338). In Koch's chronology for south-western Germany rosette brooches with filligree decoration are dated mainly to Phase 6 (A.D. 550–580).
Polyhedral decorative gold bead from an ear-ring (38): re-used on a necklace and of an earlier date than the brooches (ibid., 337f.).

Howletts A (Fig. 3.21):
Massive copper-alloy shield-on-tongue buckle with three shoe-shaped rivets (3): comparable to buckles from graves dated by Martin (1989, 133f.; Fig. 3.10, 4,6) to the second quarter and middle third of the 6th century (see discussion on shield-on-tongue buckle in grave 74 at Bifrons [above]).
Necklace with cylindrical millefiori bead (4): cylindrical millefiori beads are dated mainly to Schretzheim Phase 3 but occur as early as Phase 2 (A.D. 545–565/70; Koch 1990, 123).

Mersham A (Fig. 3.22):
Square brooch (e): a common type which was generally dated to Böhner's phase III (e.g. Böhner 1958, 98 Pl. 13, 12; Neuffer-Müller and Ament 1973, 73 Pl. 2, 12, 13) but more recent work suggests that they were already in production during phase II (see Bierbrauer 1985, 16 Fig. 7, a). In Koch's chronology for south-western Germany the type is mostly found in SD Phase 4 but was still worn in SD Phase 5 (A.D. 510–550).
Decorated club-shaped buckle (c): dated by Koch (pers. comm.) to the second and third quarters of the 6th century.

Mill Hill I 25 (double grave; Fig. 3.24):
Bird brooch (e) with Skeleton A: Vorges type dated to SD Phase 5 (A.D. 530–550; Koch, in press).
Radiate-head brooch with heart-shaped ornament on the headplate (e) with Skeleton B: the Mill Hill brooch has its closest parallel in a pair from grave 5 at St. Suzanne/Champ-la-Cave in France (Raveaux 1992, 107ff.; l. III; 77), associated with beads and buckle-types datable to SD Phase 6 (A.D. 550–580).
Buckle (l) and plate (h) with Skeleton B: the massive ?tinned shield-on-tongue buckle was associated with a triangular-shaped plate, if only a copper-alloy sheet set with rivets and decorated with knobs punched into the metal from back. The plate dates the belt set to the third quarter of the 6th century.
S-brooch (m) with Skeleton B: a close parallel, also with zigzag decoration, found at Sontheim is dated to SD Phase 6 (A.D. 550–80; Koch, in press)

Mill Hill I 61 (Fig. 3.12):
Bird brooches (c) and (d): not cast from the same mould but both belong to the Aubing type (Werner 1961, 43 list 7, map 7). The grave-goods associated with an Aubing-type brooch in grave 643 at Krefeld-Gellep, Stadt Krefeld, have been dated by Pirling (1966, 178 Fig. 20, 6) to Böhner's phase II. In Koch's chronology for south-western Germany, bird brooches are common types in phases 3–5 (A.D. 480–550; Koch, in press).
Buckle (h): an early 6th-century buckle with a club-shaped tongue worn with a heart-shaped plate, inlaid with ?bone and garnet or glass. The plate obviously is a variant of the kidney-shaped plate typical of Böhner's phase II. A heart-shaped Continental example is known from Harmignies and dated A.D. 475–500 (Nimy, Prov. Hainaut; Vanhaeke 1996, 845f.; Fig. 3.10).

Mill Hill I 64 (Fig. 3.31):
Belt set (d): an unusual type of shield-on-tongue buckle because of the small step around the edge of its loop leaving a raised platform; both loop and tongue are relatively flat and the tongue was originally fastened with an iron loop. The set of three shoe-shaped rivets give a date earliest in the second quarter of the 6th century.
Reticella bead (f8): see comment on bead in Bifrons grave 74 (above).

Mill Hill I 73 (Fig. 3.14):

Animal brooch (f): ungilded low-tin bronze, Herpes-type. With one exception all the Continental finds of this type are made of gilt silver and are decorated with chip-carved lines. The Mill Hill brooch is the only one with ring-and-dot decoration on its body and without a garnet-set eye. A brooch found at Altenerding, Lkr. Erding was dated to the first quarter of the 6th century (Bierbrauer 1985, Fig. 7, a). A slightly different design from grave 472 at Schretzheim was dated to phase 1 (A.D. 525–545/50) by Koch (1977, 452) and a brooch from Weingarten, Kr. Ravensburg, to phase C1 (A.D. 490–510) by Roth and Theune (1988, 30 no. 17). The type is generally dated to SD Phase 4 (A.D. 510–30; Koch, in press).

Bow brooch (e): cast imitation of a cross-bow brooch, the head-plate imitating the spring and bow of such a brooch. It seems likely that brooch (e) was copied from cross-bow brooches of the Ozengell type dated to the last quarter of the 5th century and the first quarter of the 6th (Schulze-Dörrlamm 1986, 619 f.; 714 list 6, 1). The closest parallel amongst the true cross-bow brooches was found at Altenerding and was buried in the first quarter of the 6th century (Bierbrauer 1985, 15 Fig. 11). As an imitation, the Mill Hill brooch cannot have been produced before the first prototypes were in use, but it must have been cast at a time when they were still in fashion.

Belt set (c), (j), (q) and (r): a buckle with club-shaped tongue (*Kolbendornschnalle*) and a set of three narrow shoe-shaped rivets. A set of decorated shield-on-tongue buckle and three shoe-shaped rivets in grave 102 at Rittersdorf was dated by Böhner (1958, 182; pl. 36 a–d) to his phase II. It therefore possible that the shoe-shaped rivets in grave 73 were added to the buckle with club-shaped tongue as early as in Böhner's late phase II.

Mill Hill I 92 (Fig. 3.25):

Brooch (g): very worn and incomplete. It belongs to the Douai type, although little remains of two of the five head-plate knobs and most of the lozenge-shaped foot had been broken or cut off before the remaining end was worked to a straight edge. The type was dated to the first half of the 6th century by Werner (1961, 21). Grave 14 at Mühlhausen, with a close parallel to the Mill Hill brooch, was dated to *circa* A.D. 500 by Böhme (1988, 60).

Pair of brooches (h) and (i): Hahnheim type of Martin's Eastern variant (Martin 1976, 77f.). Koch (1990, 146 and 235) has dated the type to the second quarter of the 6th century, contemporary with grave 10 at Klepsau, Hohenlohekreis (Schretzheim phase 2). In Koch's chronological sequence for south-western Germany the type is dated to Phase 5 (A.D. 530–550; Koch, in press).

Disc brooch (e): has no broken lines and only one concentric partition, a type generally dated to Böhner's phase II and early phase III (late 5th and first half of the 6th century; cf. Koch 1977, 58ff.; 1990, 140f.).

Buckle with club-shaped tongue (j): there are two straps attached to the loop which originally probably held a plate. The narrow oval shape of the loop may indicate that it was once part of a buckle with a hinged kidney-shaped plate similar to a buckle from Lyminge (Warhurst 1955, pl. X, 2). The faceted shape of the loop and the iron loop to the tongue, however, indicate a 6th-century date for the buckle, which probably was as old as the Douai-type brooch.

Mill Hill I 94 (Fig. 3.32):

Massive ?tinned shield-on-tongue buckle (c) and two copper-alloy rivets (d): for dating evidence see the buckle in Bifrons grave 74 [above]).

Reticella bead (i27): see comment on bead in Bifrons grave 74 (above).

Rust-red short barrel-shaped beads (single or double) with either yellow or white dots and crossing trails (i32,33,35): introduced at Schretzheim in Phase 3 (Koch 1977, Colour Pl. 2, 20:1–7, especially 1 and 4).

Mill Hill I 105C (Fig. 3.27):

Garnet disc brooch (d; not illustrated): the brooch is small and has one central and six wedge-shaped inlays. For dating evidence (late 5th and first half of the 6th century) see disc brooch (2) from Buckland grave 92 (above).

Copper-alloy shield-on-tongue buckle and three shoe-shaped rivets (h): for dating evidence on fully developed copper-alloy shield-on-tongue buckles see the discussion of the buckle from Bifrons grave 74 (above).

The Continental objects in these graves mostly have a date range which varies by about a quarter of a century, as is also found in Continental graves. The pair of brooches in grave D3 at Finglesham, for instance, was older at the time of burial than the associated buckle; the Douai brooch and the buckle in grave 92 at Mill Hill older than associated pair of Hahnheim-type brooches. The only major difference in the date range of associated objects is found in grave 203 at Finglesham, where the buckle is of significantly earlier date than the associated brooches. The Finglesham grave is also unusual in that it contained a combination of an unusually late type of Continental brooch, garnet-set pendants and necklace pins (Hawkes and Pollard 1981, Fig. 4, 24,24,52,53) with Kentish square-headed brooches but not with a Kentish disc brooch. It is likely that this was one of the last Kentish women to be buried with Continental brooches (see below).

It would seem that in general Kentish women kept up with Continental brooch- and bead-fashion for women and buckle-fashion for men, and it would therefore appear

	1	2	3	4	5	6	7	8	9	10	11
Mill Hill I 71	b,c	a,i									
Mill Hill I 61	c,d	e-g									
Lyminge 16	1		2								
Bifrons 63		1	2								
Bifrons 42	2.3	1		4							
Bekesbourne 19	1			2							
Finglesham D3	2	1	5	4							
Bifrons 41	3	1,2,4		5							
Mill Hill I 73	e,f			c,j,q,r							
Buckland 20	2	1	3	9							
Bifrons 29	1.2		5	6	3.4						
Lyminge 44		1			2				phase		
Sarre 4		1.2	6	5	3				II-IV		
Howletts A		2		3	1						
Sarre 158	2					1					
Mersham A	d,e	a				b					
Mill Hill I 105C		i,j		h,p	f	a					
Mill Hill I 25B	e,m	i,j		l		r					
Finglesham E2		a,b		e,g,h		c					
Bekesbourne 22				2		1					
Buckland 92	2				1		d				
Mill Hill I 102		a				i	b				
Bifrons 64		2	5		3		4				
Bifrons 74				2		1	3				
Mill Hill I 64				d,e	j		f				
Mill Hill I 94				c,d	h		i				
Finglesham 203	22-23	20-21	35, 41						25, 52		
Buckland 30					2				5		
Gilton 41						1					2
Buckland 1								1	2		
Buckland 35								1	2		
Gilton 27								1	2		
Sarre 115								1	2		
Kingston 14/205								1	2		
Buckland 29									2-5	1	
Chartham 5									2	1	
Kingston 298/299								4	1	3	2
Kingston 161								1	2		3
Buckland 67									1		2
Chartham 14				phase					1		2
Sarre A				V					1	2	3
Sibertswould 172									1		2
Sibertswould 101										1	2
Monkton 3										1	2
Upchurch A										1	2
Gilton XIII										1	2
Gilton XV										1	2

Table 3.1. Association of types of dress-equipment in women's graves in Kent. Only graves in which at least two of the listed objects were associated are represented (numbers given to objects relate to figures and Appendix 1).
Fundkombination von Trachtbestandteilen in kentischen Frauengräbern. Aufgeführt sind nur Gräber mit mindestens zwei Typen, die Numerierung entspricht den Nummern in den Abbildungen und im Anhang 1.

1 Continental brooch-types; 2 Kentish square-headed brooches; 3 D-bracteates; 4 Shield-on-tongue buckles/shoe-shaped rivets; 5 Kentish keystone-garnet disc brooches of Class 2; 6 Kentish keystone-garnet disc brooches of Class 1; 7 Reticella beads; 8 Kentish keystone-garnet disc brooches of Class 3; 9 Metal pendant-types other than D-bracteates; 10 Kentish disc brooches of Class 4ff. and further developments; 11 Amethyst beads.

that these types of objects can be used to back up the chronology for Kent.

Table 3.1 gives details of the principal objects found in 6th- and 7th-century women's graves that illustrate the general change in fashion which led Leeds (1913) to distinguish between a Frankish and a Kentish phase in the chronology of Kent (see above). Kentish brooch-types (bow brooches and disc brooches of Avent's Classes 1 and 2; columns 2, 5 and 6) were at first worn in combination with Continental brooch-types (column 1), D-bracteates (column 3), shield-on-tongue buckles and/or shoe-shaped rivets (column 4) and reticella beads (column 7). These items were replaced by a fashion including a single Kentish disc brooch (disc brooches of Avent's Classes 3ff. and composite brooches; columns 8 and 10) at the neck, other types of pendant (column 9) and amethyst beads (column 11).

The change in fashion is remarkably clear. The two groups are linked by only three graves, shown in italics. One of them is grave 203 at Finglesham, which has already been noted for the association of a late type of Continental square-headed brooch with a 'heirloom' buckle (see above).

The more detailed chronological framework for 6th-century Kentish women's graves discussed in what follows is based on observations made by Bakka (see above), and on the dating sequence provided by Continental imports.

Bakka's Kentish phase III began with the introduction of the Kentish disc brooch, which was not just another type of Kentish brooch but a replacement of Continental types of small brooch. Though most of the Kentish square-headed brooches are smaller than Continental radiate brooches, they were not normally worn at neck but an additional type used in the lower chest or pelvis area. Other than square-headed brooches, Kentish disc brooches were regularly made of silver and the little space was used to show the same decorative elements found on square-headed brooches: inlays, 'chip-carving', niello, hourglass-shaped punchmarks and gilding. The high quality of this type of brooch suggests that it was not just another type of brooch but that wearing it was the privilege of women who had access to silver rather than merely copper-alloy brooches.

The only type of Kentish small brooch known apart from disc brooches are pairs of bird brooches from grave D3 at Finglesham (Fig. 3.15, 3) and grave 30 at Bekesbourne (1), and a single bird brooch of the same type as the Bekesbourne pair found at Buckland (Parfitt 1995, 461 colour photo). The bird brooches from Bekesbourne and Buckland are decorated with the same elements and techniques as Kentish disc brooches of Avent's Classes 1 and 2: garnet settings, a 'chip-carved' animal with a head and two limbs framed in niello zigzag. Only the bird brooches from grave D3 at Finglesham are different in that they are not decorated with niello, but with circular punchmarks instead of hourglass-shaped ones, similar to the decorative elements found on some of the early Kentish small square-headed brooches. It seems possible, therefore, that the bird brooches from Finglesham were an early Kentish small brooch-design that was not as successful as the disc brooches which were further developed to single disc brooches at neck.

Table 3.2 combines details of Bakka's stages II and III with dating evidence from Continental types. Columns A–C represent brooch-type combinations, columns 1–8 Continental types of object and columns 9–16 Kentish types of object. Only graves with objects fulfilling at least three of the diagnostic criteria are listed.

Column A represents Bakka's definition of stage II, which begins with Kentish square-headed brooches and ends with the introduction of Kentish disc brooches. Only graves with combinations of at least four brooches including one or more Kentish square-headed brooches but excluding Kentish disc brooches are listed. This definition was designed to exclude graves whose brooch-combinations are likely to exclude Kentish disc brooches not for chronological but for cultural reasons. Grave 92 at Mill Hill I, for instance, with a combination of four Continental brooches suggests that the woman buried was an immigrant from the Continent. In grave 86 at this site a woman was buried with three annular brooches and a set of three Continental radiate-headed brooches, and may have been of Anglian origin (Parfitt and Brugmann 1997).

Column B represents a slightly different definition of Bakka's stage III 'starting with the beginning of the keystone garnet disc brooches and finishing with the end of Kentish square-headed brooches' (see above). Only graves in which Kentish disc brooches were combined with other types of brooches, though not necessarily Kentish bow brooches are listed. Instead, stage II ends with the *Mehrfibeltracht*, the fashion of wearing combinations of brooches instead of a single brooch at neck. In chronological terms there is no significant difference between these definitions. Additionally, Leigh's idea that the first Kentish disc brooches were produced and worn in pairs (Leigh 1980, 493ff.; 1984, 73) is represented: 'X' stands for the combination of two Kentish disc brooches with other types of brooch and '&' for a single Kentish disc brooch combined with other brooch-types. The 'pair' of Kentish disc brooches which is not a true pair is the one from grave 105 at Mill Hill (Fig. 3.27, a, f).

Column C represents the beginning of the period following Bakka's stage III, in which a single Kentish disc brooch of Avent's Classes 1 or 2 was worn at neck (*Einfibeltracht*). Table 3.1 shows that it is followed by phases in which Kentish disc brooches of Avent's Classes 3ff. were used.

The datings of Continental brooch-types, buckles and beads are based on the following evidence:

Column 1 (Continental radiate-head and small brooch-types of Böhner's late phase II and early phase III; SD Phases 4–5):
Bifrons 29 (Fig. 3.16, 1–2): see above.

| | brooch comb | | | continental types of brooches | | | buckles | | | | beads | | Kentish types of bow brooches | | | disc brooches | | | skimmers | | |
	A	B	C	1	2	3	4	5	6	7	8	9	10	11	12	13	14	15	16	phase
Bifrons 51	X			2								1								
Mill Hill I 61	X			c,d		h,i						f,g								
Mill Hill I 73	X			e,f		j		***												II
Bifrons 41	X			3		5	2.9	*				1-2								
Bifrons 42	X			2.3		4	2.4	-				1						5		
Finglesham D3	X			2			3.7	**	4											
Lyminge 44		X										1			2			3		
Sarre 4		X					4.0	**	5			1.2			3			4		
Bifrons 29		X		1.2			4.0	-	6						3, 4					
Mersham A		&		e					c			a				b				
Buckland 92		&		2						d					1					III
Mill Hill I 105C		X		d			3.4	***	h					j	a	f				
Howletts A		&					3.6	***	3	4				2	1					
Finglesham 203	X				22.23	16-19														
Mill Hill I 25B		&			e,m		3.6		1						1	r			b	
Bekesbourne 22			X				3.0	***	2								1	3		
Bifrons 74			X				3.5	**	2	3							1			
Buckland 38			X							g					1					
Mill Hill I 64			X				3.3	***		f					j					IV
Mill Hill I 94			X				3.8	**	c	i	i				h					
Buckland 30			X							s-t	p-r				2					
Mill Hill II A			X								2				1					
Buckland 59			X							3l-o					1					

Table 3.2. Finds from graves of Kentish phases II–IV. Only graves are listed which fulfil at least three of the following criteria. The numbers given to the objects relate to the figures and Appendix 1.
Funde aus Gräbern der kentischen Phasen II–IV. Die Numerierung der Funde entspricht den Nummern in den Abbildungen und in Anhang 1. Aufgenommen wurden nur Gräber, die mindestens drei der nachfolgend aufgeführten Kriterien erfüllen.

A. Set of at least four brooches including Kentish brooch-types other than disc brooches; B. Kentish disc brooch(es) combined with other type(s) of brooch: X = two Kentish disc brooches, & = one Kentish disc brooch. C. Single Kentish disc brooch. 1 Continental brooch-types of Böhner's late phase II and early phase III; SD Phases 4–5. 2. Continental brooch-types of SD Phase 6. 3. Early types of buckle with club-shaped tongue and shield on tongue buckles up to 2.9 cm in width (Böhner's late phase II/early phase III; SD Phases 3–5). 4. Width of shield-on-tongue buckles given in centimetres. 5. Number of shoe-shaped rivets. 6. Late types of buckle with club-shaped tongue of the second and third quarters of the 6th century (Böhner's early and middle phase III; SD 5 and later). 7. Reticella, mosaic and millefiori beads. 8. Rust-red barrel-shaped beads (single or double) with either yellow or white dots and crossing trails. 9. Kentish square-headed brooch, type Åberg 131. 10. Kentish square-headed brooches with drop-shaped garnets. 11. Kentish square-headed brooch, type Åberg 132. 12. Kentish disc brooches of Avent's Class 2. 13. Kentish disc brooches of Avent's Class 1.2. 14. Kentish disc brooches of Avent's Class 1.1. 15. Skimmer, type B1a. 16. Skimmer, type B1b.

Bifrons 41 (Fig. 3.8, 3): see above.
Bifrons 42 (Fig. 3.9, 2–3): see above.
Bifrons 51: *pair of small bird brooches with inset garnet eyes (2):* For the general date-range of bird brooches see above (Bifrons 41).
Buckland 92 (Fig. 3.18, 2): see above.
Mersham A (Fig. 3.22, e): see above.
Mill Hill I 61 (Fig. 3.12, c, d): see above.
Mill Hill I 73 (Fig. 3.14, e, f): see above.
Mill Hill I 105C: see above.

Column 2 (Continental bow and small brooch-types dated to SD Phase 6 (A.D. 550–580; Koch, in press):
Finglesham 203 (Figs. 3.20, 3.22, 3.23): see above.
Mill Hill 25B (Fig. 3.24, e, m): see above.

Column 3 (early types of buckles with club-shaped tongue and shield-on-tongue buckles up to 2.9 cm in width (Böhner's late phase II/early phase III; SD Phases 3–5):
Bifrons 41 (Fig. 3.8, 5): see above.
Bifrons 42 (Fig. 3.9, 4): see above.
Finglesham 203 (Fig. 3.20, 16–19): see above.
Mill Hill I 61 (Fig. 3.12, h, i): see above.
Mill Hill I 73 (Fig. 3.14, j): see above.

Column 4: width of shield-on-tongue buckles in centimetres.

Column 5: number of shoe-shaped rivets worn with the belt. Shield-on-tongue buckles without shoe-shaped rivets are marked with a hyphen.

Column 6 (late type of buckles of the second and third quarters of the 6th century [Böhner's early and middle phase III; SD 5 and later]):

Bekesbourne 22: *copper-alloy shield-on-tongue buckle (2):* the buckle is the smallest in this group and the rivets are likely to be later additions because of their large size.
Bifrons 29 (Fig. 3.16, 6): see above.
Bifrons 74 (2): see above.
Finglesham D3 (Fig. 3.15, 4): see above.
Howletts A (Fig. 3.21, 3): see above.
Mersham A (Fig. 3.22, c): see above.
Mill Hill I 25B (Fig. 3.24, h, l): see above.
Mill Hill I 64 (Fig. 3.31, d): see above.
Mill Hill I 94 (Fig. 3.32): see above.
Mill Hill I 105C (Fig. 3.27, h): see above.
Sarre 4 (Fig. 3.28, 5): *copper-alloy shield-on-tongue buckle:* compare buckles in graves 25B and 94 at Mill Hill (above).

Column 7 (reticella, mosaic and millefiori beads):
Bifrons 74 (3): see above.
Buckland 38 (Fig. 3.30, g): *globular millefiori bead*: globular millefiori beads are rare in the first half of the 6th century and occur in larger numbers mainly in Schretzheim phase 3 (Koch 1977, 218; 1990, 123).
Buckland 30 (Fig. 3.29, s–t): see above.
Buckland 92 (Fig. 3.18, d): see above.
Howletts A (Fig. 3.21, 4): see above.
Mill Hill I 64 (Fig. 3.31, f8): see above.
Mill Hill I 94 (Fig. 3.32, i27): see above.

Column 8 (rust-red disc or barrel-shaped beads [single or double] with either yellow or white dots and crossing trails [Schretzheim type 20; Mill Hill type E4–6]):
Buckland 30 (Fig. 3.29, p–r): see above.
Buckland 59 (3l–o) *three beads with yellow decoration and two with white decoration*: at Schretzheim this type occurs from phase 3 onwards (Koch 1977, Colour Pl. 2, 20:1–7, especially 20:1 and 4).
Mill Hill I 94 (Fig. 3.32, i32–35): see above.
Mill Hill II A: *35 rust-red short barrel-shaped beads (single or double) with either yellow or white dots(2)*: see above.

Column 9: square-head brooch-type Åberg 131 named after the figure in Åberg's book *The Anglo-Saxons in England* (1926) which shows a square-head brooch with raised decoration on the bow and foot in the shape of a cross (Fig. 3.6).

Column 10: brooches decorated with one or more drop-shaped garnets on the foot. The brooches, from grave 41 at Bifrons (Fig. 3.8, 1–2) and grave A at Mersham (Fig. 3.22, a), are manifestly of the same series. The same applies to the brooches from grave 42 at Bifrons (Fig. 3.9, 1) and grave 4 at Sarre (Fig. 3.28, 2). The pair from grave 44 at Lyminge (Fig. 3.23, 1) has no exact parallels but all five brooches are likely to have been made in the same workshop.

Column 11: square-head brooch-type Åberg 132 named after a figure in Åberg (1926). The scrolled foot terminal and the diamond shaped central part of the foot are typical of all brooches of this series. The brooch shown on the distribution map in figure 3.7, however, is the only one with a garnet inlay (Fig. 3.27, j).

Columns 12–14 represent Kentish disc brooches of Avent's Classes 2, 1.2 and 1.1.

Columns 15 and 16 represent skimmers of Martin's 'English group B1' (Martin 1984, Fig. 57). They are made of silver, with a handle of octagonal section and decorated with niello. Two sub-types can be distinguished amongst these skimmers. Sub-type B1a (column 15; Figs. 3.9, 5; 3.23, 3; 3.28, 4) has nine holes in a cross, with flattened terminals set with garnets on the handle. The flat part of the handle of sub-type B1b (column 16; Fig. 3.24, b) is narrower and longer than that of sub-type B1a and decorated with punchmarks.

Phase II in table 3.2 is shown with five graves. According to its brooch-type combination A, grave D3 at Finglesham is also of Phase II but the bird brooches and the relatively late type of shield-on-tongue buckle indicate that it fits in both Phase II and Phase III (see above). Phase III is represented by eight graves and grave 203 at Finglesham, which is unusual in the type-A combination of brooches including late types of Continental brooch (column 2). Phase IV is shown with eight graves.

Bifrons and Mill Hill I are the only cemeteries represented in all three phases. The basis of the three phases is evidently slender, and includes some types of objects which are represented in no more than two graves. Few types of objects were used in just one phase:

Phase II: Kentish square-head brooches of type Åberg 131 (column 9). Early types of Continental buckle (column 3) excluding the 'heirloom' from grave 203 at Finglesham (see above).

Phase III: Later Continental types of brooch (column 2), Kentish square-head brooch, type Åberg 132 (column 11), and Kentish disc brooches of Avent's Class 1.2 (column 13).

Phase IV: Rust-red beads with white or yellow crossing trails and dots (column 8) and Kentish disc brooches of Avent's Class 1.1 (column 14). However, in grave E2 at Finglesham a brooch of Avent's Class 1.1 was associated with two Kentish square-headed brooches (see below).

The table would list both more graves and more types of objects if the combination of only two instead of three criteria was admitted but this would give brooch-type combinations too much weight as chronological criteria (columns A–C) in comparison with stylistic evidence.

However, dating some other graves according to some of the criteria listed in table 3.2 seems possible:

Phase II:

Buckland 13 (Fig. 3.10): is dated by Evison (1987) to Buckland Phase 1 (A.D. 475–525) on the stylistic evidence of two Jutish-Kentish square-headed brooches (1–2) and the position of the grave on the cemetery. It contained a combination of four brooches, none a Kentish disc brooch.

Buckland 20 (Fig. 3.11): is dated to Buckland phase 1 (Evison 1987) on the same criteria. The shield-on-tongue buckle (width 2.4 cm) and two disc-shaped rivets (9) fit the definition of 'early' buckle-types (column 3) in table 3.2.

Mill Hill I 71 (Fig. 3.13): was positioned beside grave 73 and cut by the same grave (Fig. 3.5). The combination of Continental types of brooch[9] (c–d) and two Kentish square brooches (a, i) makes it an almost 'typical' grave of Phase II.

Phase III:

Bifrons 64 (Fig. 3.17): combination of two Kentish square-headed brooches (1, 2) with a Kentish disc brooch of Avent's Class 2 (3) and a reticella bead (4).

Finglesham E2 (Fig. 3.19): combination of two Kentish square-headed brooches with a Kentish disc brooch of Avent's Class 1.1. The size of the shield-on-tongue buckle (e) (width 3.5 cm) worn with two rivets, one disc (g) and the other shoe-shaped (h) make it a 'late' one (cf. column 6) in table 3.2.

Mill Hill I 102 (Fig. 3.26): combination of three Kentish bow brooches (a, d), a Kentish disc brooch of Avent's Class 1.2 (j) and a reticella bead (b12).

Phase IV:

The women buried in graves ***62*** and ***81*** at ***Gilton*** had a single brooch of Avent's Class 2 at neck and may have also worn datable bead-types with their necklaces. The beads however are not described in detail in the inventory.

Unfortunately, the layouts of most Kentish cemeteries can provide very little additional information in terms of horizontal stratigraphy, either because their plans were never drawn up, there were too few graves, or the evidence remains unpublished. The few sites providing useful evidence are listed below:

Bifrons: Godfrey-Faussett (1876) drew no plan. In one case the description of graves provides additional dating evidence: grave 42 (Fig. 3.9) lay close by the side of grave 41 (Fig. 3.8), so that the two may have been under one tumulus. Both seemed more carefully and regularly dug than most of their neighbours. Table 3.2 shows similar grave goods dating to Phase II with neither grave containing a Kentish disc brooch.

Buckland (Fig. 3.3): The layout of the graves has been discussed in detail by Evison (1987). Only the phases following her phase 2 (ibid. Fig. 101ff.) show clearly in the cemetery layout. Phases 1 and 2 correspond roughly to Kentish Phases II–IV.

Lyminge: Evison (1987, 162ff.; Fig. 3.32) published a plan of the site on which she mapped three chronological phases. Further graves are shown by Härke (1992, 267 Fig. 64) but the evidence does not add to the discussion.

Mill Hill I (Fig. 3.5): Phases II–IV show very well on the site plan; however, the plan is a result of this study and some of the graves were assigned to a certain phase *because* of their position (Parfitt and Brugmann 1997).

Table 3.3 attempts a synchronization of Kentish and some well established Continental chronological frameworks. It has already been noted that Bakka equated his Stufe II with Böhner's phase II and his Stufe III with AM II (Tab. 3.3).

Although a number of 5th- to 6th-century objects were found in the Kentish graves discussed above, none of these graves date to the 5th century and the phase may have begun in the early 6th century. It remains to be seen whether a preceding 5th-century Kentish Phase I can be defined, which ought to cover graves or groups of objects contemporary with the original 'Jutish immigration' recorded in the written sources.

Contrary to what Bakka assumed, Kentish Phase III seems to have started later than Böhner's phase II. Most belt sets found in graves associated with Bakka's phase II are stylistically earlier types than most belt sets found in graves associated with Bakka's phase III (Tab. 3.2). However, the change is gradual and the line drawn between the two groups in table 3.2 (columns 3 and 6) is artificial. A general dating of all belt sets to Böhner's phase II (ending A.D. 525 at latest according to Martin 1989) would be too early. The line between Bakka's phases II and III may correspond to the line drawn between SD Phases 4 and 5 (*circa* A.D. 530) but it seems safer, at present, to allow for a wider margin, e.g. *circa* A.D. 530/40.

In phase III, Continental brooch-types mainly of SD Phase 5 but also of Phase 6 (columns 3 and 8) were worn, and buckles of the second and third quarters of the 6th century (SD phases 5 and 6) and reticella beads (column 12) are introduced (in SD phase 5). In phase IV, shield-on-tongue buckles went out of use and rust-red beads with white or yellow crossing trails and dots (column 14) were introduced (Schretzheim Phase 3).

The use of Continental objects went out of fashion in Kent before buckles with plates became fashionable (during Schretzheim phase 3; SD Phase 7). With Kentish phase IV and Schretzheim phase 3 (Koch 1977, 62) fashion changed to a single brooch at the neck. It is therefore likely that Kentish phase III ended at about the same time as Schretzheim phase 2 and the evidence thus confirms Bakka's notion that Kentish Phase III ended approximately with AM II. Phase IV ends with Kentish disc brooches of

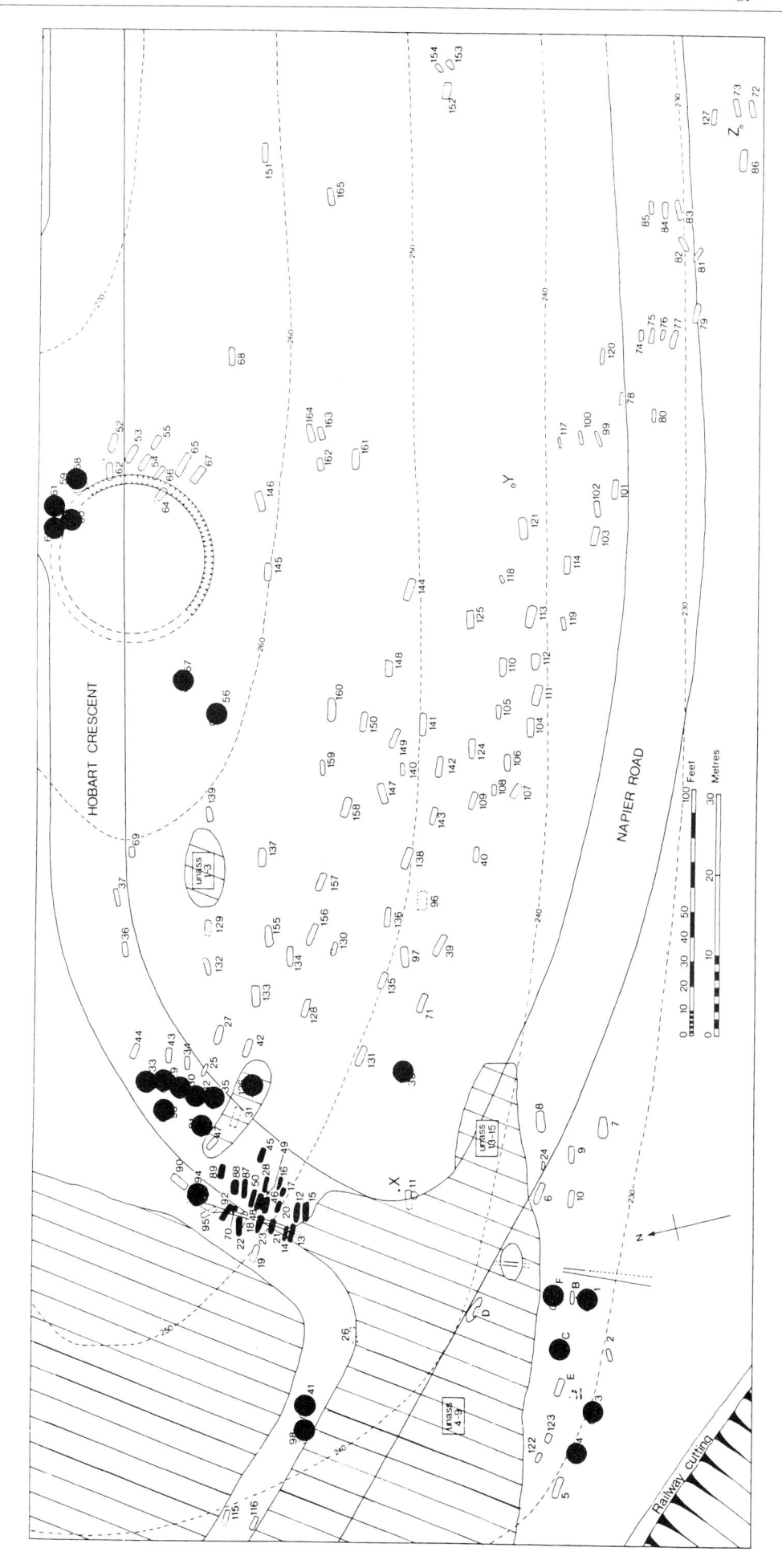

Fig. 3.3 Site plan of Buckland cemetery showing phases 1–2 (graves filled in black) and phase 3 (black dots). After Evison (1987, Figs. 101–103). Plan des Gräberfeldes von Buckland mit Eintragung der Phasen 1–2 (Gräber schwarz gefüllt) und Phase 3 (schwarze Punkte).

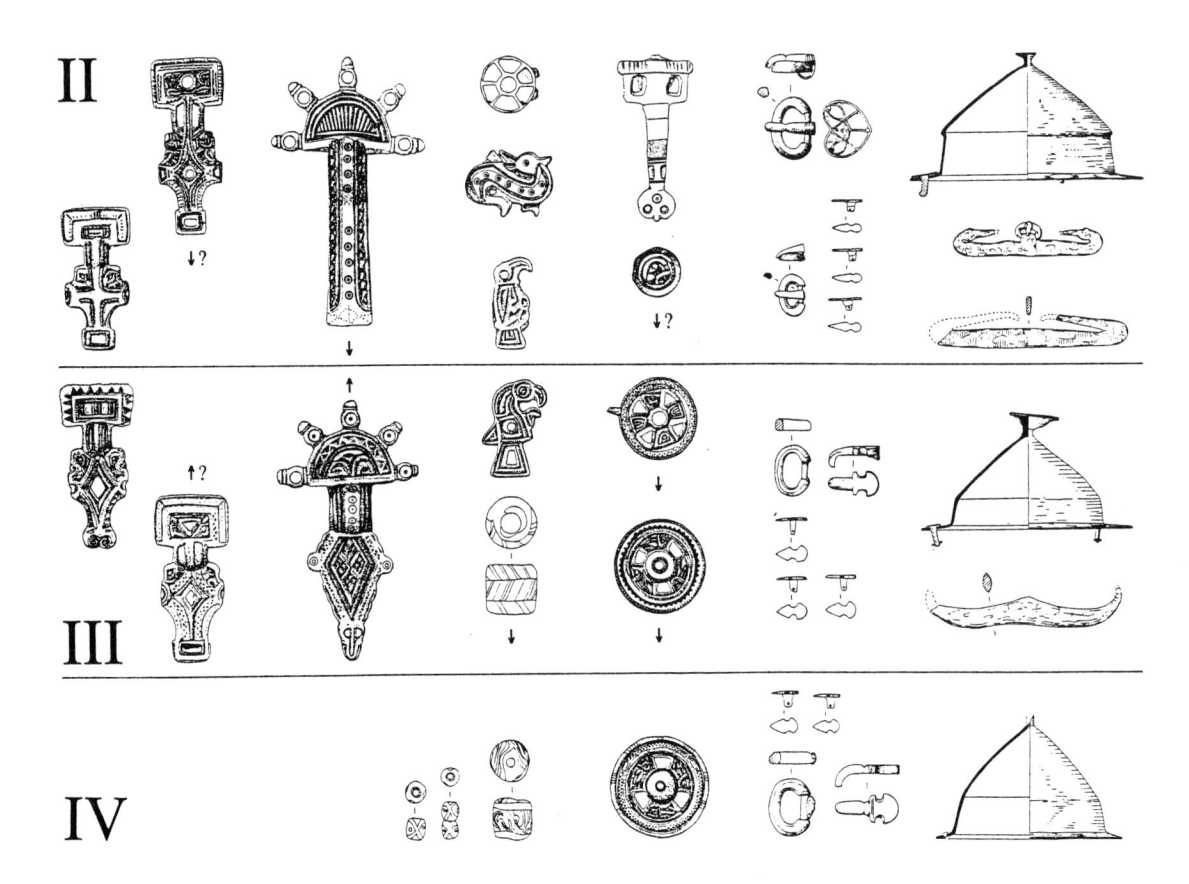

Fig. 3.4 Phases II–IV at Mill Hill I. After Parfitt and Brugmann 1997. Objects not to scale.
Phasen II–IV in Mill Hill I.

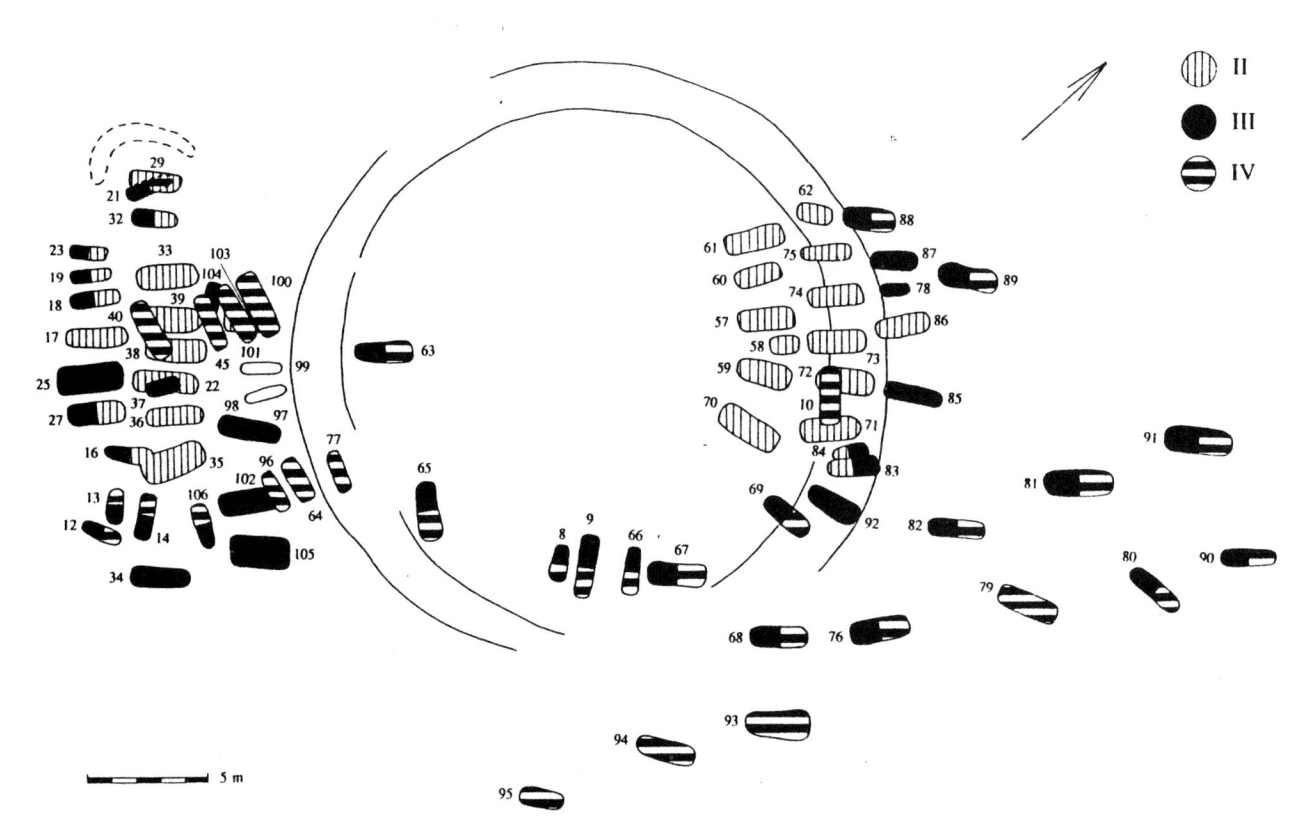

Fig. 3.5 Site plan of Mill Hill I cemetery showing phases II–IV. After Parfitt and Brugmann 1997.
Plan des Gräberfeldes von Mill Hill I, Phasen II–IV.

Trier region (Böhner 1958)	Ament 1977	Kent (Bakka 1981)	Buckland (Evison 1987)	Kent (Mill Hill)	Schretzheim (Koch 1977)	South-West Germany (Koch in press)
II (-> 525)	AM I (450/80-520/30)					SD 1 (430-60)
						SD 2 (460-80)
		II	1 (475-525)	II (?500-530/40)	Irlmauth	SD 3 (480-510)
III (525-600)	AM II (520/30-560/70)	III	2 (525-575)			SD 4 (510-30)
(IIIa nach Bakka 1981)				III (530/40-560/70)	1 (525-545/50)	SD 5 (530-550)
					2 (545-565/70)	SD 6 (550-580)
	AM III (560/70-600)			IV (560/70-580/90)	3 (565-590/600)	
(IIIb nach Bakka 1981)			3 (575-625)	V	4 (590-620/30)	SD 7 (580-600)
IV	JM I (600-630/40)					SD 8 (600-620)

Table 3.3. Provisional synchronization of Kentish phases II–IV with Continental chronological frameworks.
Vorläufige Synchronisierung der kentischen Phasen II–IV mit kontinentalen Chronologiesystemen.

Avent's Classes 1 and 2 and therefore approximately with phase 2 at Buckland (Tab. 3.3). There are no indications that Kentish phase IV lasted longer than SD Phase 6.

Absolute dates for Kentish chronology have up to now been transferred from Continental frameworks. Coin-dating of Kentish graves of phases II–IV is almost impossible. In the graves listed in table 3.1 (Kentish Phases II–V) only two 6th-century coins were found. The *terminus post quem* for Buckland 29 is A.D. 560–570 but the coin is looped and worn and therefore dated to Buckland phase 3 by Evison (1987, 136). The other coin was found in Gilton 41 and was struck in the third quarter of the 6th century and pierced for use as pendant when still in fresh condition (Hawkes, Merrick and Metcalf 1966, 103). In table 3.1, Gilton 41 is listed as one of the graves linking Kentish Phase II–IV to Phase V because of the single Kentish disc brooch of class 1.2 associated with an amethyst bead, an iron chatelaine and slip-knot rings. The coin was not worn at neck but found in or near a glass vessel at the foot of the grave outside the coffin. Chatelaines, slip-knot rings and amethyst beads are usually found in graves of phase V and it therefore seems possible that the brooch was a heirloom buried in phase V. In the absence of further coin-dated Kentish graves, absolute dates for the Kentish chronology are mostly derived from Continental evidence and therefore are bound to shift according to Continental updatings such as proposed by Martin (1989).

In due course the chronological framework proposed in this paper will have to be tested against the substantial 6th-century evidence excavated at Buckland in 1994 (see above). It may not only be possible to make corrections and improvements to datings of individual objects and the Kentish phasing but also to reduce the dependency of Kentish chronology on Continental frameworks.

DEUTSCHE ZUSAMMENFASSUNG

Obwohl die Funde aus kentischen Gräberfeldern des 6. Jahrhunderts für ihren Reichtum bekannt sind, ist es bisher nicht möglich, das Material in einer Weise für ein eigenständiges regionales Chronologieschema zu verwenden, wie es für merowingerzeitliches Material auf dem Kontinent bereits geschehen ist. Dies liegt vor allem daran, daß die meisten Funde aus Altgrabungen stammen und daher verhältnismäßig wenige geschlossene Fundkomplexe vorliegen (Tab. 3.1; Abb. 3.1–3.3). Mit Hilfe einiger kontinentaler Fibel-, Schnallen-, und Perlentypen aus kentischen Frauengräbern, deren Datierungen aus kontinentalen Chronologieschemata übernommen werden können (Tab. 3.2; Abb. 3.8–3.32), und einiger neuer Funde läßt sich die relative und absolute Chronologie Kents jedoch weiter verfeinern (Tab. 3.3; Abb. 3.4–3.5).

All illustrations of finds from Dover-Buckland are reproduced from Evison 1987 with kind permission of English Heritage.

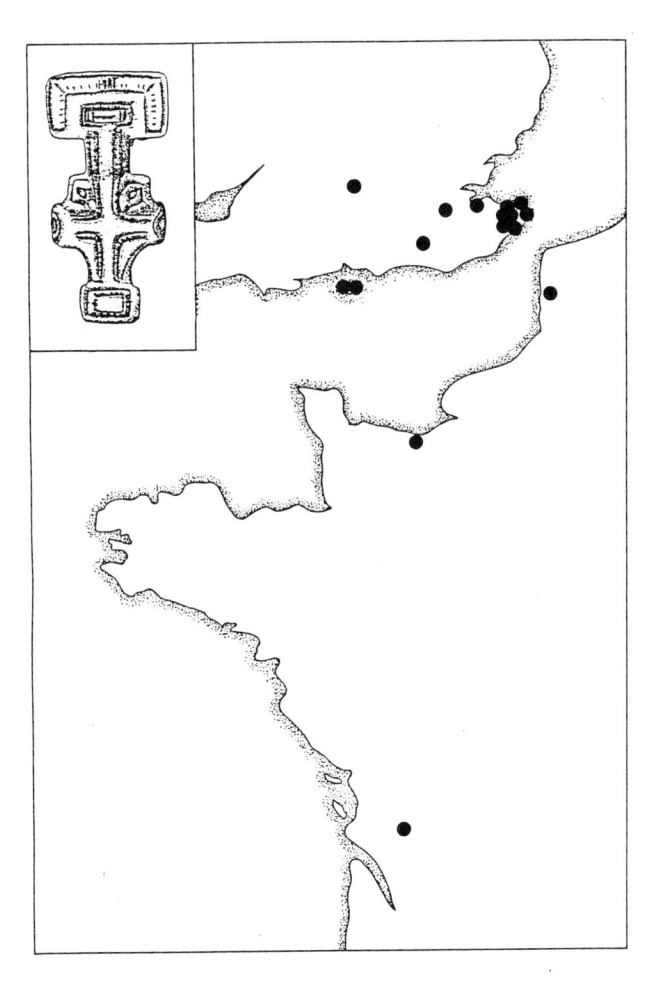

Fig. 3.6 Distribution map of Kentish square-headed brooch type Åberg 131. After Parfitt and Brugmann (1997) with list of sites. Not to scale.
Verbreitung der kentischen Bügelfibeln mit rechteckiger Kopfplatte vom Typ Åberg 131.

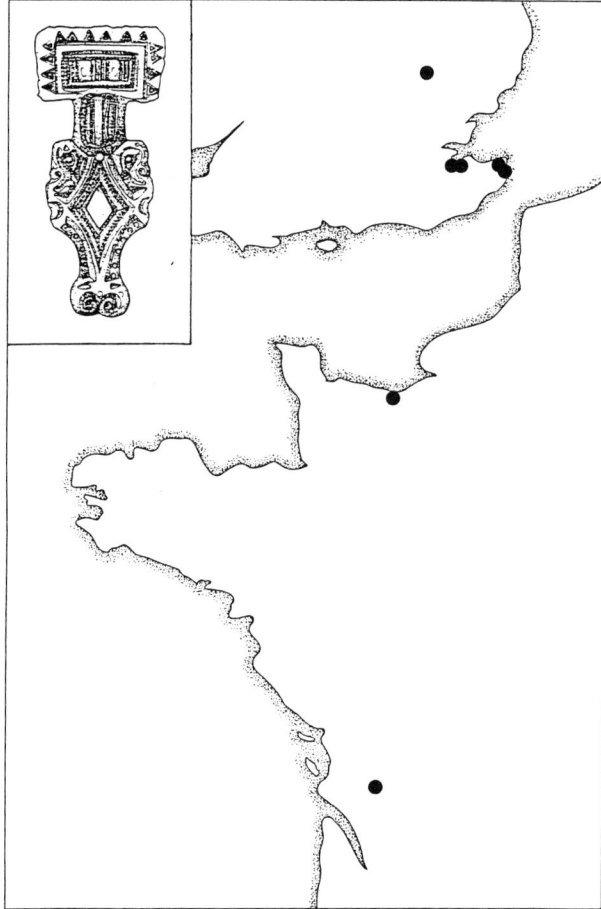

Fig. 3.7 Distribution map of Kentish square-headed brooch type Åberg 132. After Parfitt and Brugmann (1997) with list of sites. Not to scale.
Verbreitung der kentischen Bügelfibeln mit rechteckiger Kopfplatte vom Typ Åberg 132.

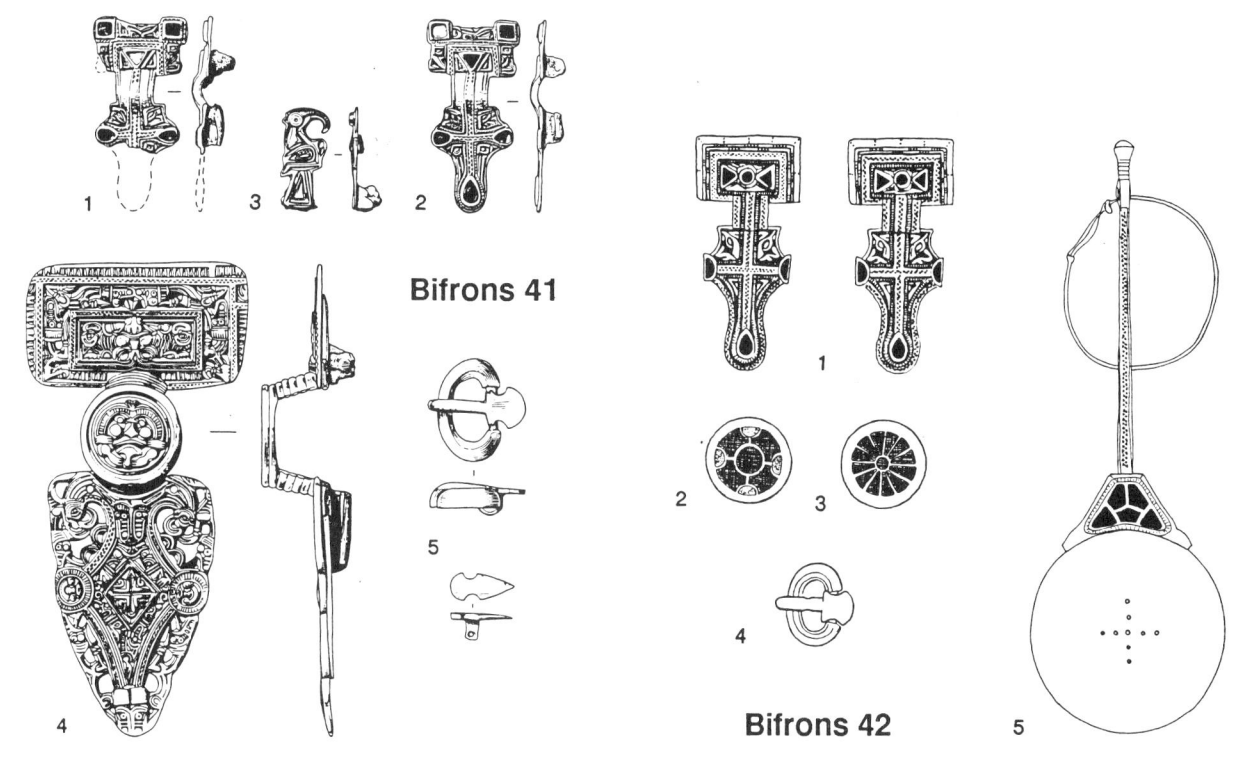

Fig. 3.8 Phase II: Brooches and buckle set from Bifrons grave 41. After Haseloff (1981, Fig. 90). Scale approx. 1:2.
Phase II: Fibeln und Schnallengarnitur aus Bifrons Grab 41.

Fig. 3.9 Phase II: Brooches, buckle and skimmer from Bifrons grave 42. After Godfrey-Faussett (1876, 314f.) and Bakka (1958, Fig. 53). Not to scale.
Phase II: Fibeln, Schnalle und Sieblöffel aus Bifrons Grab 42.

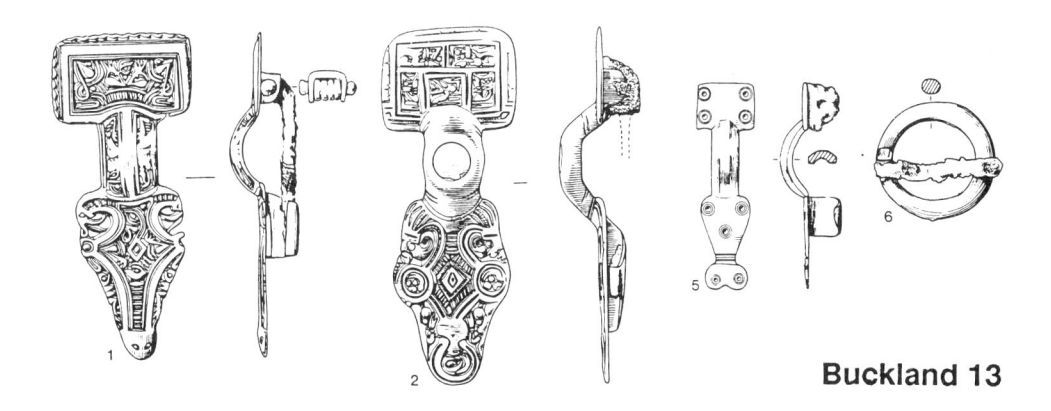

Fig. 3.10 Phase II: Brooches from Buckland grave 13. After Evison (1987, Fig. 9). Scale 1:2.
Phase II: Fibeln aus Buckland Grab 13.

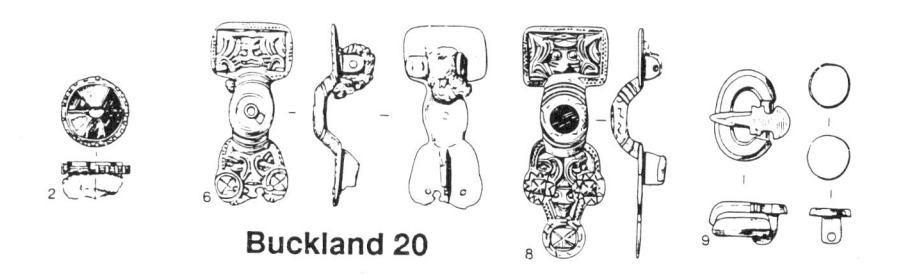

Fig. 3.11 Phase II: Brooches and buckle from Buckland grave 20. After Evison (1987, Fig. 12). Scale 1:2.
Phase II: Fibeln und Schnalle aus Buckland Grab 20.

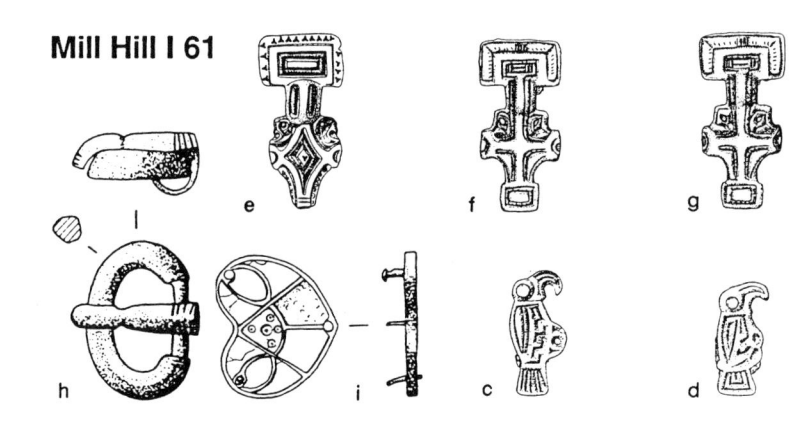

Fig. 3.12 Phase II: Brooches and buckle-set from Mill Hill I grave 61. After Parfitt and Brugmann 1997. Scale 1:2.
Phase II: Fibeln und Schnallengarnitur aus Mill Hill I Grab 61.

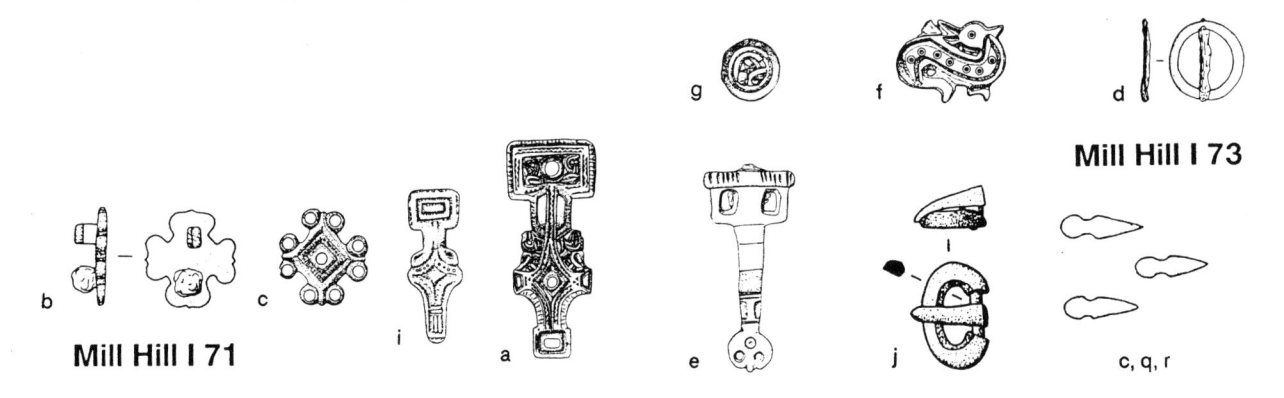

Fig. 3.13 Phase II: Brooches from Mill Hill I grave 71. After Parfitt and Brugmann 1997. Scale 1:2.
Phase II: Fibeln aus Mill Hill I Grab 71.

Fig. 3.14 Phase II: Brooches and buckle from Mill Hill I grave 73. After Parfitt and Brugmann 1997. Scale 1:2.
Phase II: Fibeln und Schnalle aus Mill Hill I Grab 73.

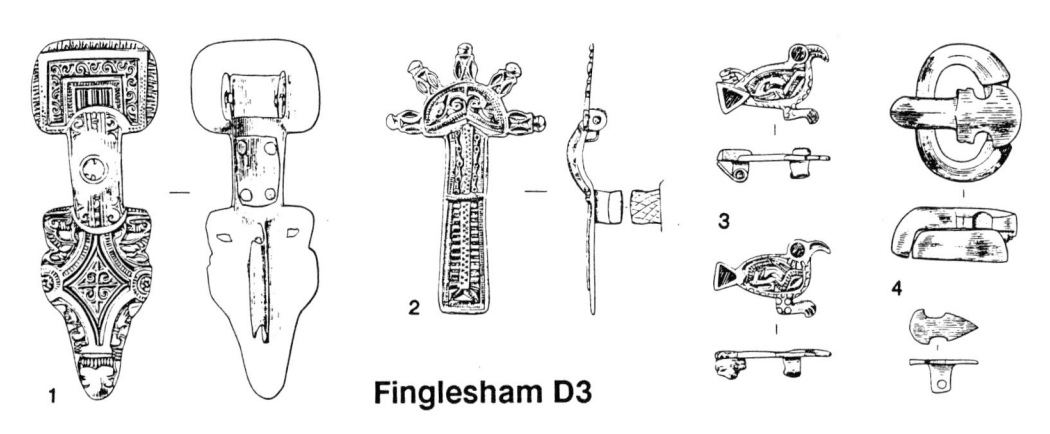

Fig. 3.15 Phase II/III: Brooches and buckle from Finglesham grave D3. After Chadwick (1958, Fig. 6; 9). Scale 1:2.
Phase II/III: Fibeln und Schnalle aus Finglesham Grab D3.

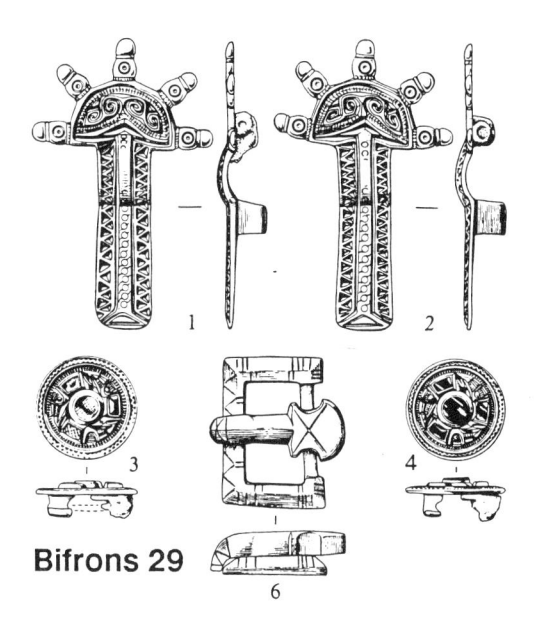

Fig. 3.16 Phase III: Brooches and buckle from Bifrons grave 29. After Hawkes and Pollard (1981, Fig. 7). Scale 1:2.
Phase III: Fibeln und Schnalle aus Bifrons Grab 29.

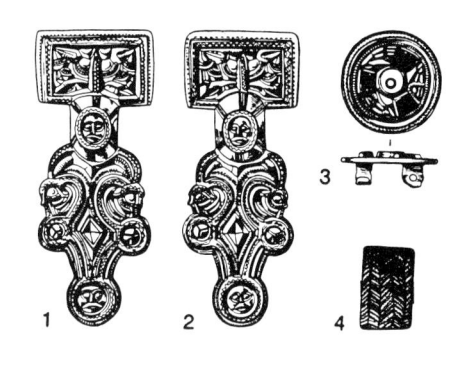

Fig. 3.17 Phase III: Brooches and bead from Bifrons grave 64. After Hawkes and Pollard (1981, Fig. 11). Scale 1:2.
Phase III: Fibeln und Perle aus Bifrons Grab 64.

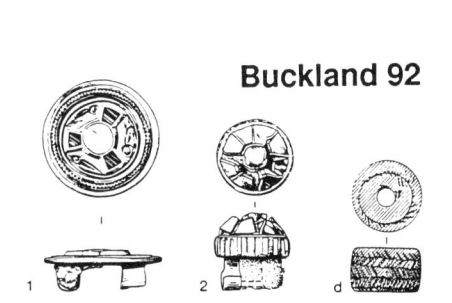

Fig. 3.18 Phase III: Brooches and reticella bead from Buckland grave 92. After Evison (1987, Fig. 42). Scale 1:2.
Phase III: Fibeln und Reticella-Perle aus Buckland Grab 92.

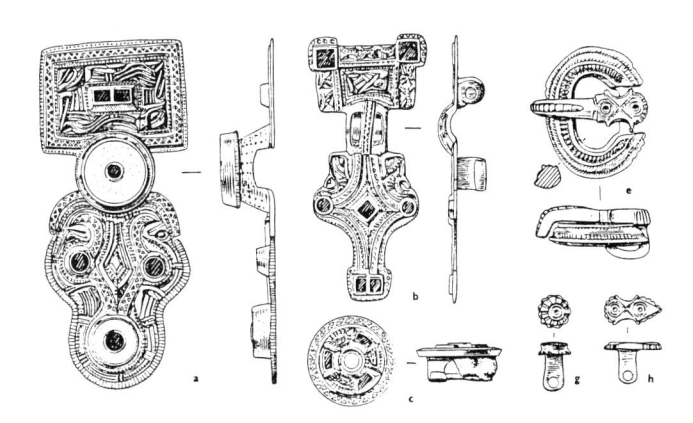

Fig. 3.19 Phase III: Brooches and buckle-set from Finglesham grave E2. After Chadwick (1958, Figs. 6 and 11). Scale 1:2.
Phase III: Fibeln und Schnallengarnitur aus Finglesham Grab E2.

Finglesham 203

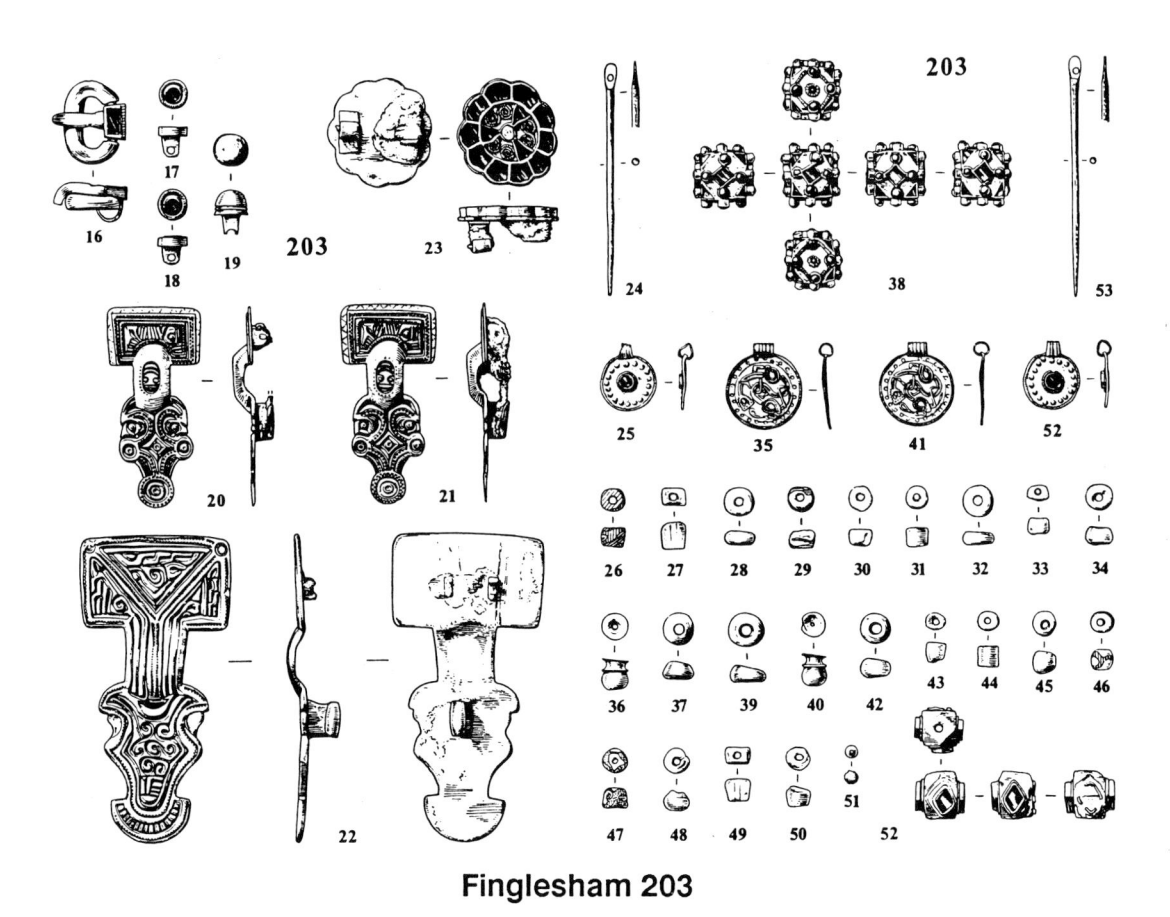

Fig. 3.20 Phase III. Brooches, buckle and necklace from Finglesham grave 203. After Hawkes and Pollard (1981, Figs. 4 and 6). Scale 1:2.
Phase III: Fibeln, Schnalle und Halskette aus Finglesham Grab 203.

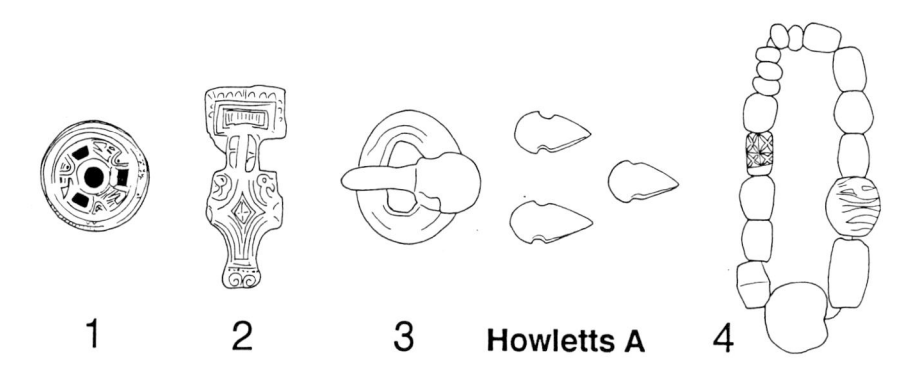

Fig. 3.21 Phase III. Brooches, buckle and beads from Howletts grave A. After Smith (1917/18, Pl. 1). Not to scale.
Phase III: Fibeln, Schnalle und Perlen aus Howletts Grab A.

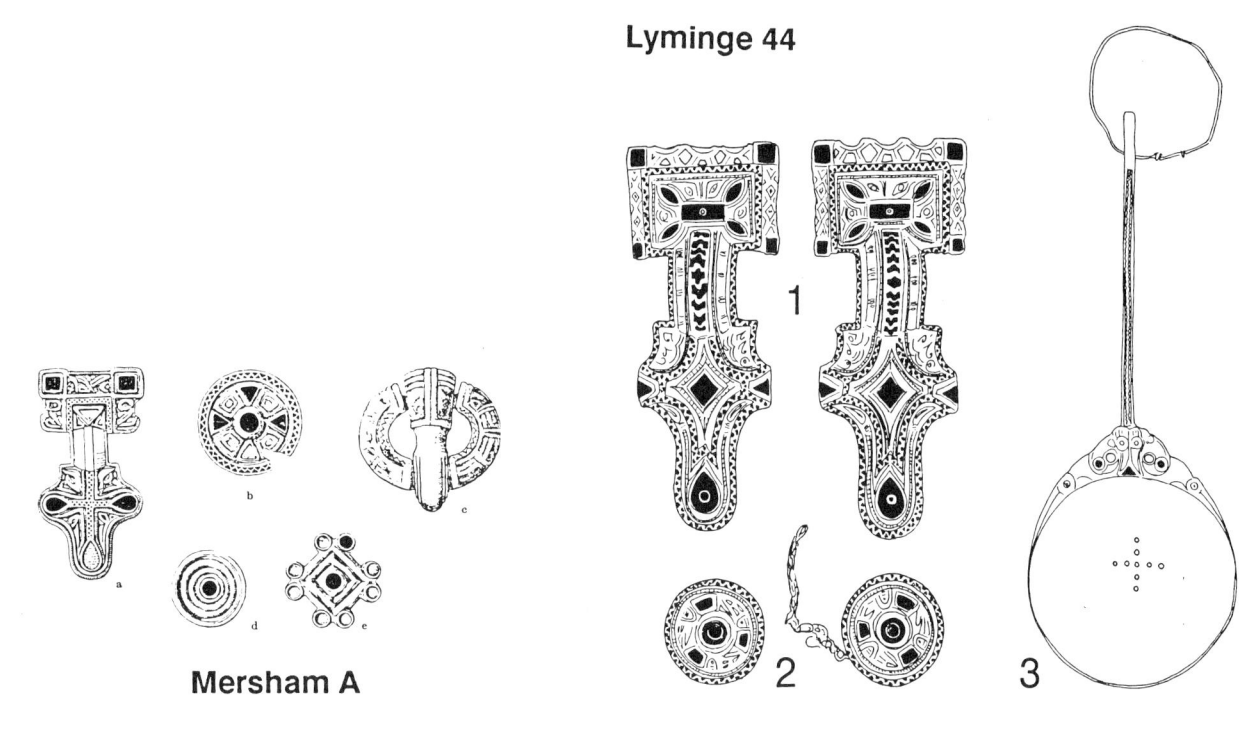

Fig. 3.22 Phase III: Brooches and buckle from Mersham grave A. After Bakka (1958, Fig. 52). Scale 1:2.
Phase III: Fibeln und Schnalle aus Mersham Grab A.

Fig. 3.23 Phase III: Brooches and skimmer from Lyminge grave 44. After Warhurst (1955, Pls. 12 and 13).
Phase III: Fibeln und Sieblöffel aus Lyminge Grab 44.

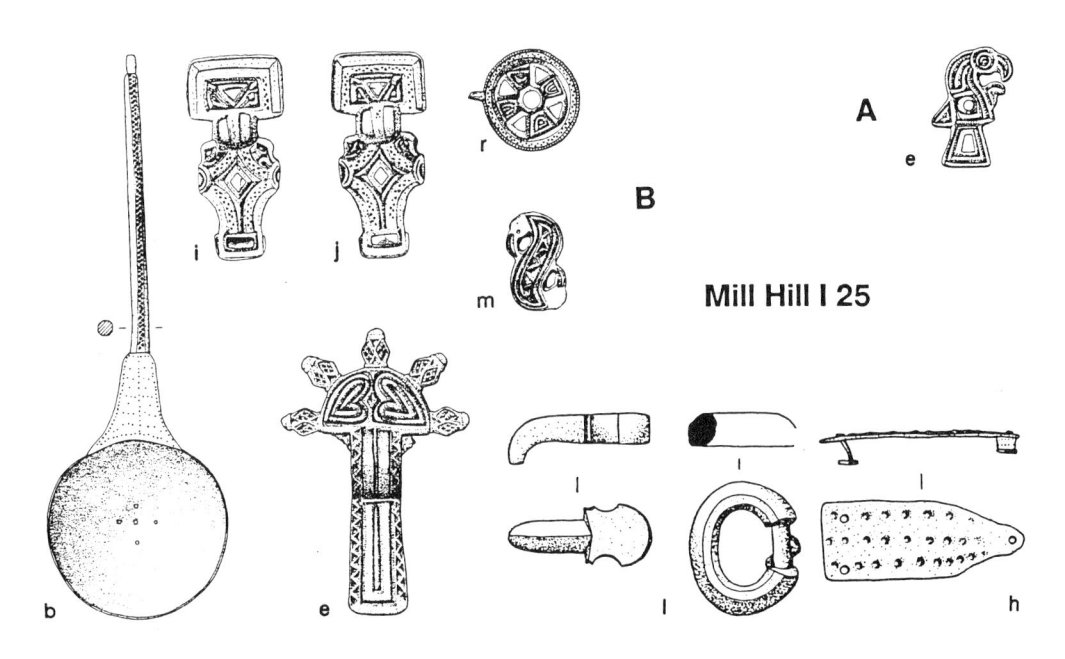

Fig. 3.24 Phase III: Brooches, buckle-set and skimmer from Mill Hill I graves 25 A and B. After Parfitt and Brugmann 1997. Scale 1:2.
Phase III: Fibeln, Schnallen und Sieblöffel aus Mill Hill I Gräber 25A und 25B.

58 · *Birte Brugmann*

Mill Hill I 92

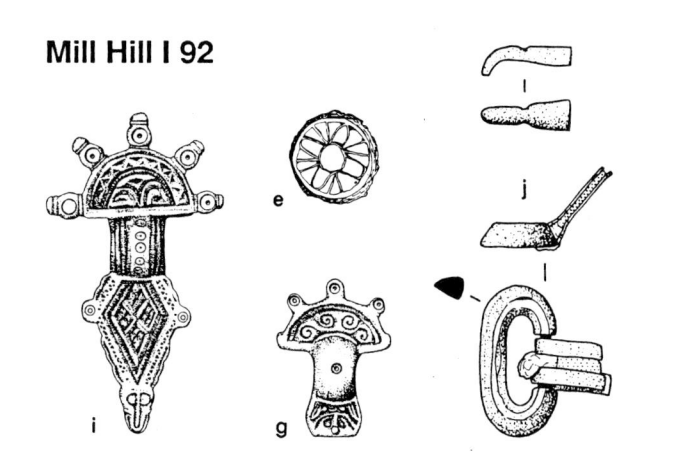

Fig. 3.25 Phase III: Brooches and buckle from Mill Hill I grave 92. After Parfitt and Brugmann 1997. Scale 1:2.
Phase III: Fibeln und Schnalle aus Mill Hill I Grab 92.

Mill Hill I 102

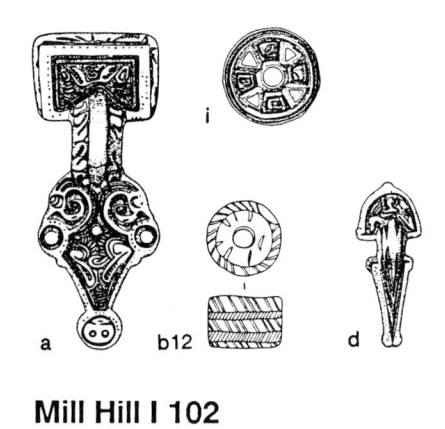

Fig. 3.26 Phase III: Brooches and reticella bead from Mill Hill I grave 102. After Parfitt and Brugmann 1997. Scale 1:2.
Phase III: Fibeln und Reticella-Perle aus Mill Hill I Grab 102.

Mill Hill I 105C

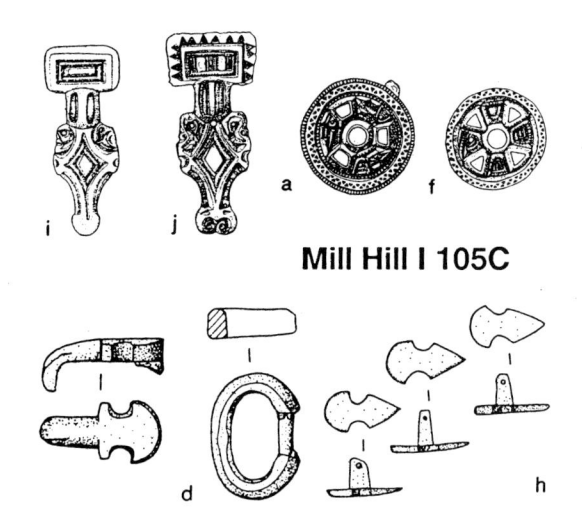

Fig. 3.27 Phase III: Brooches and buckle from Mill Hill I grave 105C. Garnet disc brooch (d) not illustrated. After Parfitt and Brugmann 1997. Scale 1:2.
Phase III: Fibeln und Schnalle aus Mill Hill I Grab 105C; Almandinscheibenfibel nicht abgebildet.

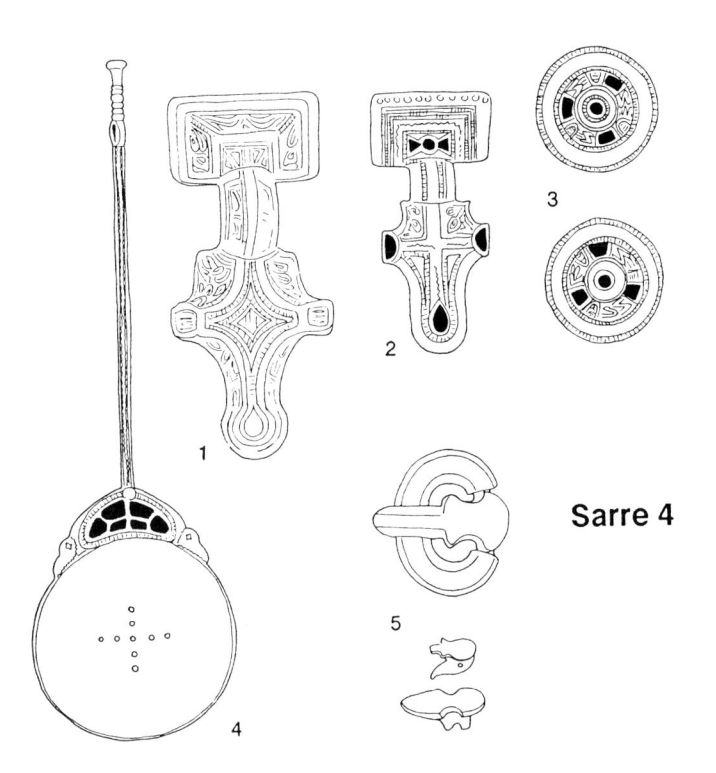

Fig. 3.28 Phase III: Brooches, buckle-set and skimmer from Sarre grave 4. After Brent (1862/63, 313; pl. 1; 2). Scale c. 1:2.

Phase III: Fibeln, Schnalle und Sieblöffel aus Sarre Grab 4.

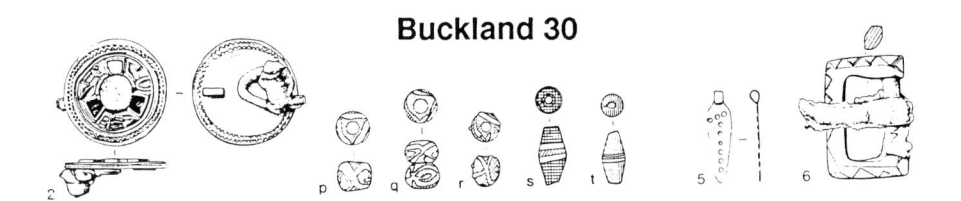

Fig. 3.29 Phase IV: Brooch, selected beads, pendant and buckle from Buckland grave 30. After Evison (1987, Fig. 18). Scale 1:2.

Phase IV: Fibel, ausgewählte Perlen, Anhänger und Schnalle aus Buckland Grab 30.

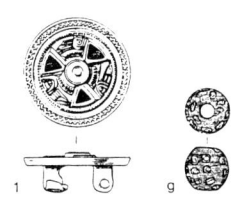

Fig. 3.30 Phase IV: Brooch and mosaic bead from Buckland grave 38. After Evison (1987, Fig. 22). Scale 1:2.

Phase IV: Fibel und Mosaikperle aus Buckland Grab 38.

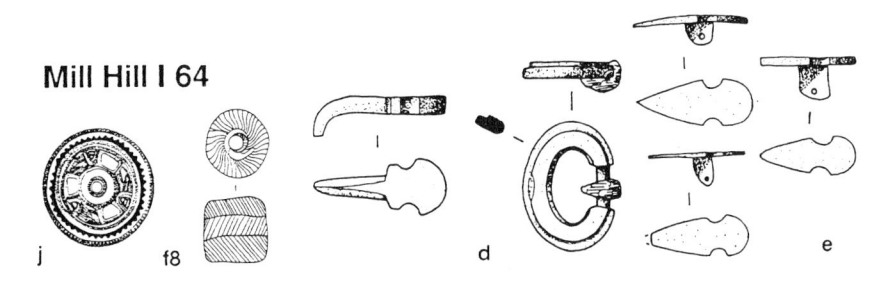

Fig. 3.31 Phase IV: Brooch, buckle-set and reticella bead from Mill Hill I grave 64. After Parfitt and Brugmann 1998. Scale 1:2.
Phase IV: Fibel und Reticella-Perle aus Mill Hill I Grab 64.

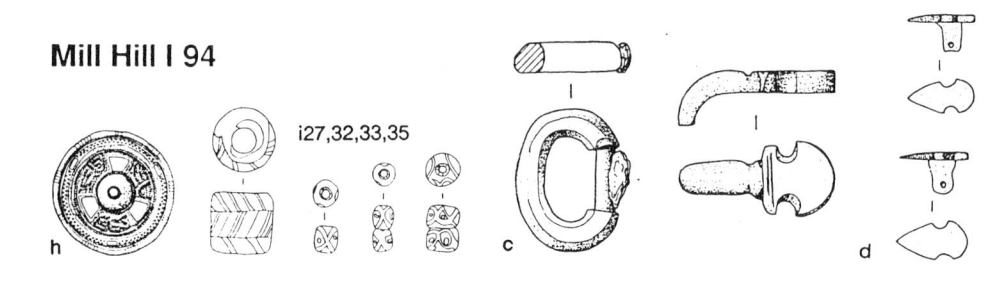

Fig. 3.32 Phase IV: Brooch, reticella bead and buckle-set from Mill Hill I grave 94. After Parfitt and Brugmann 1997. Scale 1:2.
Phase IV: Fibel, Reticella-Perle und Schnallengarnitur aus Mill Hill I Grab 94.

APPENDIX 1

Graves and grave goods mentioned in the text. Grave inventories are not necessarily complete.

Abbreviations:

BM   British Museum
CM   Canterbury Museum
MM   Maidstone Museum

Bekesbourne 19: CM.
1. pair of Continental radiate-head brooches (Jenkins 1957, 294; Kühn 1974, Pl. 84, 22.96)
2. shield-on-tongue buckle; 3 shoe-shaped rivets

Bekesbourne 22: CM.
1. Kentish disc brooch Class 1.1 (Avent 1975, Pl. 1, 1)
2. shield-on-tongue buckle; 3 shoe-shaped rivets
3. skimmer

Bekesbourne 30: CM.
1. pair of Kentish bird brooches (Jenkins 1957, 295)

Bifrons 29: Fig. 3.16; Godfrey-Faussett 1876, 309f.
1–2. pair of radiate-head brooches (Godfrey-Faussett 1876, 309 Fig.; Brown 1915, Pl. 35, 1; Kühn 1974, Pl. 252, 61.22; Hawkes and Pollard 1981, Fig. 7, 1, 2)
3–4. pair of Kentish disc brooches Class 2.1 (Hawkes and Pollard 1981, Fig. 7, 3, 4; Avent 1975, Pl. 4, 23, 24)
6. shield-on-tongue buckle (Hawkes and Pollard 1981, Fig. 7, 6; Godfrey-Faussett 1876, 309f. Fig.)
5. bracteates (Godfrey-Faussett 1876, 309f. Fig.; Hawkes and Pollard 1981, Fig. 7, 7–10)

Bifrons 41: Fig. 3.8; Godfrey-Faussett 1876, 313f.
1–2. pair of Kentish bow brooches (Haseloff 1981, Fig. 90, 1, 1, 2; Godfrey-Faussett 1876, 313 Fig.)
3. bird brooch (Godfrey-Faussett 1876, 313; Brown 1915, Pl. 35, 11; Haseloff 1981, Fig. 90, 1.3)
4. Kentish bow brooch (Godfrey-Faussett 1876, 313f., Fig.; Haseloff 1981, Fig. 90, 1.4)
5. shield-on-tongue buckle; shoe-shaped rivet (Haseloff 1981, Fig. 90, 2.9, 2.10; Godfrey-Faussett 1876, 313)

Bifrons 42: Fig. 3.9; Godfrey-Faussett 1876, 314f.
1. pair of Kentish bow brooches (ibid., 315 Fig.)
2. Continental disc brooch (ibid., 314f. Fig.)
3. Continental disc brooch (ibid.)
4. shield-on-tongue buckle (ibid., 315)
5. skimmer (ibid., Fig.)

Bifrons 51: Godfrey-Faussett 1880, 552
1. pair of Kentish bow brooch (ibid, 552; Åberg 1926, Fig. 131; Leigh 1980, Pl. 103)
2. pair of bird brooches (Thiery 1939, Pl. 11, 165; Godfrey-Faussett 1880, 552)

Bifrons 63: Godfrey-Faussett 1880, 553
1. Kentish bow brooch (ibid.; Hawkes and Pollard 1981, Fig. 8, 5)
2. bracteate (ibid., Fig. 9, 6; Godfrey-Faussett 1880, 553)

Bifrons 64: Fig. 3.17; Godfrey-Faussett 1880, 553
1–2. pair of Kentish bow brooches (ibid.; Hawkes and Pollard 1981, Fig. 10, 1, 2)
3. Kentish disc brooch Class 2.1 (ibid., Fig. 10, 3; Avent 1975, Pl. 3, 22; Godfrey-Faussett 1880, 553)
4. beads (ibid., Fig. 11, 7; Godfrey-Faussett 1880, 553)
5. bracteate (ibid.; Hawkes and Pollard 1981, Fig. 11, 6)

Bifrons 74: Godfrey-Faussett 1880, 554
1. Kentish disc brooch Class 1.1 (ibid.; Avent 1975, Pl. 1, 2)

2. shield-on-tongue buckle; 2 shoe-shaped rivets (Godfrey-Faussett 1880, 554)
3. beads (ibid.)

Buckland 1: Evison 1987, 215f.; Fig. 66
1. Kentish disc brooch (ibid., Fig. 4, 2; Avent 1975, Pl. 15, 73)
2. pendant (Evison 1987, Fig. 4, 1)

Buckland 13: Fig. 3.10; Evison 1987, 218f.; Fig. 67
1. Kentish square-headed brooch (ibid., Fig. 9, 13/2)
2. Kentish square-headed brooch (ibid., Fig. 9, 13/1)
5. small-long brooch (ibid., Fig. 9, 13/5)
6. annular brooch (ibid., Fig. 9, 13/6)

Buckland 20: Fig. 3.11; Evison 1987, 220f.; Fig. 69
2. cloisonné disc brooch (ibid., Fig. 12, 20/2)
3. bracteate (ibid., Fig. 12, 20/4)
6,8. pair of Kentish square-headed brooches (ibid., Fig. 12, 20/6, 20/8)
9. shield-on-tongue buckle and disc-shaped rivets (ibid., Fig. 12, 20/9)

Buckland 29: Evison 1987, 223f.; Fig. 69
1. Kentish disc brooch Class 6.1 (ibid., Fig. 17, 29, 1; Avent 1975, Pl. 34, 114)
2. bracteate (Evison 1987, Pl. 17, 29/8)
3. 3 pendants (ibid., Fig. 17, 29/5)
4. pendant (ibid., Fig. 17, 29/6)
5. looped coin (ibid., Fig. 17, 29/7)

Buckland 30: Fig. 3.29; Evison 1987, 224f.; Fig. 69
2. Kentish disc brooch Class 2.1 (ibid., Fig. 18, 30/2; Avent 1975, Pl. 5, 26)
5. pendant (Evison 1987, Fig. 18, 30/5)
6. buckle (ibid., Fig. 18, 30/6)
p–t. beads (ibid., Fig. 18, 30/4p-t)

Buckland 35: Evison 1987, 226; Fig. 70
1. Kentish disc brooch Class 3.1 (ibid., Fig. 21, 35/1; Avent 1975, Pl. 15, 74)
2. pendant (Evison 1987, Fig. 21, 35/3)

Buckland 38: Fig. 3.30; Evison 1987, 226ff.; Fig. 70
1. Kentish disc brooch Class 2.5 (ibid., Fig. 22, 38/1; Avent 1975, Pl. 10, 56)
g. beads (ibid., 21, 38/4)

Buckland 59: Evison 1987, 232f.; Fig. 33
1. Kentish disc brooch Class 2 (ibid., Fig. 33, 59/1)
3l–o. beads (ibid., Fig. 33, 59/3l-o)

Buckland 67: Evison 1987, 234f.; Fig. 73
1. 3-4 pendants (ibid., Fig. 37, 67/2-4)
2. amethyst bead (ibid., Fig. 37, 67/1a)

Buckland 92: Fig. 3.18; Evison 1987, 237f.; Fig. 76
1. Kentish disc brooch Class 2 (ibid., Fig. 42, 92/1)
2. cloisonné disc brooch (ibid., Fig. 42, 92/2)
d. reticella bead (ibid., Fig. 42, 92/3d)

Chartham 5: Douglas 1793, 20ff.
1. Kentish composite disc brooch Class 5 (ibid., Pl. 5, 1; Avent 1975, Pl. 57, 162)
2. pendant (Douglas 1793, Pl. 5, 2.6)

Chartham 14: Faussett 1856, 170
1. pendant (ibid., Pl. 7, 8)
2. amethyst bead (ibid., 170)

Finglesham D3: Fig. 3.15; Chadwick 1958, 11ff.
1. Kentish square-headed brooch (ibid., Fig. 9, b; Pl. 2)
2. pair of radiate-head brooches (ibid., Fig. 9, a; Pl. 2; Kühn 1974, Taf. 259, 11.53)
3. pair of Kentish bird brooches (Chadwick 1958, Fig. 9, c–d; Pl. 2)

4. shield-on-tongue buckle; shoe-shaped rivets (ibid., Fig. 6, a, j)
5. bracteates (ibid., Fig. 9, e, f; Pl. 3)

Finglesham E2: Fig. 3.19; Chadwick 1958, 18ff.
a. Kentish square-headed brooch (ibid., Fig. 11, a; Pl. 4, A)
b. pair of Kentish square-headed brooches (ibid., Fig. 11, b; Pl. 4, B)
c. Kentish disc brooch Class 1.1 (ibid., Fig. 11, c; Pl. 4, B; Avent 1975, Pl. 1, 8)
e,g,h. shield-on-tongue buckle; shoe-shaped rivet (Chadwick 1958, Fig. 6, e, g, h)

Finglesham 203: Fig. 3.20; Hawkes and Pollard 1981, 333ff.
16–19. buckle; 3 disc-shaped studs (ibid., Fig. 4, 16–19)
20,21. pair of Kentish bow brooches (ibid., Fig. 4, 20, 21)
22. Continental square-headed brooch (ibid., Fig. 4, 22)
23. Continental rosette brooch (ibid., Fig. 4, 23)
25,52. pendant (ibid., Fig. 5, 25, 52)
35,41. bracteates (ibid., Fig. 5, 35, 41)

Gilton XIII: Douglas 1793, 37f.
1. Kentish disc brooch Class 6.2 (ibid., Pl. 9, 2; Avent 1975, Pl. 40, 125)
2. amethyst beads (Douglas 1793, Pl. 9, 1)

Gilton XV: Douglas 1793, 48ff.
1. Kentish composite disc brooch Class 2 (ibid., Pl. 12, 1; Avent 1975, Pl. 52, 153)
2. amethyst beads (Douglas 1793, Pl. 12, 2, 3)

Gilton 27: Faussett 1856, 12f.
1. Kentish disc brooch Class 3.3 (ibid., Pl. 3, 7; Avent 1975, Pl. 26, 97)
2. pendant (Faussett 1856, Pl. 4, 23)

Gilton 41: Faussett 1856, 15f.
1. Kentish disc brooch Class 1.2 (ibid., Pl. 2, 7; Avent 1975, Pl. 2, 15)
2. amethyst bead (ibid., 15)
3. perforated coin (ibid., 16 Pl. 11, 2; cover)

Gilton 62: Faussett 1856, 21; Fig. 6; pl. 3
1. Kentish disc brooch Class 2 (ibid., 21; Fig. 6; pl. 3; Avent 1975, corpus no. 45)
2. beads (Faussett 1865, 21)

Gilton 81: Faussett 1856, 26
1. Kentish disc brooch Class 2 (ibid. 26; Fig. 4; pl. 3; Avent 1975, corpus no. 53)
2. beads (Faussett 1856, 26)

Howletts A: Fig. 3.21; Smith 1917/18, 104
1. Kentish disc brooch Class 2.2 (ibid., Pl. 1, 3; Avent 1975, Pl. 8, 41)
2. Kentish square-headed brooch (Smith 1917/18, Pl. 1, 2; Åberg 1926, Fig. 132)
3. shield-on-tongue buckle; 3 shoe-shaped rivets (Smith 1917/18, Pl. 1, 5, 6)
4. mosaic bead with other types (ibid., Pl. 1, 4; BM)

Kingston 14/205: Faussett 1856, 77ff., Gr 205; Douglas 1793, 37ff., tumulus XIV
1. Kentish composite disc brooch Class 3.2 (ibid., Pl. 10, 6, 7; Faussett 1856, Pl. 1, 1; Avent 1975, Pl. 68, 179)
2. pendant (Faussett 1856, Pl. 1, 2; Douglas 1793, Pl. 10, 1)

Kingston 161: Faussett 1856, 71
1. Kentish disc brooch Class 3.1 (ibid., Pl. 3, 9; Avent 1975, Pl. 19, 81)
2. pendants (Faussett 1856, Pl. 11, 22)
3. 5 amethyst beads (ibid., 71)

Kingston 298/299: Faussett 1856, 91ff.; double grave
1. 289, pendant (ibid., 91 Fig. 1)
2. 299, 2 amethyst beads (ibid., 91)

3. 299, Kentish composite disc brooch unclassified (ibid., Pl. 2, 1; Avent 1975, Pl. 58, 166)
4. 299, Kentish composite disc brooch Class 3.2 (ibid., Pl. 24, 93; Faussett 1856, Pl. 3, 9)

Lyminge 16: Warhurst 1955, 13ff.
1. c. radiate-head brooch (ibid., Pl. 8, 5; Kühn 1974, Taf. 269, 12, 52; Pl. 15, 21/1)
2. bracteate (Warhurst 1955, Pl. 7, a, 1)

Lyminge 44: Fig. 3.23; Warhurst 1955, 28ff.; Pl.
1. pair of Kentish square-headed brooches (ibid., Pl. 12, 1A, 1B)
2. pair of Kentish disc brooches Class 2.1 (ibid., Pl. 12, 2A, 2B; Avent 1975, Pl. 6, 31, 32)
3. skimmer (Warhurst 1955, Pl. 13; Fig. 11, 1)

Mersham A: Fig. 3.22; Bakka 1958, Fig. 52
a. Kentish square-headed brooch (Bakka 1958, Fig. 52, a)
b. Kentish disc brooch Class 1.2 (ibid., Fig. 52, b; Avent 1975, Pl. 3, 17)
e. square brooch (Bakka 1958, Abb. 52, e)
c. buckle (ibid., Fig. 52, c)
d. disc brooch (ibid., Fig. 52, d)

Mill Hill I 25A: Fig. 3.24; Parfitt and Brugmann 1997
1. Continental bird brooch

Mill Hill I 25B: Fig. 3.24; Parfitt and Brugmann 1997
b. skimmer
e. radiate-head brooch
h,l. buckle and plate
i–j. Kentish square-headed brooches
m. Continental S-shaped brooch
t. Kentish disc brooch Class 1.2

Mill Hill I 61: Fig. 3.12; Parfitt and Brugmann 1997
c–d. Continental bird brooches
e–g. Kentish square-headed brooches
h–i. buckle and plate

Mill Hill I 64: Fig. 3.31; Parfitt and Brugmann 1997
d. buckle
f8. reticella bead
j. Kentish disc brooch Class 2

Mill Hill I 71: Fig. 3.13; Parfitt and Brugmann 1997
a,i. Kentish square-headed brooches
b–c. Continental square brooches

Mill Hill I 73: Fig. 3.14; Parfitt and Brugmann 1997
d. annular brooch
e. bow brooch
f. Continental animal brooch
g. button brooch
j. buckle

Mill Hill I 92: Fig. 3.25; Parfitt and Brugmann 1997
e. Continental disc brooch
g. Continental radiate-head brooch
h–i. pair of Continental radiate-head brooches
j. buckle

Mill Hill I 94: Fig. 3.32; Parfitt and Brugmann 1997
c–d. buckle-set

h. Kentish disc brooch Class 2
i. beads

Mill Hill I 102: Fig. 3.26; Parfitt and Brugmann 1997
a. Kentish square-headed brooch
b12. reticella bead
d. Kentish radiate-head brooch
i. Kentish disc brooch

Mill Hill I 105C: Fig. 3.27; Parfitt and Brugmann 1997
a. Kentish disc brooch Class 2
d. disc brooch
f. Kentish disc brooch Class 1.2
h. buckle and rivets
i–j. Kentish square-headed brooches

Mill Hill II A: Woodruff 1904, 9
1. Kentish disc brooch Class 2 (ibid.; Avent 1975, 17; No. 58; Dover Museum, formerly Deal Library Exhibition)
2. beads (Avent 1975, 17; Dover Museum)

Monkton 3: Hawkes and Hogarth 1974, 55ff.
1. Kentish composite disc brooch (ibid., Fig. 4; Pl. 1)
2. amethyst beads (ibid., Fig. 5, a-l)

Sarre A: Roach Smith 1860
1. pendant (ibid., Pl. 2)
2. Kentish composite disc brooch Class 2 (ibid., Pl. 3; Avent 1975, Pl. 66, 177)
3. amethyst beads (ibid., Pl. 2)

Sarre 4: Fig. 3.28; Brent 1862/63, 310ff.; 1868, Pl. 8, 4
1. Kentish square-headed brooch (Brent 1862/63, Pl. 2, 2)
2. Kentish square-headed brooch (ibid., Pl. 2, 1)
3. pair of Kentish disc brooches Class 2.2 (ibid., Pl. 1, 8, 9; Avent 1975, Pl. 8, 42, 43)
4. skimmer (Brent 1862/63, Pl. 2, 3)
5. shield-on-tongue buckle; shoe-shaped rivet (ibid., 313 Fig.)
6. bracteates (Brent 1863, Pl. 1, 1–6)

Sarre 115: Brent 1864/65, 175f.
1. 2 pendants (ibid., Pl. 6, 2)
2. disc brooch (ibid., Pl. 6, 6)

Sarre 158: Brent 1864/65, 180
1. Kentish disc brooch Class 1.2 (ibid., Pl. 6, 4; Avent 1975, Pl. 3, 19)
2. cloisonné disc brooch (Brent 1864/65, Taf. 6, 10)
3. buckle (ibid., 180 Fig.)

Sibertswold 101: Faussett 1856, 118
1. Kentish composite disc brooch Class 1 (ibid., Pl. 2, 6; Avent 1975, Pl. 50, 149)
2. 17 amethyst beads (ibid., 118)

Sibertswold 172: Faussett 1856, 130ff.
1. pendants and looped coins (ibid., Pl. 4, 1, 2,, 7–9, 13, 16, 17; Pl. 11, 1, 3)
2. amethyst beads (ibid., 131)

Upchurch A: Smith 1852, 161f.
1. Kentish composite disc brooch Class 1 (ibid., Pl. 37, 1; Avent 1975, Pl. 51, 150)
2. amethyst beads (Smith 1852, 162)

NOTES

1 For references to sites excavated before 1950 see Meaney (1964) s.v. 'Kent'. All sites are shown on Fig. 3.1.
2 See e.g. Smith 1852; 1854; 1857; 1860; 1868; Godfrey-Faussett 1876; 1880.
3 Continuing the 'Valetta House' excavations published by Hurd (1913).

4 The remaining parts of Ozengell.
5 See, for instance, Myres (1971) for pottery, Bode (in press) for cruciform brooches and Hawkes and Pollard (1981) for bracteates.
6 E.g. Graves 93 and 96a at Buckland (Evison 1987, Figs. 43 and 45).

7 For an overview see, for instance, Steuer (1990) and Koch (in press).

8 Dr Koch has kindly allowed me to use not only her chronological framework before publication but also the collection of data she used for her seriation.

9 For dating evidence see brooch (e) in grave A at Mersham (above).

BIBLIOGRAPHY

Åberg, N. 1926. *The Anglo-Saxons in England*. Uppsala.

Ament, H. 1977. 'Zur archäologischen Periodisierung der Merowingerzeit'. *Germania* 55, 1977, 133–40.

Avent, R. 1975. *Anglo-Saxon garnet inlaid disc and composite brooches*. BAR British Ser 11. Oxford.

Bakka, E. 1958. 'On the beginning of Salin's Style I in England'. *Univ Bergen Årb Hist-ant rekke* 3.

Bakka, E. 1981. 'Scandinavian-type gold bracteates in Kentish and continental grave-finds'. In V. I. Evison (ed.), *Angles, Saxons and Jutes. Essays presented to J. N. L. Myres*, 11–38. Oxford.

Bierbrauer, V. 1985. 'Das Reihengräberfeld von Altenerding in Oberbayern und die bajuwarische Ethnogenese. Eine Problemskizze'. *Zeitschrift Archäol des Mittelalters* 13, 7–25.

Bode, J. forthcoming. *Das kaiser- und völkerwanderungszeitliche Gräberfeld von Schmalstede, Kr. Rendsburg-Eckernförde*. Urnengräberfriedhöfe Schleswig-Holsteins. Neumünster.

Böhme, H. W. 1988. 'Les Thuringiens dans le Nord du royaume franc'. *Rev Archéol Picardie* 3–4, 57–69.

Böhner, K. 1958. *Die fränkischen Alterthümer des Trierer Landes*. Germanische Denkmäler der Völkerwanderungszeit B 1. Berlin.

Brent, J. 1862/63. 'Account of the Society's Researches in the Saxon Cemetery at Sarr'. *Archaeol Cantiana* 5, 305–22.

Brent, J. 1864/65. 'Account of the Society's Researches in the Anglo-Saxon cemetery at Sarr'. *Archaeol Cantiana* 6, 157–85.

Brown, G. Baldwin 1915. *The Arts in Early England. Vol. 3: Saxon Art and Industry in the Pagan Period*. London.

Brugmann, B. 1994. *Das angelsächsische Gräberfeld von Mill Hill, Deal, Kent*. Doctoral Thesis Kiel Univ. Microfiche.

Chadwick, S. E. 1958. 'The Anglo-Saxon Cemetery at Finglesham, Kent: A Reconsideration'. *Medieval Archaeol* 2, 1–71.

Christlein, R. 1966. *Das alamannische Reihengräberfeld von Marktoberdorf im Allgäu*. Materialheft zur Bayrischen Vorgeschichte 21. Kallmünz.

Douglas, J. 1793. *Nenia Britannica*. London.

Evison, V. I. 1987. *Dover: The Buckland Anglo-Saxon Cemetery*. London.

Evison, V. I. 1994. *An Anglo-Saxon Cemetery at Great Chesterford, Essex*. CBA Res Rep 91. London.

Faussett, B. 1856. *Inventorium Sepulchrale*. London.

Godfrey-Faussett, T. G. 1876. 'The Saxon Cemetery at Bifrons'. *Archaeol Cantiana* 10, 298–315.

Godfrey-Faussett, T. G. 1880. 'The Saxon Cemetery at Bifrons. Concluded'. *Archaeol Cantiana* 13, 552–6.

Härke, H. 1992. *Angelsächsische Waffengräber des 5.-7. Jahrhunderts*. Cologne.

Haseloff, G. 1981. *Die germanische Tierornamentik der Völkerwanderungszeit*. 3 vols. Berlin.

Hawkes, S. C. and Hogarth, A. C. 1974. 'The Anglo-Saxon cemetery at Monkton, Thanet'. *Archaeol Cantiana* 89, 49–89.

Hawkes, S. C. and Pollard, A. M. 1981. 'The Gold Bracteates from Sixth-Century Anglo-Saxon Graves in Kent, in the Light of a New Find from Finglesham'. *Frühmittelalterliche Stud* 15, 316–70.

Hawkes, S. C., Merrick, J. M. and Metcalf, D. M. 1966. 'X-ray fluorescent analysis of some Dark Age coins and jewellery'. *Archaeometry* 9, 98–138.

Hogarth, C. 1973. 'Structural Features in Anglo-Saxon Graves'. *Archaeol J* 180, 104–19.

Hurd, H. 1913. *Some Notes on Recent Archaeological Discoveries at Broadstairs*. Broadstairs.

Jenkins, F. 1957. 'The Bekesbourne Excavations'. *Kent and Sussex J* 2, 11, 294 f.

Koch, U. 1968. *Die Grabfunde der Merowingerzeit aus dem Donautal um Regensburg*. Germanische Denkmäler der Völkerwanderungszeit A 10. Berlin.

Koch, U. 1977. *Das Reihengräberfeld bei Schretzheim*. Germanische Denkmäler der Völkerwanderungszeit A 13. Berlin.

Koch, U. 1990. *Das fränkische Gräberfeld von Klepsau im Hohenlohekreis*. Stuttgart.

Koch, U. in press: *Das alamannisch-fränkische Gräberfeld bei Pleidelsheim*. Forschungen und Berichte Vor- und Frühgeschichte Baden-Württemberg. Stuttgart.

Kühn, H. 1965. *Die germanischen Bügelfibeln der Völkerwanderungszeit in der Rheinprovinz*. Graz.

Kühn, H. 1974. *Die germanischen Bügelfibeln der Völkerwanderungszeit in Süddeutschland*. Graz.

Leeds, E. T. 1913. *The Archaeology of the Anglo-Saxon Settlements*. Oxford.

Leeds, E. T. 1936. *Early Anglo-Saxon Art and Archaeology*. Oxford.

Leigh, D. 1980. *The Square-Headed Brooches of Sixth-Century Kent*. University College, Cardiff Ph. D. thesis. Unpublished.

Martin, M. 1976. *Das fränkische Gräberfeld von Basel-Bernerring*. Basel.

Martin, M. 1984. 'Weinsiebchen und Toilettgerät'. In A. Cahn and A. Kaufmann-Heinimann (eds), *Der spätrömische Silberschatz von Kaiseraugst*, 97–132. Derendingen.

Martin, M. 1989. 'Bemerkungen zur chronologischen Gliederung der frühen Merowingerzeit', *Germania* 67, 121–41.

Meaney, A. L. 1964. *A Gazetteer of Early Anglo-Saxon Burial Sites*. London.

Möller, J. 1987. *Katalog der Grabfunde aus der Völkerwanderungs- und Merowingerzeit im südmainischen Hessen (Starkenburg)*. Germanische Denkmäler der Völkerwanderungszeit B 11. Stuttgart.

Myres, J. N. L. 1971. 'The Angles, the Saxons and the Jutes'. The Raleigh Lecture on History. British Academy 1970 (Oxford 1971).

Neuffer-Müller, C. and Ament, H. 1973. *Das fränkische Gräberfeld von Rübenach*. Germanische Denkmäler der Völkerwanderungszeit B7. Berlin.

Nissen-Fett, E., 1941. 'Relieffibeln vom nordischen Typus in Mitteleuropa'. *Bergen Mus Årb Hist-ant rekke* 5.

Parfitt, K. 1995. 'The Buckland Saxon cemetery'. *Current Archaeol* 144, 459–64.

Parfitt, K. and Brugmann, B. 1997. *The Anglo-Saxon Cemetery on Mill Hill, Deal*. Soc Medieval Archaeol Monogr Ser 14. Leeds.

Perkins, D. J. R., n. d. *The Thanet Archaeological Unit. Interim Excavation Reports 1977–1980*. Broadstairs.

Pirling, R. 1966. *Das römisch-fränkische Gräberfeld von Krefeld-Gellep*. Germanische Denkmäler der Völkerwanderungszeit B 2. Berlin.

Raveaux, J.-P. (ed.) 1992. *La Collection archéologique de Mme Perrin de la Boullaye*. Châlons-en-Champagne.

Roth, H. and Theune, C. 1988. *SW ♀ I–V: Zur Chronologie merowingerzeitlicher Frauengräber in Südwestdeutschland. Ein Vorbericht zum Gräberfeld von Weingarten, Kr. Ravensburg*. Stuttgart.

Schulze-Dörrlamm, M. 1986. 'Romanisch oder Germanisch? Untersuchungen zu den Armbrust- und Bügelknopffibeln des fünften und sechsten Jahrhunderts n. Chr. aus den Gebieten westlich des

Rheins und südlich der Donau'. *Jahrb Römisch-Germanischen Zentralmus* 33, 593–720.

Smith, C. Roach 1852. *Collectanea Antiqua 2*. London.

Smith, C. Roach 1854. *Collectanea Antiqua 3*. London.

Smith, C. Roach 1857. *Collectanea Antiqua 4*. London.

Smith, C. Roach 1860. 'On Anglo-Saxon Remains discovered recently in various places in Kent'. *Archaeol Cantiana* 3, 35–46.

Smith, C. Roach 1868. *Collectanea Antiqua 6*. London.

Smith, R. 1917/18: 'On Prehistoric and Anglo-Saxon Remains Discovered by Captain L. Moysey at Howletts, near Brigde, Kent'. *Proc Soc Ant* 30, 102–13.

Steuer, H. 1990. Review of H. Benmann und W. Schröder (eds), *Die transalpinen Verbindungen der Bayern, Alemannen und Franken bis zum 10. Jahrhundert. Fundberichte Baden-Württemberg* 15, 494–506.

Theune-Vogt, C. 1990. *Chronologische Ergebnisse zu den Perlen aus dem alamannischen Gräberfeld von Weingarten, Kr. Ravensburg. Kleine Schriften aus dem Vorgeschichtlichen Seminar der Phillips-Univ Marburg* 33. Marburg.

Thiery, G. 1939. *Die Vogelfibeln der Germanischen Völkerwanderungszeit*. Bonn.

Vanhaeke, L. 1996: 'Harmignies, Nimy' (Prov. Hainaut – B). In Wieczorek, Périn, von Welck and Menghin (eds) 1996, *q.v.* 845–6.

Warhurst, A. 1955. 'The Jutish Cemetery at Lyminge'. *Archaeol Cantiana* 69, 1–40.

Werner, J. 1961. *Katalog der Sammlung Diergard (Völkerwanderungszeitlicher Schmuck) 1*. Berlin.

Wieczorek, A. 1987. 'Die frühmerowingischen Phasen des Gräberfeldes von Rübenach'. *Berichte RGK* 68, 356–492.

Wieczorek, A., Périn, P., von Welck, K., and Menghin, W. (eds) 1996. *Die Franken, Wegbereiter Europas*. Mainz.

Woodruff, C. 1904. 'Further Discoveries of Late Celtic and Romano-British Interments at Walmer'. *Archaeol Cantiana* 26, 9–16.

4. The sixth-century transition in Anglian England: an analysis of female graves from Cambridgeshire

John Hines

INTRODUCTION:
THE SIXTH-CENTURY TRANSITION

The archaeology of the Early Anglo-Saxon Period in England is characterized above all by the presence of human graves containing grave goods – so much so, that there has in the past been some unease over the dominance of early Anglo-Saxon archaeology as a discipline by cemetery studies alone (Dickinson 1983, esp. p. 39; for a mature review of the relationship between cemetery and settlement archaeology, see Hinton 1990, 1-41). One may indeed define the 'Early Anglo-Saxon Period' as the period of furnished burial, extending from some time in the 5th century to the early 8th (Geake 1997, 1). Within this period, a division between an earlier sub-phase and a later one came clearly into focus in the inter-war years, primarily as a result of Lethbridge's excavations of larger examples of the otherwise relatively rare – or at least less easily recognized – later cemeteries (Lethbridge 1931; 1936) and Leeds's synthesizing discussions, in particular his definition of the 'Final Phase' of Anglo-Saxon furnished burial (Leeds 1936, 96-114).

The state of chronological studies and understanding at that time saw no particular problem in the relationship between these two major phases in the early Anglo-Saxon burial sequence. The earlier material, characterized by types classically assigned to the Migration Period (*Völkerwanderungszeit, folk[e]vandringstid*) on the Continent and in Scandinavia, was conventionally dated to the 5th and 6th centuries, and the Final Phase predominantly to the 7th century. The boundary line, consequently, was assigned to 'around 600'. Despite the lack of any very compelling evidence for this particular absolute date, it was all too easy to associate the material-cultural watershed with a major historical event that it apparently coincided with, the beginning of the conversion of England to Christianity, dating from Augustine's mission to Kent in A.D. 597. The supposed historical context offered an apparently satisfactory explanation: the change in burial practices and other material culture in the Final Phase could be argued to represent the influence, direct or indirect, of Christianity (Hyslop 1963; Boddington 1990).

The problematic nature of the Migration Period/Final Phase transition has become increasingly evident as the accepted absolute dates for comparative material and phase boundaries on the Continent and in Scandinavia have moved earlier. Style I, the Germanic zoomorphic art style characteristic of the later Migration Period, has moved from the 6th century back to a starting date some time in the second half of the 5th century (Haseloff 1974; 1981), and Style II, its successor which characterizes the following phase – i.e. the Swedish Vendel Period, Danish Late Germanic Iron Age etc. – has been re-assigned from the 7th century to dates starting earlier and earlier in the 6th: ca. 560/570; ca. 550; now even ca. 520 (Haseloff 1981, 647-73; Wilson 1976, 10; Høilund Nielsen 1987; Lund Hansen 1992; Martin 1989). Of especial importance has been Birgit Arrhenius' study of the chronology of the Vendel graves (1983), which effectively aligned the start of the Swedish Vendel Period with the developments that characterize the beginning of Ament's AMIII on the Continent, ca. 560/570. It has, in consequence, been rather difficult to sustain the conventional dating of the division between the Migration Period and the Final Phase in England to a point at least a generation later, at ca. 600 (Hines 1984, 16-32). The issue is not simply one of accepting a revised dating. Rather the redating seems to stretch the Anglo-Saxon chronological sequence in an awkward way, and thus to highlight the uncertainties in our perception of what happened as the Migration Period gave way to the Final Phase, and thus what the implications of the transition may be in our reconstructed cultural history.

A cemetery site which is central to the analysis which underlies this paper provides a simple but effective illustration of the problems involved here. Recent excavations (1989-1991) at Edix Hill, Cambridgeshire, produced the skeletal remains of 147 indentifiable individuals from 113 grave cuts of the Early Anglo-Saxon Period (Malim and Hines 1998). By conventional chronology we would estimate that the sequence of burials here covers a timespan from ca. A.D. 500 to the mid-7th century – i.e. a period of some 150 years. Amongst the female graves, at least 29 are datable to the Migration Period while only 6 at most are demonstrably datable to the Final Phase. A further 11 female graves cannot be assigned to one phase rather than the other. If all these poorly furnished and thus 'undatable' graves were assumed to belong to the Final Phase, the earlier phase burials would still outnumber the later ones by more than 3:2. The cemetery has not been excavated in its entirety, and it is possible that a higher proportion of Final-phase graves may be concentrated in some as yet unexcavated area, but the distribution of datable burials within the areas and trial trenches that have been dug gives no reason to believe that this is likely to be the case. Even at the narrowest ratio that can be produced – which implies that *all* excavated Migration-period burials were sufficiently furnished now to be datable – an even rate of burial in the cemetery would lead us to estimate ca. 590 (three-fifths of the way through its life-span) as the approximate date of the transition from Migration-period to Final-phase burials at Edix Hill. With any earlier date than this, we must infer that the rate of burial here, and thus by implication the size of the burying population, was lower in the Final Phase than earlier. Thus the establishment of what changes took place and of their date are inextricably interlinked, and both are vital to our understanding of the community represented in the cemetery of Edix Hill and its culture.

Even though, as we have noted, it is in broad terms perfectly evident that the sequence of art styles, technical practices and artefact-types in England has much in common with the material-cultural sequences found on the Continent and in Scandinavia, we cannot safely assume that changes were so closely concurrent in all areas that their chronologies can be directly correlated on this basis (see Brugmann, this volume). One check on such relationships would be through absolute dates for the material phenomena, but at present we only have these in sufficient measure in parts of the Continent (see Theune, this volume). In England we have very few reliable absolute datings for deposits before the 7th century, and prospects for much improvement in this respect for the earlier Anglo-Saxon centuries are virtually nil. The problem of correlation should not, however, merely be dismissed as a fundamental flaw in the whole enterprise of attempting to obtain both finer and more reliable dates. It is fundamental to accurate reconstruction of trans-European cultural developments in the post-Roman centuries, without which no proper explanatory model of the context of any archaeological evidence of this period can be possible.

Even at a purely local level, the need to take the problem seriously is clear enough. There was evidently a massive material-cultural change between the Migration Period and the Final Phase in England. But the precise nature of the change remains obscure. Was it abrupt, or was it gradual – or did it, perhaps, take place at different rates, and indeed different times, in different areas? Until such relative chronological and geographical questions can be answered there is no basis for any general absolute date or date-bracket for the change, and thus no foundation for any detailed assessment of its circumstances.

Before proceeding further, a word of explanation of the terminology used here is appropriate. I shall use the term Early Anglo-Saxon Period (capitalized) as the proper name for the period of furnished burial (5th to 8th century). The still frequent use of the term 'pagan Anglo-Saxon' for this period on the false basis that the provision of grave goods is definitely non-Christian is deplorable. I propose also to retain the terms Migration Period and Final Phase for the two major subdivisions of the Early Anglo-Saxon Period, as already introduced. Both terms are generally familiar, and the former term accords with the German, Swedish and Norwegian terms for the cultural phase with which the Migration Period in England clearly overlaps and shares a wide range of material characteristics within the 5th and 6th centuries, even though there is room for keen controversy over precisely how the Migration Period is defined and dated in those areas (see, for instance, Kristoffersen, and Axboe, this volume).

BACKGROUND TO THE PROJECT

A programme of analytical research attempting to answer the questions outlined above began with the post-excavation analysis of finds from the Edix Hill cemetery. The existence of this Anglo-Saxon cemetery had been known since the mid-19th century, as large quantities of finds were collected from the site during coprolite digging and a limited amount of more scholarly excavation up to the 1860's. The site came to be known by various names – as Orwell, Malton, or (most commonly) Barrington A. It was rediscovered by metal-detector users in the later 1980's, and subjected to an exploratory excavation by the Cambridgeshire Archaeology unit, directed by Tim Malim, between 1989 and 1991.

It was soon appreciated that the time-span of the cemetery rendered it particularly important as a site with continuity across the Migration Period/Final Phase transition. Chronological analysis was therefore given a high priority in post-excavation research project funded by English Heritage. The quantity of data from this site alone, however, was relatively small, especially as the burials have to be divided for analytical purposes into three major categories: adult male, adult female, and children's graves. (Culturally, the child-adult boundary seems to lie around 12-14 years of age.) Here, as elsewhere (see Theune, this volume), the male and female sets have to be analysed quite separately. It is very rare for children's graves to be sufficiently furnished for inclusion within the male and female categories.

There are, however, several other Early Anglo-Saxon burial sites in the neighbourhood of Edix Hill (Fig. 4.1). Recently excavated sites are Great Chesterford (Evison 1994), where the range of burials excavated begins and ends earlier than on Edix Hill (approximately from some time in the second half of the 5th century into the second half of the 6th), and Melbourn (Wilson 1956), a site

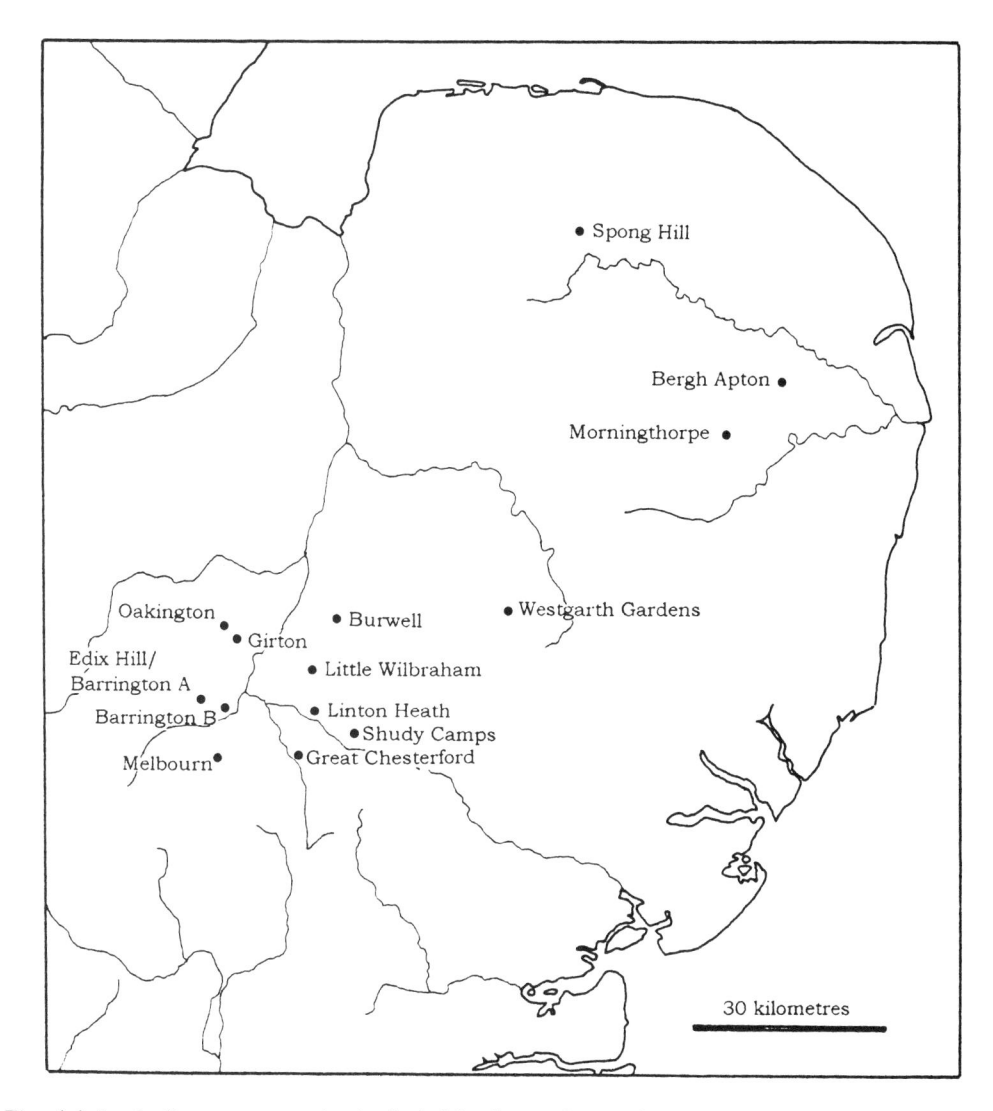

Fig. 4.1 Anglo-Saxon cemeteries included in the analyses of grave goods discussed in the text.
Im Text erwähnte und in der Analyse benutzte angelsächsische Gräberfelder.

contemporary only with the latest burials on Edix Hill. The classic Final-phase cemeteries excavated by Lethbridge (1931; 1936) at Burwell and Shudy Camps lie by the eastern borders of Cambridgeshire. North of Cambridge 24 graves have recently been excavated from the Oakington cemetery (Taylor *et al.* 1998), not far from the site of the cemetery at Girton College (Hollingworth and O'Reilly 1925). From 19th-century excavations we have records of grave assemblages, and a lot of surviving finds, both from Barrington A and from the closely adjacent site of Barrington B (Foster 1883), which seems to belong to exactly the same date-range as Barrington A/Edix Hill. There are also sufficiently thorough records of the burials in the substantial cemeteries of Little Wilbraham and Linton Heath for much of this material to be used for comparative and analytical purposes (Neville 1852; 1854). There are several further cemetery-sites in the area, for instance at Grantchester, Haslingfield, Shelford and Trumpington, but we have no usable grave assemblages from these sites. With, in particular, the date-range covered by Great Chesterford, Barrington A/Edix Hill, Barrington B, and Melbourn within the clearly defined locality of South Cambridgeshire, it was hoped to be able to use the aggregated evidence of these sites to produce a clear relative chronology for the period from the later 5th to the mid-7th century.

ANALYSIS AND RESULTS

The basic method by which we sought to establish a fine-chronological sequence was by typological classification of the artefactual material and then attempting to order the graves by means of a seriation of the individual grave groups as assemblages of classified types. For most of the recurrent Early Anglo-Saxon artefact-types there are established typological schemes, although these vary markedly in their level of detail. In several cases in this analysis typological differences between sub-types of a particular class of artefacts had to be ignored because otherwise certain types would appear too infrequently – normally only once – in the data set and thus be useless for purposes of seriation. A complete review of the typology and modifications to it used here is unnecessary; significant deviations from normal classification are noted as appropriate below.

The particular method of seriation used in this case was correspondence analysis, using the KVARK programme written by Torsten Madsen. The initial analyses relevant to this account were carried out by Karen Høilund Nielsen, the latter by the author. A number of different data sets were examined in turn, the analysis of each one naturally affected by the results obtained and questions posed by previous analyses. It was recognized at an early stage that the more copiously and variously furnished female graves were far more likely to yield substantial and detailed chronological results than the male graves, and it is only the results of examination of this set of graves that are reported here.

The first data set analysed was the graves from Edix Hill and Barrington A (i.e. material from both recent and 19th-century excavations in the one cemetery) alone; the second the 'South Cambridgeshire' cemeteries (Edix Hill/Barrington A, Barrington B, Great Chesterford and Melbourn); thirdly, for comparative and control purposes, the second set plus the finds from a group of recently excavated and published cemeteries in East Anglia (Bergh Apton, Morningthorpe, Spong Hill inhumations and Westgarth Gardens: Green and Rogerson 1978; Green, Rogerson and White 1987; Hills, Penn and Rickett 1984; West 1988); and fourthly the full set of Cambridgeshire cemeteries, including also Burwell, Girton, Linton Heath, Little Wilbraham, Oakington and Shudy Camps. The only reason for investigating the latter set last was the practical one of the time required to establish the classification of artefacts within the grave groups by reference to the museum collections.

The results of the analyses are informative and valuable in many ways, not least in the form of one major and largely (though not entirely) unexpected discovery of a pattern within the data that is one of the obstacles in the way of a fine chronology of the Anglo-Saxon Migration Period. Within the Migration-period graves alone it proved relatively easy to produce the sort of parabola that represents a smooth sequence which could, in turn, be a chronological order, whichever of the four sets one examines (Fig. 4.2). In such diagrams, however, established typological chronology – the validity of which is cautiously accepted here – indicates that the chronological sequence from one end of the curve to the other is not one of early-to-late but rather early/late/early. This led to the conclusion – of which a detailed account is given in the Edix Hill report (Malim and Hines 1998) – that the Migration Period, in the 6th century at least, in Cambridgeshire and East Anglia, was characterized by the presence of alternative 'costume groups', which we have labelled A, B, C and D. These were characterized by the recurrent association of particular types of brooch and wrist-clasp (see below for a subsequent modification of this range):

A – Cast saucer brooches. Applied disc or saucer brooches. Great square-headed brooches.
B – Type-B cruciform brooches. Cross-headed small long brooches and their derivatives. Anglian equal-armed brooches. Copper-alloy pins. Form B 17a and b wrist clasps.
C – Type-Bb and Type-C cruciform brooches. Square-headed small long brooches and their derivatives. Form B12, B 13b, B 14a and B 18, and Class C 1 (Barrington Type) wrist clasps.
D – Type-D and Type-Z cruciform brooches. Trefoil-headed small long brooches. Annular brooches. Penannular brooches. Disc brooches. Form B 7 clasps. Iron pins.

Of course, the fact that these groups fall into a smooth sequential pattern indicates that they merge and overlap rather than standing totally apart, but the associative tendencies here are still strong enough to produce a pattern stronger than any chronological sequence yet detected.

When the Edix Hill report was written, this phenomenon could be described and discussed on the basis of the South Cambridgeshire plus East Anglia data set noted above. It has subsequently been tested by adding in the complete Cambridgeshire data. This largely supports the conclusions previously drawn, albeit with one substantial modification of a kind that can fairly be regarded as a refinement of our understanding rather than a change implying some profound error in the whole model. Firstly, it proved necessary to remove artefact-types characteristic of costume group A (cast saucer brooches and great square-headed brooches) in order to work towards a basic parabola. Subsequently

B17ABCL	:	Form B17a or b clasps
ANGEQUA	:	Anglian equal-armed brooch
XFORMB	:	Type-B cruciform brooch
SLBCPO	:	Cross-headed small long brooch
SKRINGS	:	Copper-alloy slip-knot ring
SLBXDER	:	Cross-head derivative small long brooch
AEPIN	:	Copper-alloy pin
OWKDISC	:	Openwork disc brooch
B13ACL	:	Form B 13a clasps
XFORMZ	:	Type-Z cruciform brooch
PENANN	:	Penannular brooch
B7CL	:	Form B7 clasps
DISCB	:	Disc brooch
B13CCL	:	Form B13c clasps
SLBTRE	:	Trefoil head small long brooch
AGFRING	:	Silver finger ring
XFORMD	:	Type-D cruciform brooch
AELARGRI	:	Large copper-alloy ring
ANN	:	Annular brooch
FEPIN	:	Iron pin
AEGH	:	Copper-alloy girdle hanger
SMALLSHB	:	Small square-headed brooch
B20CL	:	Form B20 clasps
B18CL	:	Form B18 clasps
B1214ACL	:	Form B12 or B14a clasps
SLBSQH	:	Square-headed small long brooch
SHB23	:	Phase 2 or Phase 3 great square-headed brooch
ACL	:	Class A clasps
XFORMC	:	Type-C cruciform brooch
XFORMBB	:	Type-Bb cruciform brooch
SLBSQDER	:	Square-head derivative small long brooch
C1BARCL	:	Form C1 clasps, Barrington type

Key to Figure 4.2.

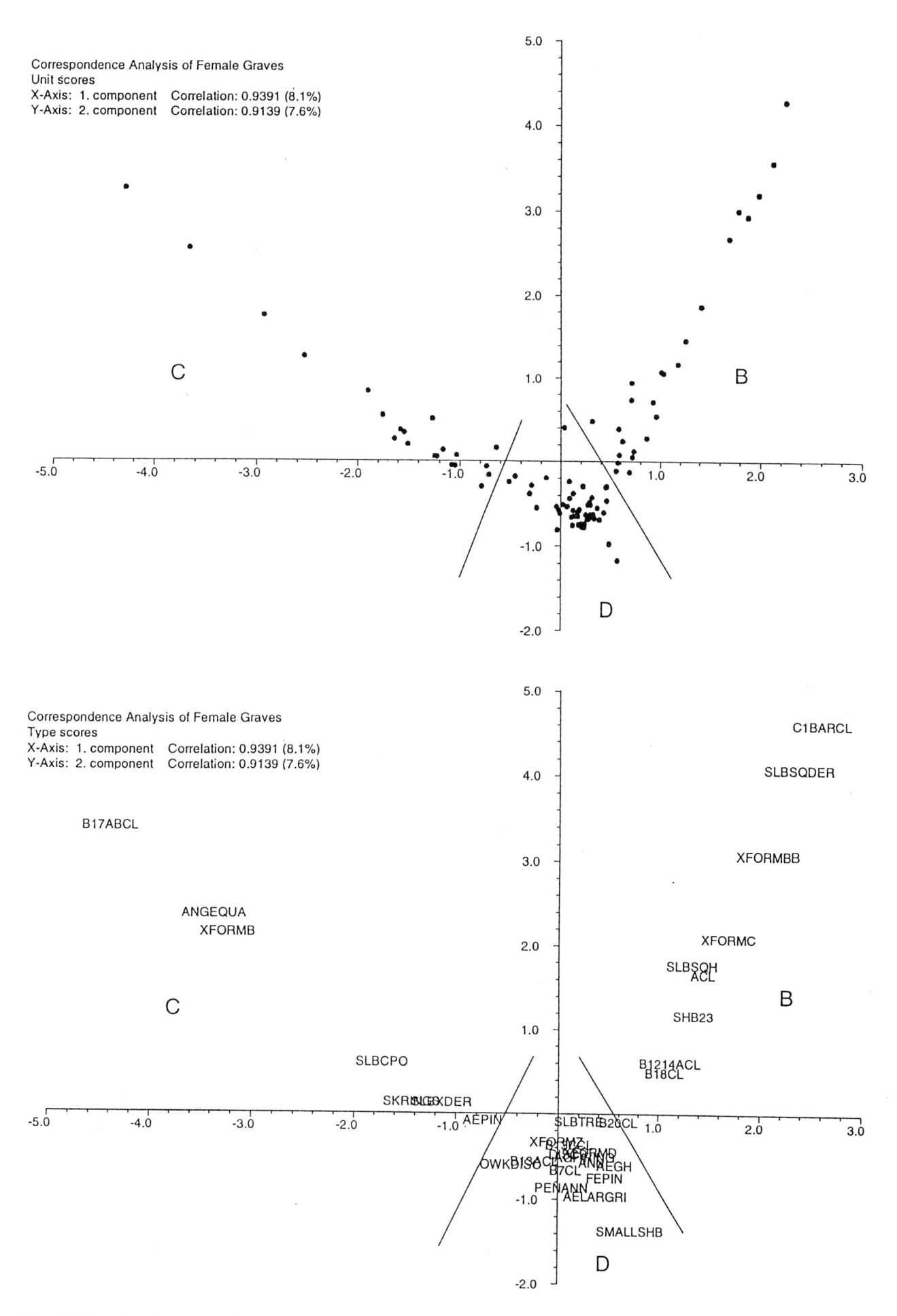

Fig. 4.2 Results of correspondence analysis of South Cambridgeshire and East Anglian Migration-period female graves, showing costume groups B, C and D.

Korrespondenzanalyse der völkerwanderungszeitlichen Frauengräber aus South Cambridgeshire und East Anglia: Trachtgruppen B, C und D.

the distinction between simple and derivative small long brooches was abandoned. The most significant change was the removal of Type-C cruciform brooches (as defined by Catherine Mortimer, 1990, and more or less the same as Åberg's more familiar group III: Åberg 1926, 28-56), which had formerly been thought to be typical of costume group C. In this data set these seem to have moved much closer to costume group D, but on the whole tended to fall in the middle of the parabola and thus to close up the ends (Fig. 4.3). With the further removal of two grave groups, Little Wilbraham grave 9 and Spong Hill grave 2, both of which, anomalously in costume-group terms, combine square-headed and cross-headed types of small long brooch, a good parabola still comprising no less than 138 grave groups was obtained (Fig. 4.4).

This sequence again reveals the distinctive grouping of

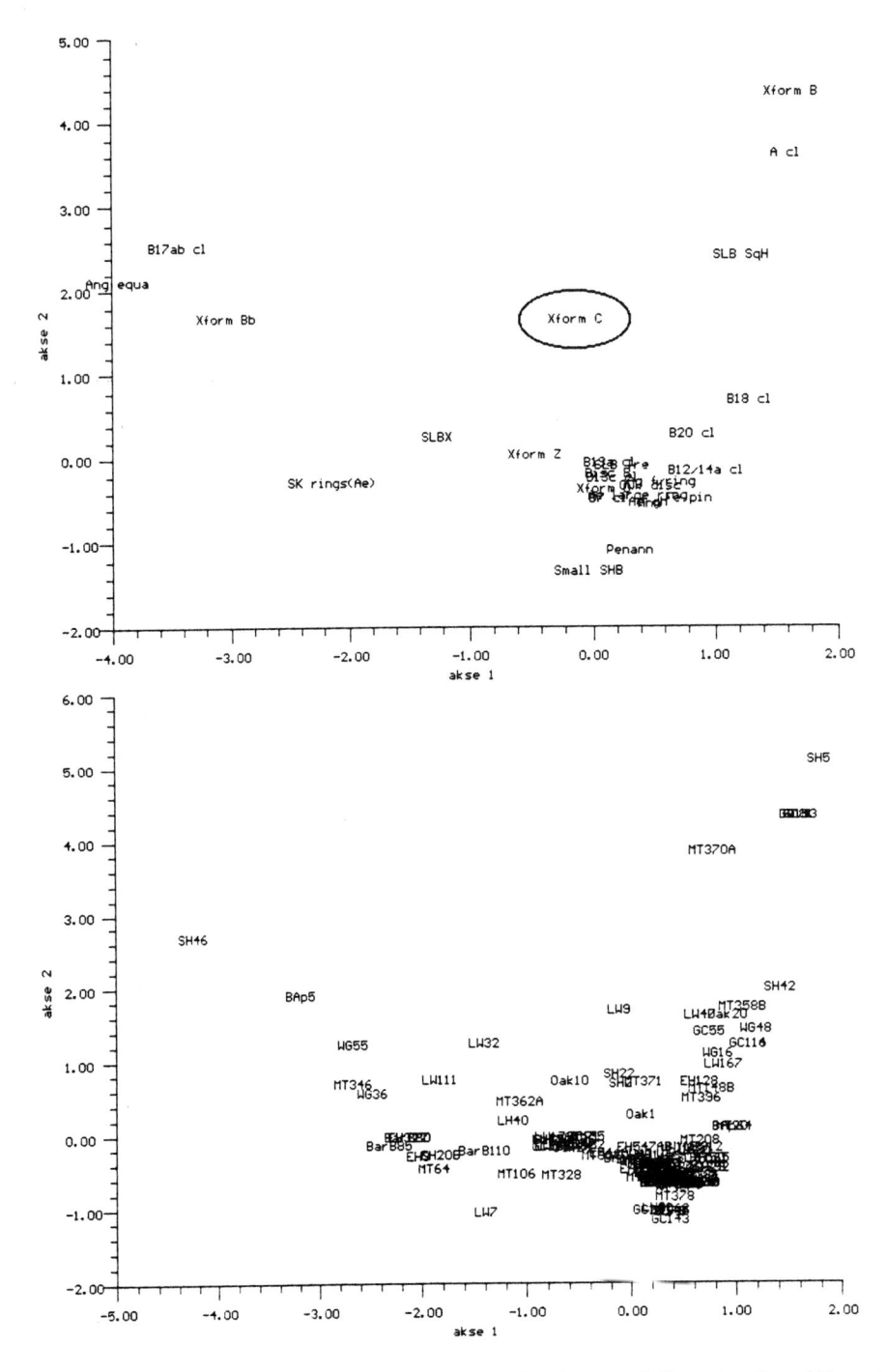

Fig. 4.3 Results of correspondence analysis of further Cambridgeshire and East Anglian Migration-period female graves, before the removal of Type-C cruciform brooches (= Xform C).
Korrespondenzanalyse der völkerwanderungszeitlichen Frauengräber aus dem weiteren Cambridgeshire und East Anglia – vor der Herausnahme der kreuzförmigen Fibeln vom Typ C.

Fig. 4.4 Results of correspondence analysis of further Cambridgeshire and East Anglian Migration-period female graves, showing revised costume groups B, C and D.
Korrespondenzanalyse der völkerwanderungszeitlichen Frauengräber aus dem weiteren Cambridgeshire und East Anglia – mit Eintragung der revidierten Trachtgruppen B, C und D.

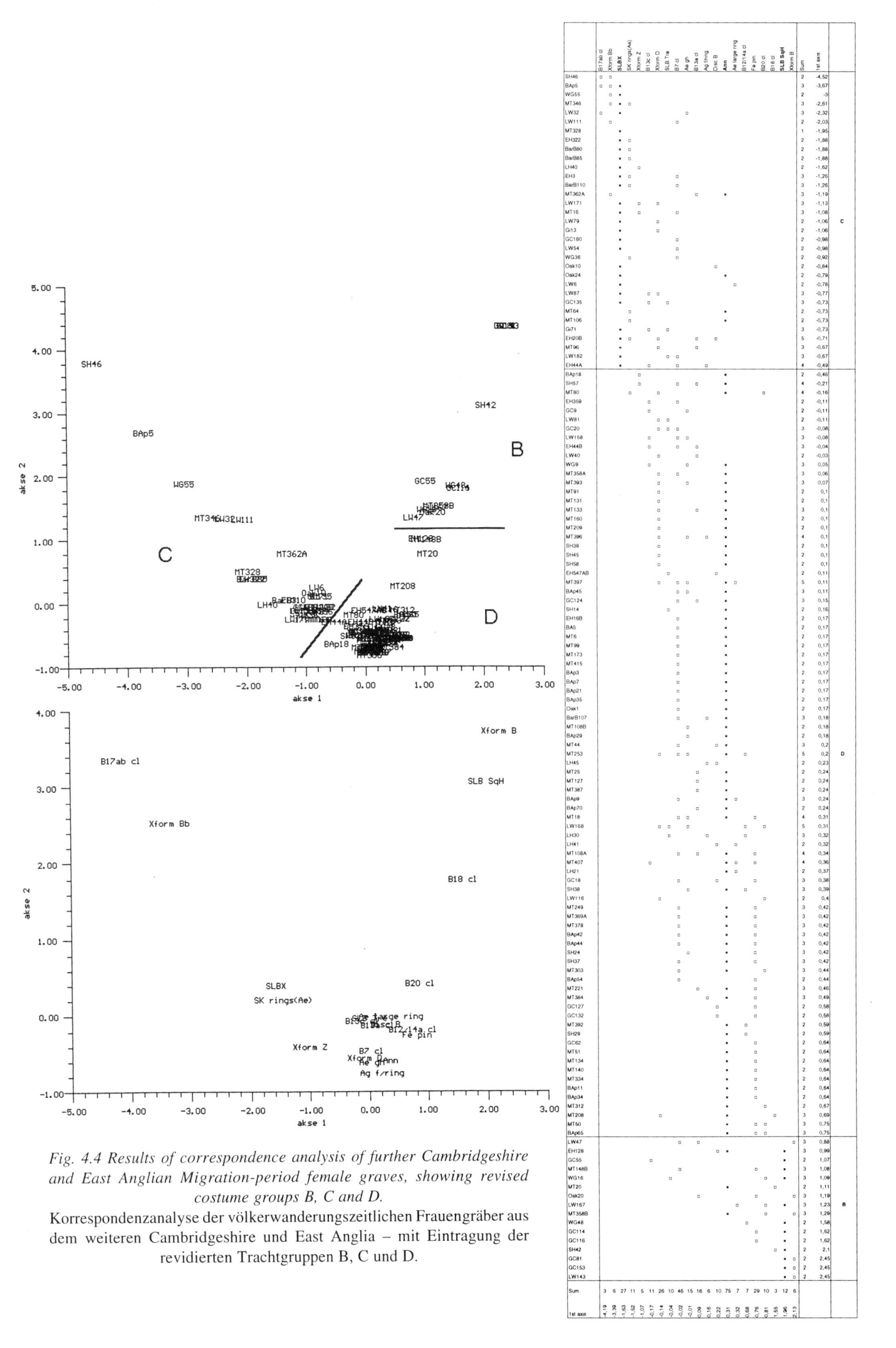

a variety of brooch- and clasp-types at the two ends, namely Type-B cruciform brooches (Xform B: 7 occurrences in the data set), square-headed small long brooches (SLB SqH: 12 occurrences), form B 18 clasps (B 18 cl: only 3 occurrences) as costume group B, and Type-Bb cruciform brooches (i.e. cruciform brooches with a spatulate foot: here Xform Bb: 6 occurrences), cross-headed small long brooches (SLBX: 27 occurrences), form B 17 clasps (B 17 ab cl: only 3 occurrences), and copper-alloy slip-knot rings (SK rings (Ae): 11 occurrences), as costume group C. Costume group D remains in the middle of this series. Interestingly, the five occurrences of zoomorphic florid cruciform brooches (Xform Z) now show a leaning towards costume group C rather than costume group D. What the sorted matrix (Fig. 4.4 right) shows most clearly, however, is the degree to which the costume groups reflect the separation of alternative types of shoulder brooch: the small long brooch-types noted immediately above in costume groups B and C and the annular brooches in costume group D. Yet there are other associated types that seem to characterize the groups in a significant way, not least form B 7 clasps (B7 cl) in costume group D, so that it appears implausible that these groupings are merely random associations with the contrasting shoulder brooch-types. Probably the most significant point here is the thorough-going contrast between costume groups B and C, which is likely to represent a stage of development preceding the predominance of costume group D in the final decades of the Migration Period.

In order to search further for a possible chronological series, and in particular to examine the problem of the Migration Period/Final Phase transition, the geographically more compact but chronologically much wider data set of the Cambridgeshire female graves was then analysed on its own. Preliminary steps taken here in order to make the data more tractable were the conjunction of Type-D and Type-Z cruciform brooches (Åberg's groups IV and V), and treating all applied disc or saucer brooches, all cast saucer brooches, all copper-alloy buckles except for the small oval type, all iron buckles, and all scutiform pendants, as single categories. There were inevitably several artefact-types which occurred only once in the original data set and which had to be omitted, including what are otherwise recognized as chronologically diagnostic forms such as the shield-on-tongue buckle and shoe-shaped studs, amethyst beads, the weaving batten, and the Kentish garnet-inlaid disc brooch.

Even then, correspondence analysis at first produced at best a triangular scatter of points rather than anything approaching a parabola. To disentangle this it was necessary to remove Phase-3 great square-headed brooches, silver bracelets and cast saucer brooches: in other words, once again to separate out costume group A. This in fact then produced the basis of a clear parabola on the first and third axes of the correspondence analysis (cf. Fig. 4.5). The pattern was subsequently clarified by dropping a number of artefact-types, the copper-alloy pin, which could not be comprehensively subclassified, the tooth pendant, of which only 2 examples were present, and copper-alloy vessel rim mounts, which if they are, as widely thought, repair mounts would inconsistently represent the deposition of turned wooden vessels in graves; followed by a number of graves which then were left with only one artefact-type still in the analysis. Finally Barrington A grave 15, which is highly unusual as an inhumation grave containing, *inter alia*, a typically Migration-period small long brooch and an amber bead in association with a fragment of a 'bone' (presumably antler) comb which is otherwise highly typical of Final-phase inhumation graves, was also removed. The result, then, is a nice parabola, convincingly early (Migration-period) on one side and late (Final-phase) on the other (Fig. 4.5).

In light of the prominence and importance of the transition between these two major sub-phases of the Early Anglo-Saxon Period as noted above, the middle of the sequence of the Cambridgeshire female graves is an area of especial interest. Here we can find a small group of five graves which seem, if any do, to stand between the Migration Period and Final Phase clusters. Burwell grave 61 and Shudy Camps grave 31 are from what are generally regarded as exclusively and definitively Final-phase cemeteries, while Great Chesterford grave 21 is from a cemetery that would otherwise appear not to extend beyond the Migration Period. The other two graves are from Edix Hill (graves 54 and 83), the cemetery where we know we have continuity between the two phases.

None of these graves is particularly richly furnished (Tab. 4.1). They fall together primarily because they all

---

Great Chesterford 21: *Pair of square-head derivative small long brooches.* Copper-alloy pin with flattened expanding head. Knife fragments. *Iron chatelaine set* including figure-of-eight chain links, large *iron ring fragments*, and looped heads from suspended iron objects, *probably latch lifters.*

Edix Hill 54: *Small oval copper-alloy buckle.* Iron knife. *Iron latch lifter. Iron ring.*

Edix Hill 83: *Shale annular brooch (first identified as iron).* 11 glass beads (3 monochrome and 8 polychrome) and 3 amber beads. *Iron buckle with oval loop.* Iron knife. Antler ring, with T-shaped iron girdle hanger. *Set of three iron latch lifters.* Set of iron belt rings and double-looped link. Two iron studs with sheet silver heads, apparently from a belt.

Burwell 61: Copper-alloy pin with discoid head. *Iron buckle.* Iron knife. *Copper-alloy loop.*

Shudy Camps 31: Iron knife. *Iron chatelaine*, with large *iron ring*, and iron and copper-alloy amulet capsule containing lignite.

Tab. 4.1 Inventory of the five grave groups falling around the boundary between the Migration Period and the Final Phase in figure 4.5. (Artefact-types included in the seriation and correspondence analysis highlighted.)
Inhalt der fünf Bestattungen an der Grenze zwischen Völkerwanderungszeit und 'Final Phase' in Abb. 4.5.

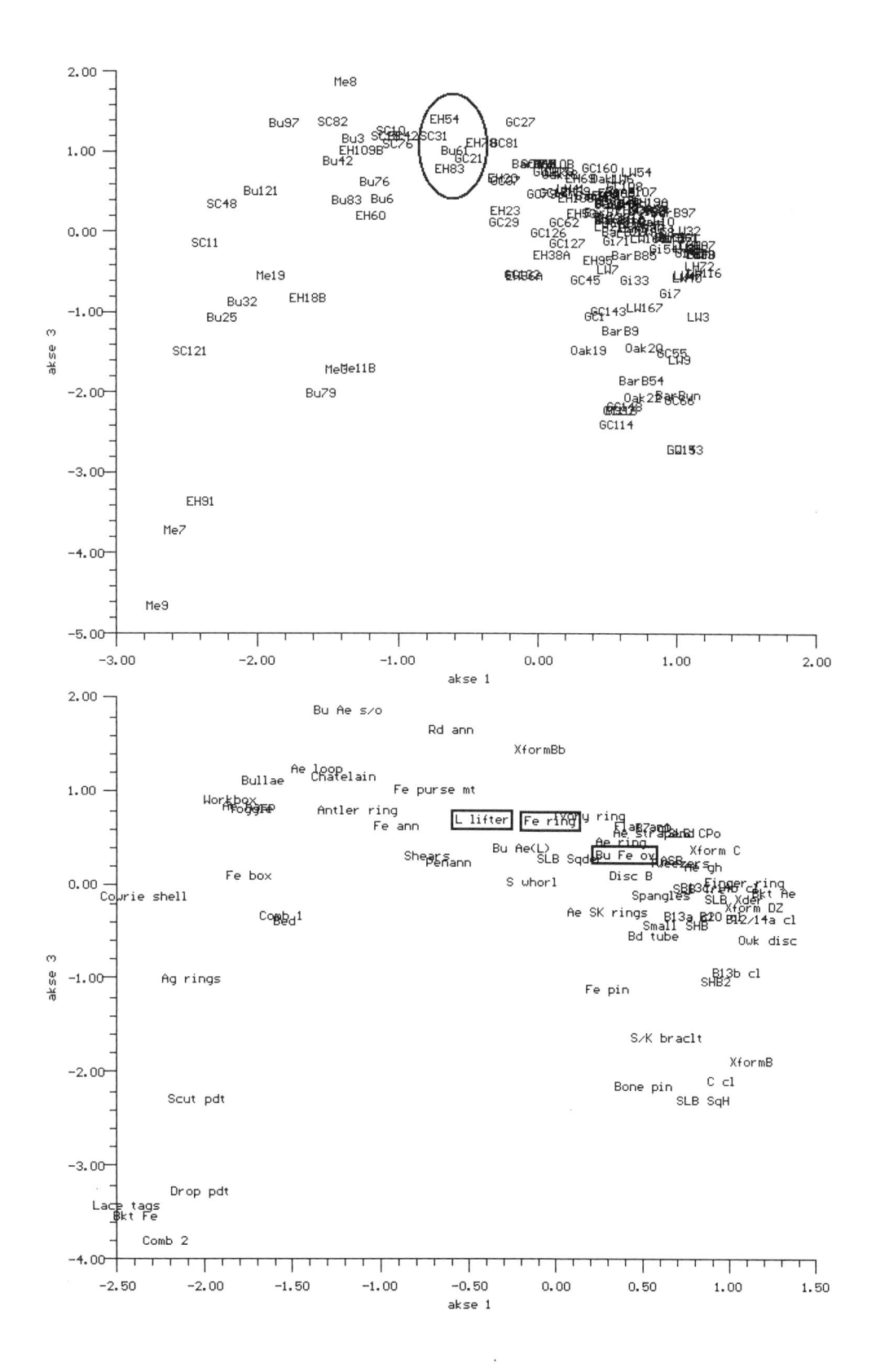

The sixth-century transition in Anglian England: an analysis of female graves from Cambridgeshire

Fig. 4.5 Results of correspondence analysis of Cambridgeshire female graves of the Early Anglo-Saxon Period, showing chronological sequence between the Migration Period and the Final Phase.
Korrespondenzanalyse der frühen angelsächsischen Frauengräber aus Cambridgeshire – mit Eintragung der chronologischen Abfolge zwischen Völkerwanderungszeit und 'Final Phase'.

contain at least one of a range of iron belt fittings which in fact are the three most common artefact-types within this analytical scheme: iron belt rings (33 occurrences in all), latch lifters (29) and iron buckles (26). Two of the graves also contained the remains of iron chatelaine chains. All of these are artefact-types which it is already believed are found in both sub-phases, though chatelaines are unquestionably more typical of the Final Phase (Geake 1997, 57-8). A correspondence analysis of this whole set of graves with these three widespread artefact-types removed does not lose the basic parabola visible in figure 4.5, although inevitably it is less clear and the separate clustering of the Migration-period and Final-phase groups is the more apparent (Fig. 4.6). Three of these grave groups in fact contain at least one item that would conventionally be regarded as characteristically Final-phase: a small oval copper-alloy buckle in Edix Hill grave 54, an antler ring in Edix Hill grave 83 and a small copper-alloy loop, conjecturally from a box or container, in Burwell grave 61, as in graves 6 and 23 at that site (Lethbridge 1931, 59). Great Chesterford grave 21, conversely, has a typically Migration-period pair of square-headed derivative small long brooches.

When we have artefact-types continuing from one period into another we can, obviously, encounter ambiguous graves, especially when the level of furnishing is relatively low. In this case, however, only one of these five graves, Edix Hill grave 83, can properly be described as being ambiguous in such a way as to be best regarded as genuinely transitional between the two sets. This grave contained not only an antler belt ring paralleled at Burwell in graves 76 and 83, but also three amber beads, a type most characteristic of the Migration Period although not exclusively so (Geake 1997, 47-8 and 218). The annular brooch it contained was initially identified as being made of iron, although subsequent laboratory examination indicates it is actually made of shale. It is nevertheless counted in this analysis as being of the same category as reportedly iron annular brooches from Burwell graves 76 and 83 and Great Chesterford grave 29 – and thus with parallels in both typically Final-phase and Migration-period cemeteries – although clearly the true material of those other brooches needs to be checked when possible. In all but one case, then, there is still evidence by which we can reasonably differentiate between the grave groups and assign them to one phase or the other. One grave out of 138 does not constitute a burial horizon, and there are thus no grounds for identifying any form of transitional phase between the Migration Period and the Final Phase in Cambridgeshire.

Although it is only the female graves that are assessed in detail here, the correlation of the female-grave chronology with that of the male graves can usefully, if briefly, be discussed with reference to the burials at Edix Hill. A set of male graves at Edix Hill belonging to a distinctly late phase of burial can be identified on the basis of the types of shield boss and/or spearhead buried there, or stratigraphic relationships with later graves thus identified. On the basis of the latter evidence it becomes clear that antler combs are as characteristic a feature of later male graves as they are of female graves. Small oval copper-alloy buckles equivalent to those found in the Final-phase female graves are also found in male graves. It is, of course, patently obvious that the latest phases of both male and female burial in a single cemetery should normally be broadly contemporary, but this does not mean that we can immediately assign the transition between earlier and later male graves to exactly the same point as the transition between Migration-period and Final-phase female graves as just described. On the other hand it is not unlikely that the profound and sudden material-cultural changes evident in the female range should have had a simultaneous counterpart in the male range, and it will be by reference to shared artefact-types that we may, eventually, be able to investigate this problem more successfully.

CORRELATION AND ABSOLUTE CHRONOLOGY

Parallels between the definitive material of the Migration Period or the Final Phase in Cambridgeshire and objects found over a much wider area, in England, on the Continent and in Scandinavia, must be the primary point of reference for aligning these two Anglo-Saxon phases with the relative and absolute chronologies in those other areas. Local brooch-types of the Migration Period are also found in northern Germany and Scandinavia (i.e. cruciform, small long, saucer and square-headed brooches), in some cases in typologically highly similar versions (see Böhme 1974; Reichstein 1975; Hines 1997). The Anglian English wrist-clasps can be fitted into a single historical sequence with the Scandinavian clasps (Hines 1993). There are certain imported or copied Continental types characteristic of the Migration Period in England, e.g. radiate-head brooches, shield-on-tongue buckles and shoe-shaped rivets (see Brugmann, this volume), while there is also some detectable Anglo-Saxon influence on square-headed brooches found in the Rhineland and Alamannia (Hines 1997, 205-22). The common Germanic animal style, Salin's Style I, is characteristic of the Anglo-Saxon Migration Period, along, towards the end, with a 'Bichrome Style' that has an intriguing parallel in Scandinavia (Hines 1997, 133). Parallels between England and overseas in the Final Phase are more matters of technique and style than of typological links, although the presence of buckles with long triangular backplates and amethyst beads is one of the latter kind. Otherwise the most conspicuous parallels are the introduction of more elaborate cloisonné work, cell patterns without cloisonné, and Style II. On this basis it is abundantly clear that the transition between the Migration Period and the Final Phase is very similar to the phase boundaries at cemeteries such as Rübenach and Schretzheim which Ament added together to define the transition from AMII to AMIII in his general chronological scheme for the Continent and dated to 560/70, and correspondingly with the features of

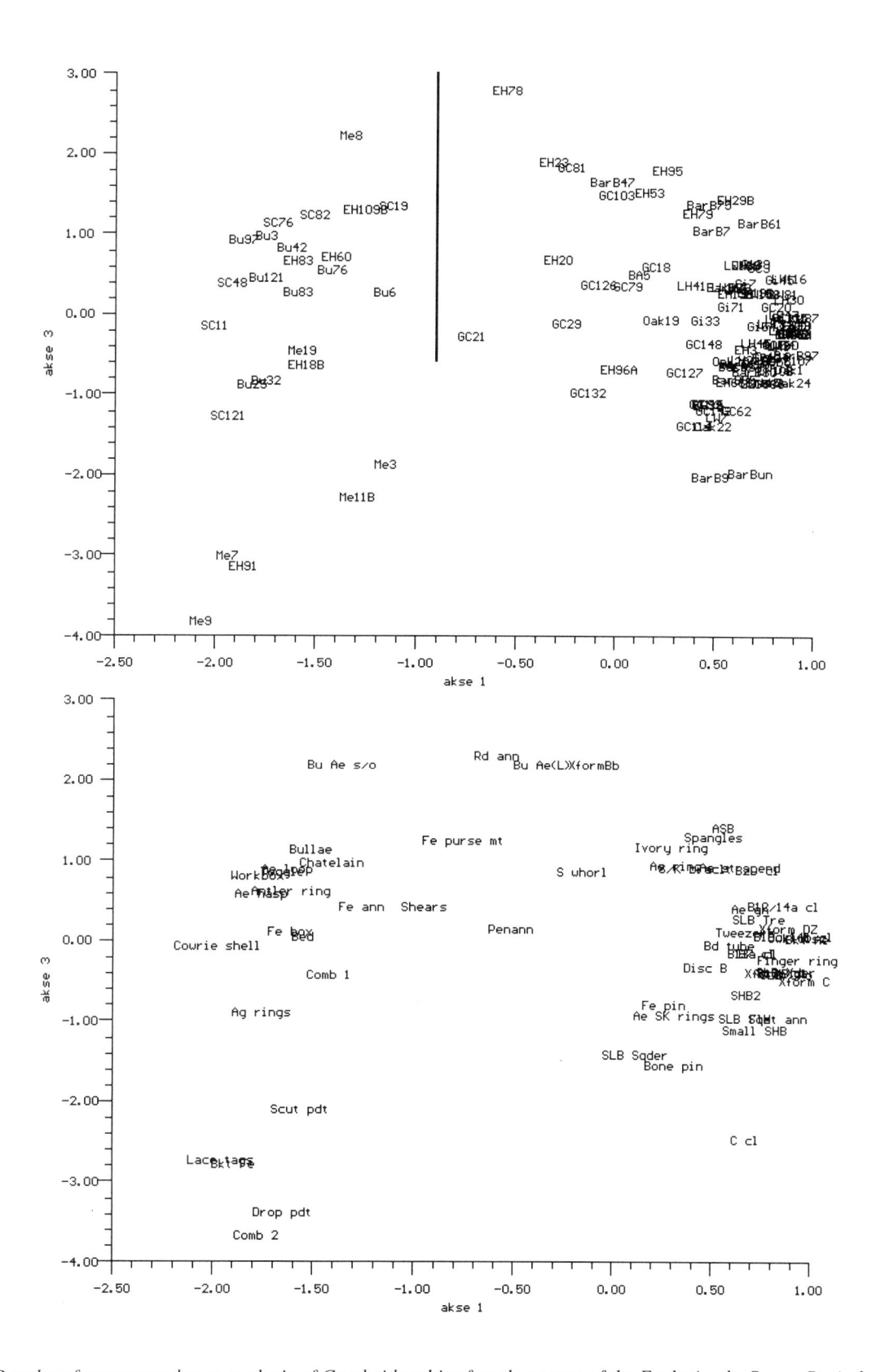

Fig. 4.6 Results of correspondence analysis of Cambridgeshire female graves of the Early Anglo-Saxon Period, with the commonest artefact-types omitted (iron belt rings, latch lifters and iron buckles), showing the chronological division between the Migration Period and the Final Phase.
Korrespondenzanalyse der frühen angelsächsischen Frauengräber aus Cambridgeshire, ohne die gemeinsam vorkommenden Typen. Die Trennung zwischen Völkerwanderungszeit und 'Final Phase' ist markiert.

marking the beginning of Phase SW ♀ III in south-western Germany as defined by Roth and Theune (1988; see Theune, this volume). Note that Siegmund (1989) could push elements of this transition a little earlier in the Rhineland, to his Phase 5 (555/65). The end of the Migration Period in England is also very similar to the end of the Norwegian Migration Period after its final Stufe IV as defined and described by Bakka (1973; cf. Kristoffersen, this volume). Around the west coast of Norway, indeed, we can see a variation between more and less conservative regions in the late Migration Period similar to that seen in England towards the end of in the Anglo-Saxon Migration Period (Hines 1997, 308-9).

It is for these reasons that not only the equivalent term 'Migration Period' is justified as a name for the earlier phase in England, but also that the 560's (= 560/570) are suggested as a 'conventional date' for its abrupt end wherever that can be corroborated. By *conventional date* is meant an approximate absolute date that is the best estimate one can make, on evidence currently available, for a clearly defined point in the archaeological sequence. Another example of an important conventional date is Günther Haseloff's estimate of *circa* A.D. 475 for the introduction of Style I art already referred to. Unlike multidimensional scaling in imaginary space, however, conventional dates are placed in real, not imaginary, time. And one wants one's conventional dates to be as realistic as possible. It is hardly surprising, then, that their essentially heuristic character is often misunderstood.

As has been consistently noted, if reliable absolute dates could be put to material characteristic of the Migration Period and the Final Phase in England they should contribute to a clearer understanding of the character of the phase-boundary, not only in terms of the local sequence of events but also in terms of correlation with external sequences. If a suitable series of such dates could be obtained within a consistent area, sharply distinct absolute dates would confirm the separation of the two phases, and might pinpoint the boundary line between them, while overlapping dates would support the view of a fuzzy boundary there. The differences in the absolute dates that we have are in fact quite clear, but regrettably leave a massive area of doubt. Before the Sutton Hoo boat grave, coin-'dated' graves are restricted to Kent. Just one of these is clearly relevant to the central issue under discussion here, Gilton grave 41 (Faussett 1856, 15-16; Hawkes, Merrick and Metcalf 1966, 103-4). This grave group, however, simply reinforces the problem rather than helping to solve it. In brief, the collection of artefacts here, including amethyst beads, silver wire necklace rings, a chatelaine, and a keystone garnet disc brooch of Avent's class 1.2, looks like a transitional one, albeit with a majority of material characteristic of the Final Phase. Brugmann (this volume) notes its ambiguous place between her phases IV and V. The coin in this grave is a Visigothic copy of a coin of Justinian (527-565) in virtually mint condition. It is reported that this coin is unlikely to have been struck *post*-575, and it could of course be considerably earlier. It is not difficult to infer or justify a plausible date of burial *circa* 560-580 for this grave group, but even within that bracket we cannot be sure whether we have dated a late Migration-period grave, an early Final-phase grave, or the boundary line itself. Nor do we yet know precisely what relevance that date in Kent has to other regions of England. To confuse matters even more, this grave group also reportedly included an openwork silver mount which has now been identified by specialists at the Department of Medieval and Later Antiquities of the British Museum as a 12th-century Romanesque piece! In respect of this item at least, then, the grave group is corrupt – but this does not mean that one can completely discount the possible integrity of the remainder of the assemblage.

CONCLUDING REVIEW

Right across the Anglo-Saxon culture area in England there is evidence of a substantial change in material culture, very broadly sometime during the second half of the 6th century. As virtually all of the other papers in this volume show, a similar change is apparent in the Rhineland and south-western Germany and in Scandinavia, and its precise description and dating cause deep archaeological problems there too. In some cases, even within what we call Anglian England, we can see that the sequence of change is of substantially different character in different areas; in the majority of cases, however, regional differences in the Migration Period make it impossible to draw even such conclusions with real confidence. We are thus not always able to tell whether this change was a sharp change, a gradual change, or a change accompanied by a distinct transitional phase. In one area that has been subjected to close examination, however, Cambridgeshire, it is now concluded that there was an abrupt watershed between the two phases, with no transitional phase. We cannot locate this inferred boundary line in absolute chronology other than by noting its similarity to the AMII/AMIII boundary on the Continent, which is coin-dated to *circa* 555/570. We do not have the evidence from which we can infer how long it took for changes in fashion on the Continent to reach here and most of the rest of England – if indeed there were any significant delay at all.

But we can nevertheless regard Anglo-Saxon chronology not only with respect but even in a positive light. Material connexions and parallels between England and the Continental and Scandinavian sequences not only give us some sort of guide to the approximate date of the Anglo-Saxon finds but can also help in correlating and dating the overseas sequences too. The continuous common development and apparent interchange between England and the more northerly parts of Scandinavia (Norway in particular), in respect of square-headed and disc-on-bow brooches for instance, up to and even beyond the end of the Migration Period, is of great importance. The fact that the latest stylistic developments of Scandinavian square-headed brooches (e.g. on the Norwegian Jorenkjøl brooch) are closely mirrored in the latest developments on Anglo-Saxon square-headed brooches supports Bakka's proposed closing date (*circa* 560) for the late and final phase, Stufe IV, of his Migration Period, not only in Norway but also in other northerly areas of Scandinavia (cf. Sjøvold 1988). This phase is not properly represented in Denmark, which is why it has become impossible to talk of a Migration Period (*alias* Early Germanic Iron Age) continuing beyond about the first quarter of the 6th century there (Lund Hansen 1992). But even within southern Scandinavia one can discern a phase (Høilund Nielsen 1987, Phase 1A), that must have been more or less coterminous with AMII and Bakka's VWZ IV, and which has something of the character of a transitional phase between the Early Germanic Period (= Migration Period) and the Late (= Vendel Period), for instance by combining the latest square-headed brooches with the earliest of the small equal-armed and disc brooches characteristic of the later phase.

Further research can and will be done on the Migration Period/Final Phase boundary in England. More detailed analyses of the East Anglian cemeteries referred to above (see Fig. 4.1) are at present underway. Large collections of material from other regions, some of it recently excavated, can be analysed too. Two areas that seem particularly ripe and appropriate for comparative analysis are Lincolnshire/Humberside and Northamptonshire/Rutland (cf. Leahy, forthcoming; Adams and Jackson 1989; Timby 1996). None of this may throw much light into the darkness that currently envelops us, but even if it gives us a more confident sense of the scope and limits of Anglo-Saxon chronology the work will be worthwhile.

It may well be that we will, in the end, have to accept the situation that the archaeology of early Anglo-Saxon England will not comprehensively divide into relatively short phases such as those of Bakka's Migration Period, the Danish Late Roman Iron Age (Lund Hansen 1977), and the Merowingerzeit in western Europe (Nieveler and Siegmund, and Theune, this volume). Even without such a complete *Stufengliederung*, we do have at our disposal some form of fine chronology within the Migration Period and Final Phase of England. This fine chronology is based on individual artefact-types, where clear and reliable sequences of typological and/or artistic development can be traced (e.g. Fig. 4.7). Such 'special' sequences can be aligned with more general relative – and thus absolute – chronologies at a number of points, so that, for instance, we can be confident that the earliest Anglo-Saxon square-headed brooches, in the upper row of the matrix in figure 4.7, are contemporary with the Continental brooches of

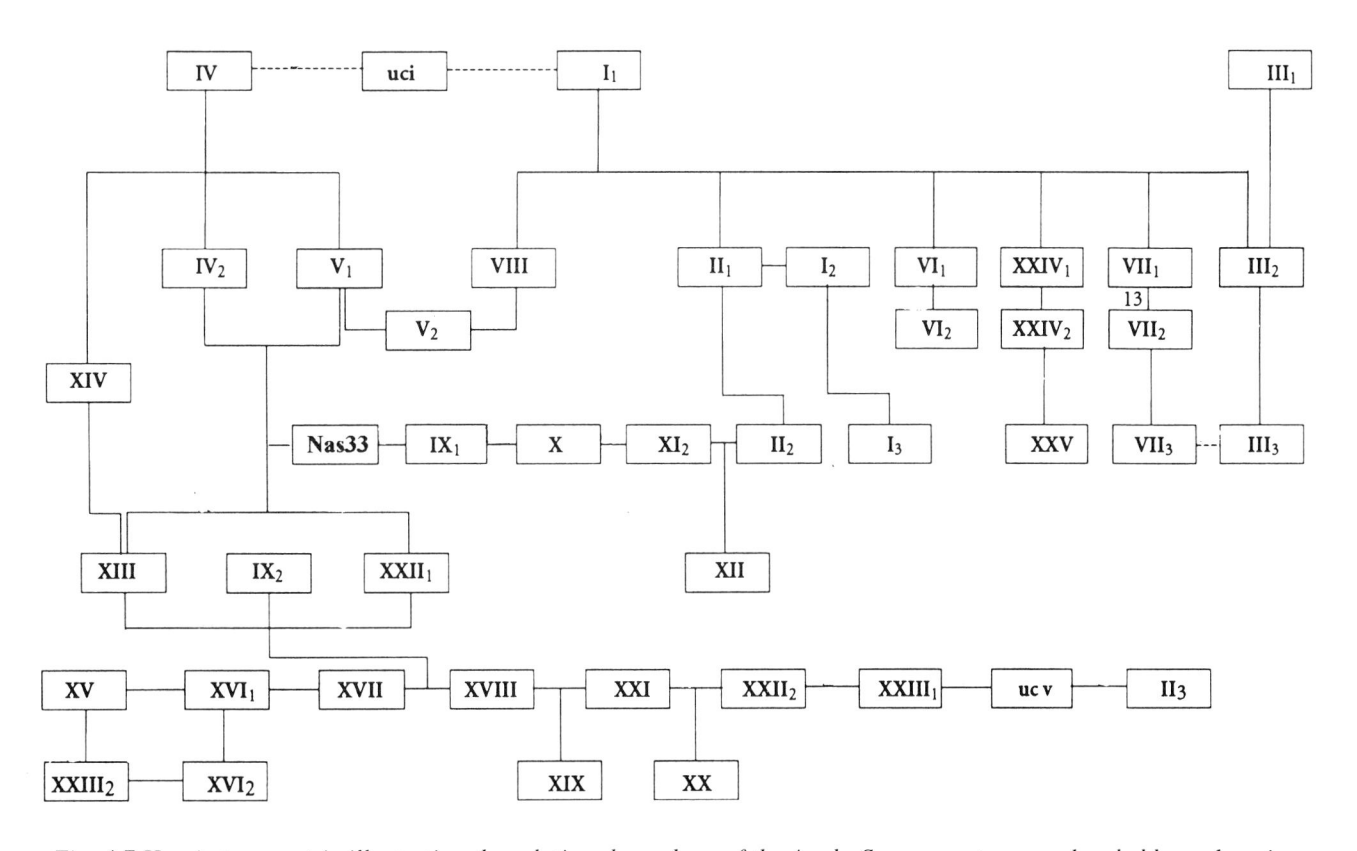

Fig. 4.7 Harris-type matrix illustrating the relative chronology of the Anglo-Saxon great square-headed brooch series. Figures in boxes refer to brooch groups. See further Hines 1997, esp. 198-204. The roman numerals refer to brooch groups and the smaller arabic numerals to sub-phases within these groups.
Harris-Matrix zur Veranschaulichung der relativen Chronologie der großen angelsächsischen Fibeln mit rechteckiger Kopfplatte.

what Haseloff called the Jutlandic brooch-group and dated, with some coin support, to *circa* 500/20. In the Final Phase, likewise, 'estimates' of absolute dates can sometimes be derived by comparing the apparently regularly declining level of gold purity in gold jewellery in the 7th century with a well-documented and well-dated fall in the standards of gold coin at this time (Hawkes, Merrick and Metcalf 1966). This allows us, then, to provide probable dates narrower than just 'Migration Period' or 'Final Phase' for a limited number of closed contexts. While it would strictly be more scientific to express the dating of those contexts in terms of a relative-chronological phase-system that encodes the evidence used, e.g. SHB2i for a grave group containing a square-headed brooch of the earlier of the two stages of Phase 2, such a system would in fact reflect no more than the relatively close datability of a small number of special contexts. Consequently the majority of Anglo-Saxon archaeologists who work with chronology are in the habit of using either directly applicable or interpolated absolute conventional dates for such contexts – i.e. *circa* 510-530 for the example cited. The problem, indeed, inheres in the apparently more general phase-systems that have been used over wide areas of England and which suggest a three-phase division within the Migration Period (Vierck 1977; Hines 1990). But of course this does not mean we may give up the search for general phases.

Altogether, Anglo-Saxon chronology has special problems which may require us to develop a special system for expressing and understanding the datings that can be made of this material. What is undoubtedly true, though, is that the special problems of Anglo-Saxon chronology derive from the peculiar character of the early Anglo-Saxon archaeological material. The peculiar character of that material in itself allows for other, extremely important, possibilities of research and interpretation – for instance in terms of identity-formation, maintenance and shift within the total community. All of the changes we can detect in Anglo-Saxon material culture we shall be able to understand best with the finest relative and absolute chronologies we can achieve. There is indeed much work still to be done, but also very good reason to undertake it.

DEUTSCHE ZUSAMMENFASSUNG

Die Trennung zwischen der Völkerwanderungszeit und der sog. 'Final Phase' ist die einzige klare chronologische Gliederung innerhalb der frühen angelsächsischen Epoche. Unklar blieb jedoch, ob hier eine rasch eintretende und durchgreifende Zäsur vorliegt, oder ob eine Übergangsperiode herausgestellt werden kann. Zudem blieb offen, wann dieser Wechsel eintrat und ob er überall in gleicher Weise vollzogen wurde. Dieses Problem wird hier anhand der Frauengräber aus der jüngst ergrabenen Nekropole von Edix Hill (Barrington A) und weiteren Plätzen in Cambridgeshire erneut beleuchtet.

Wie die Korrespondenzanalyse offenlegt, wird das Material -sogar innerhalb der Völkerwanderungszeit- durch die Existenz unterschiedlicher, konkurrierender Trachtgruppen verwirrt. Immerhin läßt sich die so gewonnene Ordnung einigermaßen mit der Völkerwanderungszeit und der 'Final phase' korrelieren. Dadurch werden einige Typen greifbar, die über beide Phasen hin andauern sowie eine geringe Zahl chronologisch ambivalenter Gräber. Insgesamt gibt es jedoch keinen Hinweis auf eine wirkliche Übergangsphase. Der Wechsel zwischen Völkerwanderungszeit und 'Final Phase' dürfte hier abrupt erfolgt sein.

Hinsichtlich der absoluten Chronologie kann nur allgemein auf die Ähnlichkeit dieses Wechsels mit jenem hingewiesen werden, den Ament für den Kontinent zwischen seinen Stufen AM II und AM III herausstellte, wonach eine absolute Datierung auf etwa 555-570 n.Chr. vorgeschlagen werden kann. Es sei betont, daß im angelsächsichen Bereich mit seinem ungewöhnlichen hohen Grad an regionaler und kultureller Variabilität keine enge Verknüpfung mit den kontinentalen Gegebenheiten erwartet werden kann.

BIBLIOGRAPHY

Åberg, N. 1926. *The Anglo-Saxons in England, during the Early Centuries after the Invasion*. Cambridge.

Adams, B. and Jackson, D. 1989. 'The Anglo-Saxon cemetery at Wakerley, Northamptonshire. Excavations by Mr D Jackson, 1968-69'. *Northamptonshire Archaeol* 22, 68-178.

Arrhenius, B. 1983. 'The chronology of the Vendel graves'. In J. P. Lamm and H.-Å. Nordström (eds), *Vendel Period Studies*, 39-70. Stockholm.

Bakka, E. 1973. 'Goldbrakteaten in norwegischen Grabfunden. Datierungsfragen'. *Frühmittelalterliche Stud* 7, 53-87.

Boddington, A. 1990. 'Models of burial, settlement and worship: the Final Phase reviewed'. In E. Southworth (ed), *Anglo-Saxon Cemeteries: A Reappraisal*, 177-99. Stroud.

Böhme, H. W. 1974. *Germanische Grabfunde des 4. bis 5. Jahrhunderts zwischen unterer Elbe und Loire*. Munich.

Dickinson, T. M. 1983. 'Anglo-Saxon archaeology: twenty-five years on'. In D. A. Hinton (ed), *25 Years of Medieval Archaeology*, 38-43. Sheffield.

Evison, V. I. 1994. *An Anglo-Saxon Cemetery at Great Chesterford, Essex*. CBA Res Rep 91, London.

Fausset, B. 1856. *Inventorium Sepulchrale*. London.

Foster, W. K. 1883. 'Account of the excavation of an Anglo-Saxon cemetery at Barrington, Cambridgeshire'. *Cambridge Antiq Soc Comm* 5, 5-32.

Geake, H. 1997. *The Use of Grave-Goods in Conversion-Period England, c.600-c.850*. BAR British Series 261. Oxford.

Green, B. and Rogerson, A. 1978. *The Anglo-Saxon Cemetery at Bergh Apton, Norfolk: Catalogue*. East Anglian Archaeol 7, Gressenhall.

Green, B., Rogerson, A. and White, S. 1987. *The Anglo Saxon Cemetery at Morning Thorpe, Norfolk*. East Anglian Archaeol 36, Gressenhall.

Haseloff, G. 1974. 'Salin's Style I'. *Medieval Archaeol* 18, 1-17.

Haseloff, G. 1981. *Die Germanische Tierornamentik der Völkerwanderungszeit*. 3 vols. Berlin.

Hawkes, S. C., Merrick, J. M. and Metcalf, D. M. 1966. 'X-ray fluorescent analysis of some Dark Age coins and jewellery'. *Archaeometry* 9, 98-138.

Hills, C., Penn, K. and Rickett, R. 1984. *The Anglo-Saxon Cemetery* at Spong Hill, North Elmham. Part III: Catalogue of Inhumations. East Anglian Archaeol 21. Gressenhall.

Hines, J. 1984. *The Scandinavian Character of Anglian England in the pre-Viking Period*. BAR British Series 124. Oxford.

Hines, J. 1990. 'Philology, archaeology and the *adventus Saxonum vel Anglorum*'. In A. Bammesberger and A. Wollmann (eds), *Britain 400-600: Language and History*, 17-36. Heidelberg.

Hines, J. 1993. *Clasps – Hektespenner – Agraffen: Anglo-Scandinavian Clasps of Classes A–C of the Third to Sixth Centuries A.D.* Stockholm.

Hines, J. 1997. *A New Corpus of Anglo-Saxon Great Square-Headed Brooches*. Reports of Res Committee of Soc Antiq 51, London.

Hinton, D. A. 1990. *Archaeology, Economy and Society: England from the Fifth to the Fifteenth Century*. London.

Høilund Nielsen, K. 1987. 'Zur Chronologie der jüngeren germanischen Eisenzeit auf Bornholm. Untersuchungen zu Schmuckgarnituren'. *Acta Archaeol* 57 (for 1986), 47-86.

Hollingworth, E. J. and O'Reilly, M. M. 1925. *The Anglo-Saxon Cemetery at Girton College, Cambridge*. Cambridge.

Hyslop, M. 1963. 'Two Anglo-Saxon cemeteries at Chamberlains Barn, Leighton Buzzard, Bedfordshire'. *Archaeol J* 120, 161-200.

Leahy, K. forthcoming. 'The formation of the Anglo-Saxon kingdom of Lindsey'. *Anglo-Saxon Stud Archaeol Hist* 10.

Leeds, E. T. 1936. *Early Anglo-Saxon Art and Archaeology*. Oxford.

Lethbridge, T. C. 1931. *Recent Excavations in Anglo-Saxon Cemeteries in Cambridgeshire and Suffolk*. Cambridge Antiq Soc Quarto Pub N.S. 3.

Lethbridge, T. C. 1936. *A Cemetery at Shudy Camps, Cambridgeshire*. Cambridge Antiq Soc Quarto Pub, N.S. 5.

Lund Hansen, U. 1977. 'Das Gräberfeld bei Harpelev, Seeland. Studien zur jüngeren römischen Kaiserzeit in der seeländischen Inselgruppe'. *Acta Archaeol* 47 (for 1976), 91-160.

Lund Hansen, U. 1992. 'Hovedproblemer i romersk og germansk jenalders kronologi i Skandinavien og på Kontinent'. In P. Mortensen and B. M. Rasmussen (eds), *Jernalderens Stammesamfund*, 21-36. Fra Stamme til Stat i Danmark 1. Aarhus.

Malim, T. and Hines, J. 1998. *The Anglo-Saxon Cemetery at Edix Hill (Barrington A), Cambridgeshire*. CBA Res Rep 112, London.

Martin, M. 1989. 'Bemerkungen zur chronologischen Gliederung der frühen Merowingerzeit'. *Germania* 67, 121-41.

Mortimer, C. M. 1990. *Some Aspects of Early Medieval Copper-Alloy Technology, as Illustrated by the Anglian Cruciform Brooch*. University of Oxford, D. Phil. thesis (unpublished).

Neville, R. C. 1852. *Saxon Obsequies*. London.

Neville, R. C. 1854. 'Anglo-Saxon cemetery on Linton Heath, Cambridgeshire'. *Archaeol J* 11, 95-115.

Reichstein, J. 1975. *Die kreuzförmige Fibel*. Neumünster.

Roth, H. and Theune, C. 1988. *SW ♀ I–V: Zur Chronologie merowingerzeitlicher Frauengräber in Südwestdeutschland*. Archäologische Informationen aus Baden-Württemberg 6, Stuttgart.

Siegmund, F. 1989. *Fränkische Funde vom deutschen Niederrhein und der nördlichen Kölner Bucht*. Universität zu Köln, Inaugural-Dissertation zur Erlangung des Doktorgrads der Philosophischen Fakultät.

Sjøvold, Th. 1988. 'The northernmost Migration Period relief brooch in the world and its family connections'. In B. Hårdh *et al.* (eds), *Trade and Exchange in Prehistory*, 213-23. Lund.

Taylor, A., Duhig, C. and Hines, J. 1998. 'An Anglo-Saxon cemetery at Oakington, Cambridgeshire'. *Proc Cambridge Antiq Soc* 86, 57-90.

Timby, J. 1996. *The Anglo-Saxon Cemetery at Empingham II, Rutland*. Oxford, Oxbow.

Vierck, H. 1977. 'Zur relativen und absoluten Chronologie der anglischen Grabfunde in England'. In G. Kossack and J. Reichstein (eds), *Archäologische Beiträge zur Chronologie der Völkerwanderungszeit*, 42-52. Bonn, Habelt.

West, S. E. 1988. *The Anglo-Saxon Cemetery at Westgarth Gardens, Bury St Edmunds, Suffolk*. East Anglian Archaeol 38, Bury St Edmunds.

Wilson, D. M. 1956. 'The initial excavation of an Anglo-Saxon cemetery at Melbourn, Cambridgeshire'. *Proc Cambridge Antiq Soc* 49, 29-41.

Wilson, D. M. 1976. 'Introduction'. In D. M. Wilson (ed), *The Archaeology of Anglo-Saxon England*, 1-22. London.

5. Dating burials of the seventh and eighth centuries: a case study from Ipswich, Suffolk

Christopher Scull and Alex Bayliss

INTRODUCTION

The archaeological chronology for Anglo-Saxon burials of the 5th to 8th centuries is based upon grave goods. Typological studies and cross-dating have established a sequence of material culture types, but for absolute dates before the late 7th century this sequence is largely dependent at second-hand upon coin-dated Continental chronologies.

The existing chronological framework, even for the relatively secure and well-dated material culture sequence for female graves of the late 5th and 6th centuries, has serious limitations. These include regionality; the probability that alternative costume traditions are represented in female graves; uncertainty as to whether sequences represent manufacture, use or deposition; and the difficulty of integrating female and male sequences (Høilund Nielsen 1997; Hines, this volume). Some of these problems are exacerbated by changes in material culture and burial practice from the later 6th century onwards when there is a marked change in the suite of material culture types deposited in graves and an overall decline in the frequency of furnished burial and in deposition of grave goods. In particular, artefact-based chronologies are of limited value when dealing with unfurnished graves or those without diagnostic material culture types.

The current basis of the chronology for 7th- and 8th-century grave finds has been summarized by Geake (1997). No detailed seriation or correspondence analysis of a large sample has yet been undertaken, such as has been used to establish detailed regional burial chronologies for areas of the Continent and Scandinavia (see, for example, contributions to this volume by Nieveler and Siegmund, and Theune), and much of the detailed typological classification necessary for such an undertaking remains to be done. Consequently, although it is possible to identify a range of material culture types as characteristic of the 7th to early 8th centuries it has not been possible for the most part to propose any more refined dating than earlier or later in the period *c.* A.D. 600–720/30 or to test whether greater precision might be achieved. Most calendrical or absolute dates are provided by coins, mostly late *thrymsas* of series Pa and Va and *sceattas*, which establish *termini post quos* for the burials in which they occur. However, the exact dating of these issues is itself subject to an element of imprecision (Grierson and Blackburn 1986; Metcalf 1993) and it is also possible that some material culture types are systematically being dated too late because of their occurrence in graves containing coins struck after *c.* A.D. 660.

This paper deals principally with evidence from the cemetery at St Stephen's Lane/Buttermarket, Ipswich (hereafter referred to as Buttermarket). As well as being of broader interest in its own right the cemetery can stand as an illustration of the potential and constraints of the framework currently available for the dating of 7th- and 8th-century burials in England. Of particular interest is the fact that burials from this site have been subject to a programme of radiocarbon dating undertaken to test and refine conclusions drawn from the conventional artefact dating. Three main sets of issues are considered here: the conventional dating and phasing of the cemetery; the results of the programme of radiocarbon dating; and the extent to which the cemetery can contribute to a broader chronological understanding as part of a local sequence. This is work in progress (Scull in prep.) and although we do not expect the final results to be radically different from those presented here these are nonetheless provisional findings and should be treated accordingly.

THE BUTTERMARKET CEMETERY

The site

The Buttermarket cemetery was excavated in 1987–88. Seventy-seven early medieval graves were excavated within an area of 4,600 sq m (Figs. 5.2–5.3). This is not, however, a complete sample. There had been considerable disturbance by later activity and it is likely that at least as many graves had been destroyed as survived. Many of the

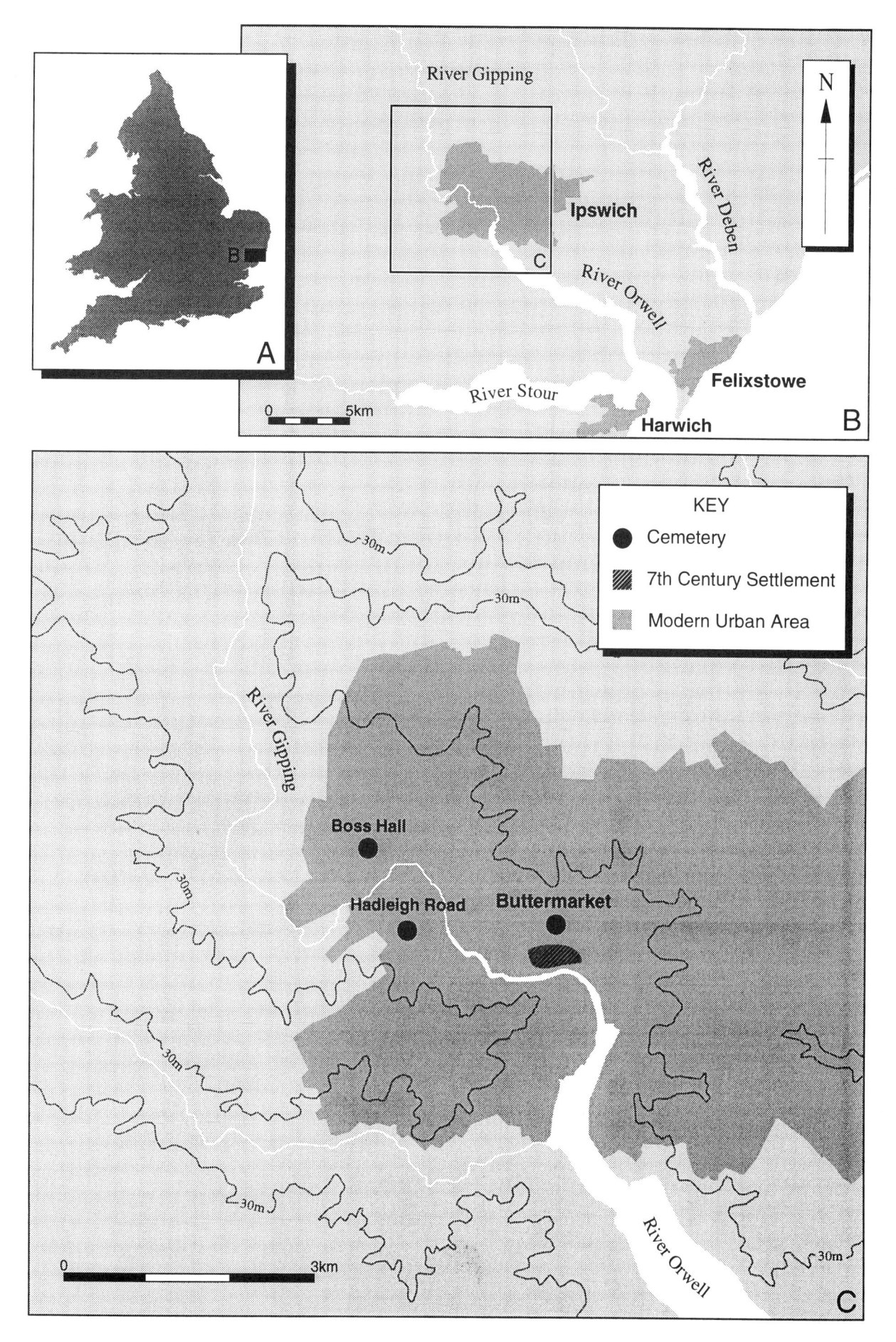

Fig. 5.1 Map showing the location of sites mentioned in the text (John Vallender).
Karte der im Text erwähnten Fundstellen.

surviving graves had been cut and damaged by later features, and it is also apparent that the cemetery extended beyond the edges of the excavation. The soil was acid and skeletal preservation was frequently poor.

This was an inhumation cemetery: no cremations were recovered. Most of the graves were aligned west–east with a few south–north. A majority of graves contained evidence of coffins or chambers, or some other form of container or structure. Annular ditches around some graves suggest that these had had mounds raised over them. Grave goods were recovered from thirty-two burials. Most of the burials with grave goods were poorly furnished, often having only a knife, but a minority of graves were more elaborately furnished. It is on these that the conventional dating and phasing of the cemetery is based.

The cemetery lay immediately north of the 7th- and 8th-century trading settlement, or *emporium*, which it served (Fig. 5.1). It was superseded by metalled streets whose frontages were developed with buildings, part of an expansion of the settlement area from 6 ha in the 7th century to 50 ha in the 9th century. The abandonment of the cemetery thus provides a *terminus post quem* for the expansion of the settlement and the excavated sequence at Buttermarket is therefore central to our understanding of the physical development of Ipswich in the pre-Viking period (Wade 1993).

Artefact dating and phasing

The excavated sequence shows that the cemetery had been abandoned by early in the 9th century at latest. Before the programme of radiocarbon dating was initiated any more refined dating and phasing of the site had to be based upon grave goods. Only a minority of burials was furnished, and of these only a minority contained assemblages or items which might allow dating to within 50 years. These fall into three groups.

1. Coin-dated burials

Coins were recovered from three graves: 1356, 4152 and 4275. A heavy issue penny of Offa from grave 4152 gave a *terminus post quem* of A.D. 792 for the burial, making this the latest dateable grave. Late *thrymsas*, or base forgeries of these, were recovered from graves 1356 and 4275. Grave 1356 contained two coins of series PaIII, one of which is a contemporary base forgery. Grave 4275 contained two contemporary base forgeries, both pierced and looped for suspension, one of series PaIII, the other provisionally identified as an unrecorded type related to the 'Constantine' and Vanimundus series (Marion Archibald pers. comm.). A *terminus post quem* of A.D. 660/680 for both graves would accommodate the main disagreements over the precise dating of the late *thrymsas*, and in both cases burial before the end of the 7th century is indicated (Grierson and Blackburn 1986, 163–4; 184; Metcalf 1993, 63–84). The assemblage from grave 1356 included a purse with a kidney-shaped lid, a buckle, two knives and a pointed tool. The forgeries from grave 4275 formed part of a necklace of biconical silver wire beads, silver 'bulla' pendants and knotted silver wire rings.

2. Burials with items or assemblages which can be tied directly into coin-dated sequences established for the Continent

Graves 2297 and 3871 each contained buckle sets belonging to Rhineland phase 8 (A.D. 610–640; Nieveler and Siegmund, this volume); grave 2297 also contained a broad sax. Grave 1306 contained an assemblage which almost certainly comes from the Continent and which belongs to Rhineland Phase 9 (A.D. 640–670; Siegmund and Nieveler, this volume): a broad sax, a shield boss, two palm cups, two spearheads, and a buckle-set of type Bern-Solothurn.

3. Burials with items or assemblages which suggest a date earlier or later in the period c. A.D. 600–720/30

Grave 3659 contained a shield boss of Dickinson and Härke's group 7. Grave 2962 contained a necklace of silver 'bulla' pendants and biconical silver wire beads similar to that from grave 4275. Both graves are likely to be later rather than earlier in the 7th century (Dickinson and Härke 1992; Geake 1997, 36–7, 42–3).

These grave finds indicated that the cemetery had been in use for around two hundred years from the beginning of the 7th century. Plotting these most closely dateable items and assemblages might suggest that the cemetery developed westwards and southwards from an early core (Fig. 5.2) but the grave finds provided little basis for a more confident phasing and constituted only the most oblique evidence for the dating of unfurnished or poorly-furnished graves.

If this suggested sequence of spatial development were to be accepted as an heuristic model then otherwise undated burials might be assigned to earlier or later phases on the basis of physical proximity to graves with chronologically-diagnostic artefacts or assemblages. Similar arguments based upon horizontal stratigraphy have been used to date unfurnished graves in other Anglo-Saxon cemeteries (Evison 1987, 134–61) and underpin the finer distinctions in some Continental chronological analyses (Nieveler and Siegmund, this volume). In these cases the number of graves involved is very much larger than at Buttermarket, where the argument rests on the assumption that a minority of furnished burials is representative of all surviving graves – a proposition which is complicated by the small number of burials in question and which cannot be tested without a dating method independent of the artefacts. This raises questions which have implications for broader models of burial practice in the 7th and 8th centuries as well as for interpretation of the site. Was the cemetery in continuous use through the 7th and 8th centuries, or was it predominantly a 7th-century cemetery which was subsequently used sporadically for burial up to the end of the 8th century? Were the unfurnished or poorly-furnished burials predominantly later rather than earlier in the lifetime of the cemetery, and should the absence of grave goods therefore

Fig. 5.2 Plan of the St Stephen's Lane/Buttermarket cemetery showing graves with closely dateable artefacts or assemblages. Graves and other cemetery features are shown in black; other surviving deposits of the 7th century or earlier are stippled (Vince Griffin and David Nuttall).
Plan des Gräberfeldes St Stephen's Lane/Buttermarket in Ipswich mit Eintragung der archäologisch datierbaren Gräber. Gräber und andere Bestattungsreste schwarz, übrige Befunde des 7. Jahrh. gerastert.

Fig. 5.3 Plan of the St Stephen's Lane/Buttermarket cemetery showing burials selected for radiocarbon dating. Graves and other cemetery features are shown in black; other surviving deposits of the 7th century or earlier are stippled (Vince Griffin and David Nuttall).
Gräberfeldplan mit Markierung der für die ^{14}C-Datierungen ausgewählten Bestattungen (vgl. Abb. 2).

be taken as a chronological indicator or a feature of variation within contemporary burial practice?

Radiocarbon dating

In the past radiometric dating techniques have not offered sufficient precision to be considered useful for this period (see, for example, Geake 1997, 10). However, the development of high-precision radiocarbon dating (Pearson 1984; McCormac *et al.* 1993; Wilson *et al.* 1996), allied with explicit mathematical modelling of chronological problems (Buck *et al.* 1991; Bronk Ramsey 1995; Buck *et al.* 1996), now has the potential to provided estimated date ranges which span less than than 50 years at 95% confidence for the period between the late 6th and early 8th centuries, and which span around a century at 95% confidence later in the 8th century. This precision was considered to be sufficiently refined for radiocarbon analysis to be able to address the outstanding questions of dating and phasing at Buttermarket and so a high-precision radiocarbon dating programme was undertaken.

Although skeletal preservation was frequently poor sufficient bone was recovered from forty-six graves to allow some assessment of age or sex but in many cases even this skeletal material was fragmentary and insufficient for high-precision radiocarbon dating. Twenty-two samples for dating have been taken from nineteen skeletons (25 per cent of all graves and 41 per cent of burials with bone surviving). So far sixteen radiocarbon measurements have been completed on bone from sixteen burials (21 per cent of all graves and 34 per cent of burials with bone surviving) (Tab. 5.1). Results of six further radiocarbon measurements are awaited. Further scientific analyses, such as $\delta^{15}N$ measurements, which will enable more detailed discussion and analysis of the radiocarbon evidence are also awaited. Consequently the following summary only presents the most basic interpretation of the evidence – more detailed analysis and modelling is to be published elsewhere (Scull and Bayliss, forthcoming; Scull in prep.).

The radiocarbon dating programme was designed to test the model of spatial development based upon artefact dating; to determine whether otherwise undatable burials belonged to the 7th or 8th centuries, and whether the cemetery was in continuous use throughout this time or saw only sporadic use during the 8th century; to establish with greater precision when the cemetery came into use; and, if possible, to refine artefactual dating. The radiocarbon measurements were undertaken by the Radiocarbon Dating Laboratory, The Queen's University, Belfast, and were processed using methods described in McCormac and Housley (1995), Pearson (1984), McCormac *et al.* (1993), Wilson *et al.* (1996), and Wilson (1997).

All radiocarbon measurements are conventional radiocarbon ages (Stuiver and Polach 1977) and have been calibrated using OxCal v2.18 (Bronk Ramsey 1995) and the data published by Stuiver and Pearson (1986). The calibrated date ranges listed in Table 5.1 have been calculated using the maximum intercept method of Stuiver and Reimer (1986) and have been rounded outwards to the nearest five years (Mook 1986). The probability

Laboratory Number	Grave	Radiocarbon Age (BP)	$\delta^{13}C$ (‰)	Calibrated date range (68% confidence)	Calibrated date range (95% confidence)
UB-4039	G3871	1441±20	-20.9±0.2	cal A.D. 600–640	cal A.D. 565–650
UB-4040	G3897	1396±20	-20.1±0.2	cal A.D. 640–660	cal A.D. 610–665
UB-4041	G249	1301±20	-20.3±0.2	cal A.D. 670–690	cal A.D. 660–770
UB-4042	G1674	1407±20	-21.3±0.2	cal A.D. 630–655	cal A.D. 605–660
UB-4043	G2365	1320±22	-20.9±0.2	cal A.D. 665–685	cal A.D. 655–760
UB-4044	G4152	1413±21	-20.3±0.2	cal A.D. 625–655	cal A.D. 600–660
UB-4045	G4307	1398±21	-19.5±0.2	cal A.D. 635–660	cal A.D. 610–665
UB-4046	G4344	1404±21	-20.2±0.2	cal A.D. 630–655	cal A.D. 605–665
UB-4047	G4431	1349±20	-19.7±0.2	cal A.D. 655–670	cal A.D. 650–680
UB-4048	G4926	1377±20	-20.1±0.2	cal A.D. 645–665	cal A.D. 635–670
UB-4049	G4979	1337±20	-20.5±0.2	cal A.D. 660–675	cal A.D. 650–685
UB-4070	G4547	1463±19	-20.3±0.2	cal A.D. 570–620	cal A.D. 550–640
UB-4074	G2297	1419±23	-20.6±0.2	cal A.D. 610–650	cal A.D. 600–660
UB-4075	G3898	1436±23	-20.0±0.2	cal A.D. 600–645	cal A.D. 565–655
UB-4076	G4269	1356±23	-20.4±0.2	cal A.D. 655–670	cal A.D. 645–680
UB-4077	G4275	1476±24	-21.0±0.2	cal A.D. 555–610	cal A.D. 540–640

Tab. 5.1 Radiocarbon determinations from burials at St Stephen's Lane/Buttermarket.
^{14}C-Messungen der Gräber von St Stephen's Lane/Buttermarket.

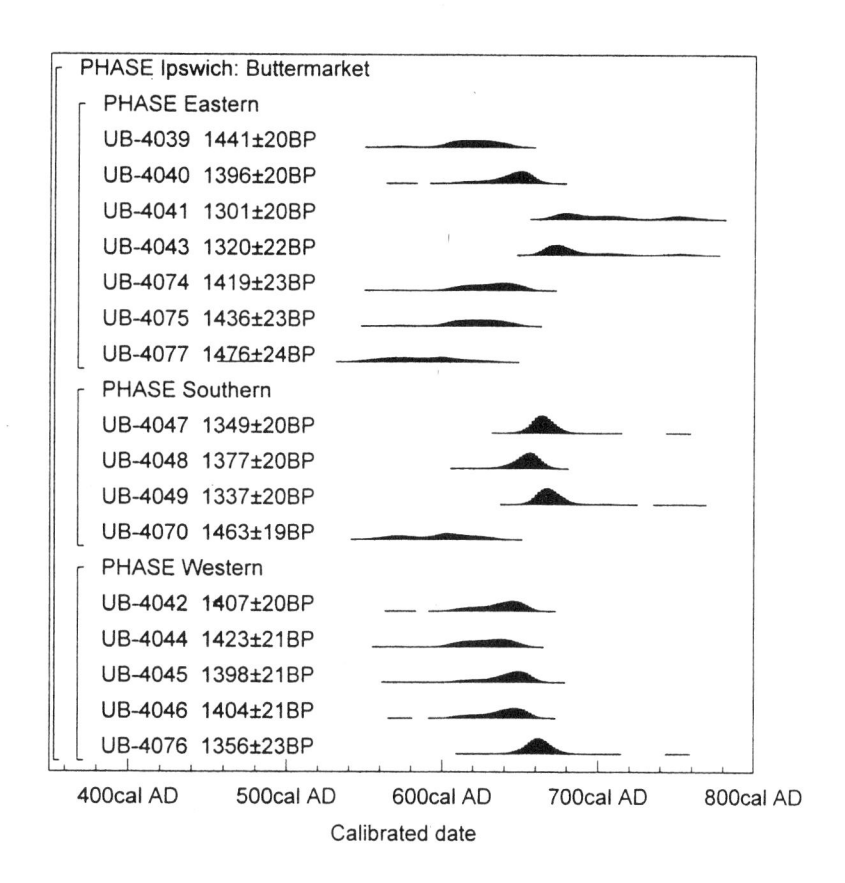

```
PHASE Ipswich: Buttermarket
  PHASE Eastern
    UB-4039  1441±20BP
    UB-4040  1396±20BP
    UB-4041  1301±20BP
    UB-4043  1320±22BP
    UB-4074  1419±23BP
    UB-4075  1436±23BP
    UB-4077  1476±24BP
  PHASE Southern
    UB-4047  1349±20BP
    UB-4048  1377±20BP
    UB-4049  1337±20BP
    UB-4070  1463±19BP
  PHASE Western
    UB-4042  1407±20BP
    UB-4044  1423±21BP
    UB-4045  1398±21BP
    UB-4046  1404±21BP
    UB-4076  1356±23BP

400cal AD   500cal AD   600cal AD   700cal AD   800cal AD
                    Calibrated date
```

Fig. 5.4 Probability distributions of dates from burials at St Stephen's Lane/Buttermarket: each distribution represents the probability that a burial occurs at some particular time (Alex Bayliss).
Wahrscheinlichkeitsverteilungen der Datierungen der Gräber: jede Kurve stellt die relative Wahrscheinlichkeit dar, mit der ein Grab zu einer bestimmten Zeit erscheint.

distributions shown in figure 5.4 have been derived from the usual probability method (Stuiver and Reimer 1993).

The results contained a number of surprises. The dates for the earliest burials in the east, south, and west of the excavated area are all very close, at the beginning of the 7th century. Figure 5.4 does not appear to show any significant difference between the start of burial in the different areas of the site. This may suggest that there is no simple sequence of spatial development within the excavated area; the evidence would instead be consistent with a more complex or haphazard spatial development, perhaps resulting from a polyfocal structure.

Only eight of the sixteen graves for which there are radiocarbon dates contained grave goods. In most cases radiocarbon dating and artefact dating are consistent. The dating of graves 2297 and 3871 according to Nieveler and Siegmund's chronology for the Rhineland is broadly supported by calibrated radiocarbon dates at the 95% confidence level (cal A.D. 600–660 and cal A.D. 565–650 respectively), and at the 68% confidence level the calibrated radiocarbon date ranges tally almost precisely with the conventional dating (cal A.D. 610–650 and cal A.D. 600–640 respectively). In two cases, however, there is a discrepancy between the radiocarbon date and the *terminus post quem* suggested by coins. Grave 4152, from which the coin of Offa was recovered, contained a skeleton which has been securely radiocarbon-dated to the 7th century (cal A.D. 600–660 at 95% confidence). Grave 4275, for which the base forgeries of late *thrymsas* suggest a *terminus post quem* of A.D. 660/80, is radiocarbon-dated to cal A.D. 540–640 (at 95% confidence). These discrepancies are considered below.

Figure 5.4 presents in graphic form the radiocarbon evidence currently available. This indicates that the use of the Buttermarket cemetery was almost entirely confined to the 7th century. Although the unconstrained probability distributions of the calibrated dates do extend slightly before A.D. 600 and after A.D. 700 this is probably caused by the inevitable statistical scatter on radiocarbon measurements. Further analysis should enable statistical approaches to be adopted which can provide quantitative estimates of the dates between which the cemetery was used and take account of this statistical scatter (Buck *et al.* 1992; Scull and Bayliss forthcoming) but the data are sufficiently strong for the main trends to be apparent visually.

Discussion

Radiocarbon dating has provided an independent chronological model for the Buttermarket cemetery which both refines and challenges conclusions drawn from the artefact dating alone. It may be objected that this model is based on only 21 per cent of the excavated graves but this is a larger sample than that provided by graves with chronologically-diagnostic artefacts. Moreover, radiocarbon dating can offer greater precision than all but a handful of the burial assemblages, and offers a quantitative estimate of the date of actual burial.

The date for the beginning of the cemetery suggested by the radiocarbon evidence is consistent with the artefact dating but that for the abandonment of burial on the site contradicts the coin date from grave 4152. The hypothesis of spatial development, based upon a small number of grave finds, is not supported by the radiocarbon dates. In this case the radiocarbon evidence indicates that the identification of horizontal stratigraphy was unreliable and, while the size and nature of the sample means that it would be unwise to extrapolate uncritically from this to other cases, it serves as a reminder that analyses based upon such arguments should be treated as heuristic models and warns of the danger of circular argument if this is not recognized. It is also important to emphasize that the radiocarbon dates show that furnished and unfurnished burial were contemporary traditions. Without this evidence there would have been a strong temptation to attribute a significant number of unfurnished or poorly-furnished graves to a period after the general abandonment of furnished burial in the earlier 8th century, especially as most of the burials in the area of grave 4152 were themselves unfurnished or had few grave goods. However, it is clear in this case that the absence of grave goods cannot in any way be considered a chronological indicator.

The discrepancy between radiocarbon dates and coin dates in two cases demands further work which is ongoing. At present, in the case of grave 4275, the conflict between the radiocarbon date and the *terminus post quem* given by the coins may be more apparent than real and it is possible that it may be explained by the statistical uncertainties of the dating method and the element of imprecision which exists over the exact chronology of the late *thrymsas*. In the case of grave 4152, however, no such explanation will serve. The radiocarbon date from this grave is wholly consistent with the other radiocarbon measurements from the cemetery and so incompatible with the coin date that, on the current balance of evidence, it appears that there are grounds for considering the possibility that the coin from grave 4152 was intrusive or represents a substitution at post-excavation.

In fact, the Offa penny from grave 4152 is the only piece of evidence to suggest that burial continued on the site later than the early 8th century. It is not yet possible to give a final assessment of the dating evidence but at present the radiocarbon dates suggest either that the coin date for grave 4152 is erroneous or that this was a late burial interred many years after the site had otherwise been abandoned as a formal cemetery. In either event the radiocarbon chronology would appear to have shifted the *terminus post quem* for the expansion of settlement over the cemetery area from the beginning of the 9th century to the early 8th century. This shift, which can be accommodated by the post-cemetery stratigraphic sequence at Buttermarket, would have important implications for our understanding of the development of the Ipswich settlement.

The local burial sequence

Buttermarket is one of three cemeteries known within 4 km of each other around Ipswich (Fig. 5.1). The others are Boss Hall (6th century, with a single grave of the late 7th or early 8th century) and Hadleigh Road (6th and 7th centuries). Between them the three groups of burials form a sequence spanning the 6th and 7th centuries and so establish a baseline against which some chronological hypotheses may be tested (Fig. 5.5).

The female grave goods from Boss Hall are typical of 6th-century Anglian inhumations: cruciform, small-long and annular brooches, wrist clasps, and amber beads are all represented. A single burial (grave 93) with silver and gold jewellery and dated by a series B1 *sceatt* to

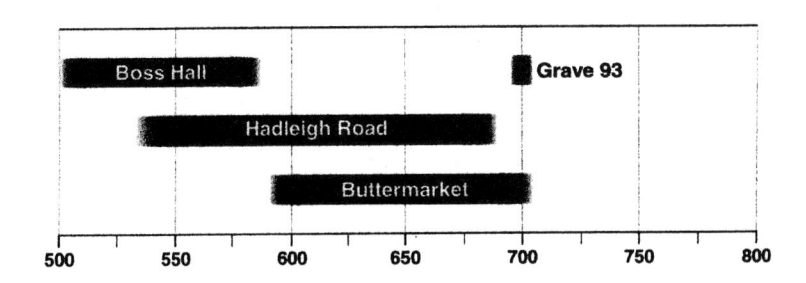

Fig. 5.5 The chronological ranges of cemeteries at Boss Hall, Hadleigh Road and St Stephen's Lane/Buttermarket (John Vallender).
Chronologische Spannen der Gräberfelder von Boss Hall, Hadleigh Road und St Stephen's Lane/Buttermarket.

A.D. 690/700 represents a single later re-use of the site (Newman 1993; Backhouse and Webster 1991, 51–53; Scull, in prep.). None of the 6th-century types is known from Buttermarket, which appears to confirm that they were no longer current by the time the cemetery was established.

Cruciform and small-long brooches are also absent from Hadleigh Road. However, 6th-century material from this site includes annular brooches and seven great square-headed brooches of Hines's groups XVI and XVII (Hines 1997). Later items include a buckle of Continental type belonging to Rhineland Phase 7 (A.D. 580/90–610; Nieveler and Siegmund, this volume) and material which indicates burial into the middle or second half of the 7th century, including palm cups and a shield boss of Dickinson and Härke's group 7 (Layard 1907; Evison 1963, fig. 24g; Plunkett 1994). The absence of cruciform and small-long brooches from Hadleigh Road cannot therefore be taken to indicate a wholly 7th-century date for this cemetery. The absence of great square-headed brooches from Buttermarket is strong evidence that at Hadleigh Road they were deposited before the Buttermarket cemetery came into use. The same pattern can be seen in the male burials at all three sites: shield bosses of Dickinson and Härke's groups 1.1, 2 and 3 are known from Boss Hall, groups 1.1, 2, 6 and 7 from Hadleigh Road, and group 7 from Buttermarket.

The Buttermarket cemetery therefore appears to have come into use after the change in material culture and cultural practice, as reflected in female burials, which may be taken archaeologically to mark the end of the Migration Period in England (Hines, this volume). The cemetery at Hadleigh Road, on the other hand, appears to span this transition, while that at Boss Hall appears to have been abandoned before it. In this area of south-east Suffolk, and by extension more widely in East Anglia, the radiocarbon date for the beginning of the Buttermarket cemetery would establish that these changes had occurred by the decades surrounding cal A.D. 600 (Fig. 5.4). As noted above further analyses are underway which will enable the provision of a quantitative estimate for the date when the Buttermarket cemetery came into use and so provide a *terminus ante quem* for the transition.

CONCLUSIONS

Taken together the three Ipswich cemeteries confirm the accepted sequence of material culture types and assemblages for the 6th and 7th centuries. The date at which the Buttermarket cemetery came into use establishes a late 6th- to early 7th-century horizon which broadly confirms the upper limit of absolute dates given to 6th-century Anglian grave goods and supports the view that the deposition, and presumably production, of this range of material culture types had been abandoned by this time. It is important to emphasize that this horizon is defined by both radiocarbon dating and conventional artefact dating.

The Insular artefact-dating of burials is weak for the late 6th and early 7th centuries. Although it is possible to identify some male grave goods of this date, most clearly shield bosses of group 6, contemporary female burials remain difficult to identify from grave goods. One important aspect of the sequence of burials around Ipswich is the presence in graves, albeit male graves, of Continental items, or items of continental type, which can be attributed securely to this period. Hines (this volume) has drawn attention to the possibility that female burials of a 'transitional phase' to be dated to this period might be discerned among his costume group D. The three Ipswich cemeteries provide exactly the chronological span of burials from the same immediate area necessary to test this hypothesis by means of a seriation by correspondence analysis; it is therefore unfortunate that the number of burials from Boss Hall is so small (twenty-two graves), that the sample from Buttermarket is damaged and incomplete, and that many of the grave associations from Hadleigh Road – excavated more than 90 years ago – are not now known.

If a 'transitional phase' is to be proposed on the basis of material culture assemblages it should now be possible to test its integrity by radiocarbon dating. The results from the Buttermarket are beginning to show the potential of high-precision radiocarbon analysis as a dating tool for burials of this period. It is able to provide deposition dates with a precision which artefact dating can match only in a very few cases and has the potential to integrate male and female sequences, regional sequences, and furnished and unfurnished burials. Its value at the level of the individual site is clear but if applied more widely, in conjunction with seriation and correspondence analysis, it would appear to have good potential to establish a robust chronological framework for furnished burials of the late 6th to early 8th centuries.

Acknowledgements

We should like to thank the organizers of the 1996 Ålborg symposium for inviting us to contribute to these proceedings, and all colleagues within and without English Heritage who have contributed to the work on which this is a preliminary report – in particular Gerry McCormac (Radiocarbon Dating Laboratory, The Queen's University, Belfast) and Marion Archibald. The cemeteries at St Stephen's Lane/Buttermarket and Boss Hall were excavated by the Suffolk Archaeological Unit and we should like to acknowledge with gratitude the help and support offered by the excavators: Tom Loader and Keith Wade (Buttermarket), and John Newman (Boss Hall). Analysis and publication of the cemeteries is being funded by English Heritage and Suffolk County Council with support from The British Museum. Figures 5.1–5.3 and 5.5 were produced by John Vallender and Vince Griffin (Central Archaeology Service, English Heritage) and David Nuttall (Suffolk County Council).

DEUTSCHE ZUSAMMENFASSUNG

Der Beitrag stellt erste, vorläufige Ergebnisse der chronologischen Analyse des frühmittelalterlichen Gräberfeldes von St Stephen's Lane/Buttermarket in Ipswich vor. Die konventionelle Datierung beruht auf Funden und Artefaktchronologien, wonach zunächst vermutet wurde, daß das Gräberfeld etwa 200 Jahre lang belegt wurde. Darauf aufbauend konnte auch ein Modell seiner räumlichen Entwicklung vorgeschlagen werden. Diese Thesen werden nun durch systematisch angelegte, hochpräzise ^{14}C-Messungen überprüft, die zu einer Neubewertung der Datierungen und des Bilds der Gräberfeldentwicklung zwingen. Die Ergebnisse haben zugleich Auswirkungen auf die Datierung weitverbreiteter Änderungen im Sachgut und in den Bestattungssitten des späten 6. Jahrhunderts. Zugleich zeigen sie exemplarisch das Potential solcher Hochpräzisionsmessungen, die man mit expliziten mathematischen Modellen verknüpft, für die Chronologie im frühen Mittelalter.

BIBLIOGRAPHY

Backhouse, J. and Webster, L. 1991. *The Making of England: Anglo-Saxon Art and Culture A.D. 600–900*. London.

Bronk Ramsey, C. 1995, 'Radiocarbon calibration and analysis of stratigraphy'. *Radiocarbon* 36, 425–30.

Buck, C. E., Cavenagh, W. G. and Litton, C. D. 1996. *Bayesian approach to interpreting archaeological data*. Chichester.

Buck, C. E., Kenworthy, J. B., Litton, C. D. and Smith, A. F. M. 1991. 'Combining archaeological and radiocarbon information: a Bayesian approach to calibration'. *Antiquity* 65, 808–21.

Buck, C. E., Litton, C. D. and Smith, A. F. M. 1992. 'Calibration of radiocarbon results pertaining to related archaeological events'. *J Archaeol Science* 19, 497–512.

Dickinson, T. M. and Härke, H. 1992. *Early Anglo-Saxon Shields*. Archaeologia 110. London.

Evison, V.I. 1963. 'Sugar loaf shield bosses'. *Antiquaries J* 43, 38–96.

Evison, V.I. 1987. *Dover: the Buckland Anglo-Saxon Cemetery*. London.

Geake, H. 1997. *The Use of Grave Goods in Conversion-period England c. 600 – c. 800*. BAR British Ser 261. Oxford.

Grierson, P. and Blackburn, M. 1986. *Medieval European Coinage, volume 1: the Early Middle Ages*. Cambridge.

Hines, J. 1997. *A New Corpus of Anglo-Saxon Great Square-headed Brooches*. Woodbridge.

Høilund Nielsen, K. 1997. 'The schism of Anglo-Saxon chronology'. In C. K. Jensen and K. Høilund Nielsen (eds), *Burial and Society: the Chronological and Social Analysis of Archaeological Burial Data*, 71–99. Århus.

Layard, N. 1907. 'An Anglo-Saxon cemetery in Ipswich'. *Archaeologia* 60, 325–42.

McCormac, F.G. and Housley, R. 1995. 'Radiocarbon analysis'. In R. M. J. Cleal, K. E. Walker and R. Montague, *Stonehenge in its Landscape: Twentieth-century Excavations*, 515–16. London.

McCormac, F. G., Kalin, R. M. and Long, A. 1993. 'Radiocarbon dating beyond 50,000 years by liquid scintillation counting'. In J. E. Noakes, F. Schonhofer and H. A. Polach (eds), *Liquid scintillation spectrometry 1992*, 125–33. Tucson.

Metcalf, M. 1993. *Thrymsas and Sceattas in the Ashmolean Museum, volume 1*. Royal Numismatic Society Special Publication 27a. London.

Mook, W. G. 1986. 'Business meeting: Recommendations/Resolutions adopted by the Twelfth International Radiocarbon Conference'. *Radiocarbon* 28, 799.

Newman, J. 1993. 'The Anglo-Saxon cemetery at Boss Hall, Ipswich'. *Bulletin Sutton Hoo Res Comm* 8, 32–5.

Pearson, G. W. 1984: *The development of high-precision ^{14}C measurements and its application to archaeological timescale problems*. Queen's University Belfast, Ph.D. thesis (unpublished).

Plunkett, S. 1994. *Guardians of the Gipping: Anglo-Saxon Treasures from Hadleigh Road, Ipswich*. Ipswich.

Scull, C. J. in prep: *Anglo-Saxon cemeteries at Boss Hall and St Stephen's Lane/Buttermarket, Ipswich*.

Scull, C.J. and Bayliss, A. forthcoming: 'Radiocarbon dating and Anglo-Saxon graves'. In U. von Freeden, U. Koch and A. Wieczorek (eds), *Völker an Nord- und Ostsee und die Franken. Beiträge des 48. Sachsensymposiums in Mannheim vom 7. bis 11. September 1997*. Berlin.

Stuiver, M. and Polach, H. A. 1977. 'Reporting of ^{14}C data'. *Radiocarbon* 19, 355–63.

Stuiver, M. and Pearson, G. W. 1986. 'High-precision calibration of the radiocarbon time scale, A.D. 1950 – 500 BC'. *Radiocarbon* 28, 805–38.

Stuiver, M. and Reimer, P. J. 1986. 'A computer program for radiocarbon age calculation'. *Radiocarbon* 28, 1022–30.

Stuiver, M. and Reimer, P.J. 1993. 'Extended ^{14}C data base and revised CALIB 3.0 ^{14}C age calibration program'. *Radiocarbon* 35, 215–30.

Wade, K. 1993. 'The urbanisation of East Anglia: the Ipswich perspective'. In J. Gardiner (ed.), *Flatlands and Wetlands: Current Themes in East Anglian Archaeology*, 144–51. East Anglian Archaeol 50.

Wilson, J. E. 1997. *Small sample high-precision ^{14}C dating*. Queen's University Belfast, M.Phil. thesis (unpublished).

Wilson, J. E., McCormac, F. G. and Hogg, A. G. 1996. 'Small sample high-precision ^{14}C dating: characterisation of vials and counter optimisation'. In G. T. Cook, D. D. Harkness, A. B. McKenzie, B. F. Miller and E. M. Scott (eds), *Liquid scintillation spectrometry 1994*, 59–65. Tucson.

Synopsis of discussion

Much of the discussion relating to the Anglo-Saxon session focussed on the problems of regional difference within England. The questions of whether any areas might be particularly good for chronological analysis, and by what criteria one could recognize them, were aired. It was noted that to be useful in any form of general or comparative studies, local schemes needed to include as much material that would allow separate schemes to be correlated as possible.

With reference to the study of the Butter Market cemetery in Ipswich, it was emphasized that it was hoped that this would be a pilot project for a comprehensive programme of high-precision radiocarbon dating of late 6th- to early 8th-century burials in England. It was also noted in this regard that regional difference appeared to become much less marked in this period than earlier.

The problems of correlation in England were not only geographical. The lack of progress in correlating male and female sequences, and the effective failure to even attempt to tackle the cremation burials alongside the inhumation graves, were admitted. It was pointed out that the situation in respect of male graves was especially unpromising in light of the virtual absence of weaponry in cremation contexts.

Several issues were raised in relation to the comparison of English sequences with overseas (particularly Continental) ones on the basis of shared types. The problem of distinguishing imported from copied artefact-types was one of these. It was agreed that much work was still to be done, but the problem was already recognized, and relevant research planned. It was also suggested that a surprisingly high proportion of possible imports seemed to be relatively poor-quality items.

The social context of foreign influence was also discussed. It was considered unlikely that the phasing of costume groups associated with particularly Continental types in Kent was in any sense more social than chronological. The question of whether the majority of foreign items came in with brides in a system of exogamy, however, was a more open one. It was clear that there was far from any agreed view on how much material indicative of their external origin women in such circumstances should be expected to have brought with them and to have retained until burial.

In a concluding review, Martin Welch described the situation as relatively encouraging compared with the period of neglect of chronological studies Anglo-Saxon archaeology had been through. He referred to the number and standard of publications of cemeteries, as well as the interest in and research on the chronological problems. However he expressed reservations about the terminology used, in particular the concept of the 'Final Phase', suggesting that it would be helpful to create a new terminology while investigating the character and appropriate definitions of phases for England without prejudice.

Further observations at this general and prospective level emphasized the need for long-term and comprehensive programmes of research, and the importance of experience in the nuances and difficulties of Anglo-Saxon archaeology in face of the danger of producing specious and false results where these were lacking. In consideration of these requirements, it was also argued that first-hand work on the material in museum collections was essential, and that one could not, for the foreseeable future, rely on published sources alone.

JH/KHN

III
SCANDINAVIA

6. Migration Period chronology in Norway

Siv Kristoffersen

This paper seeks to give an account of the history of research into the Migration Period chronology of Norway. It further attempts to establish a relative chronological framework that was designed to function in relation to a special analysis of relief decoration. Chronology was not the primary aim of that study (Kristoffersen 1997), and this paper represents the basis of an introductory chapter therefrom.[1] The work of earlier reseachers is systematically reviewed. Their results are also supported by further evidence. The foundation, however, is traditional Norwegian chronological research, mainly carried out in the period from the turn of the century to the 1930's, with all of the problems involved in that tradition.

RESEARCH HISTORY

It is generally agreed that the Migration Period begins around the year 400 A.D. or shortly before that (Montelius 1896; Shetelig 1912, 69; Nerman 1935, 119 and 121; Nissen Meyer 1935, 4 and 103; Bakka 1973, 85; Slomann 1956, 63–4; 1977, 61; Hines 1984, 17; Straume 1987, 14–15; Lund Hansen 1988, 25). The term 'Migration Period' was originally used of the period 400–800 A.D. (Montelius 1896, 7 and 15; 1906, 214; Shetelig 1912, 69; 1925), although as early as Montelius' study of 1896 a division around 600 A.D. was introduced, separating his early and late phases of the Migration Period or Periods 6 and 7. That this applied in Norway too is clear in Shetelig's work (1900, 62; 1904, 90; 1917, 76–7 and 81). The division became a conventional one somewhat later (Bøe 1931, 234; Nissen Meyer 1935, 103; 1937, 110; Fett 1940, 5; Slomann 1956, 64; Bakka 1958, 62 [somewhat earlier than 600: Tab. 6.1]). More recently, the transition to the Merovingian Period has been pushed back in time to the middle or sometime within the second half of the 6th century (Bakka 1973, 85; Hines 1984, 18 and 31; Straume 1987, 15; Tab. 6.1). Earlier studies show that there had been strong grounds for such an adjustment for a long time (Gjessing 1934; Slomann 1956, 74–80).

This holds for other Scandinavian finds too, as Nerman located the boundary between period VI and VII on Gotland between 550 and 600 (Nerman 1935, 119 and 121). Voss put the end of Style I around 575 (1954, 182). This shift back in time is based upon correlation with Continental chronology, and the relationship between Style I and Style II is a crucial issue (see below). Some scholars believe that this period boundary will come to be placed even earlier (Magnus 1975; Lund Hansen 1988, 33).

In this paper, however, the absolute chronology of the Migration Period will not be given especial attention. The definition of the transitions between the Roman Period and the Migration Period, and the Migration Period and the Merovingian Period, will be discussed. Most attention will be paid, however, to phase-divisions within the Migration Period. Especial importance is attached to local chronology: a chronological scheme which applies to the southern and western parts of Norway.

ROMAN PERIOD/MIGRATION PERIOD

This period boundary has not been particularly easy to define (Slomann 1956, 63–4; 1977, 63; Straume 1987, 14). Slomann (1977) proposed a set of artefacts as marking the transition to the Migration Period (leading types of transitional finds): the fully developed cruciform brooch; clasps with a plate and buttons and clasps with rolled spiral ends (R [=Rygh 1885] nos. 267–271); U-shaped scabbard chapes for knives and swords; and bossed pots (*Buckelurnen*). The most important element is the cruciform brooch (Slomann 1977, 63). Slomann identified problems associated with all four leading types, the most critical being the definition of a cruciform brooch. Referring to Shetelig (1906) and Stjernqvist (1961), she identified a cruciform brooch as 'fully developed' if it had a headplate wider than the bow and at an angle to it.[2] Lund Hansen has sharpened the definition up somewhat:

> A cruciform brooch must have a rectangular headplate and three knobs, two of them at either end of the spiral axis and just sticking out beneath the headplate. Later on these two knobs are placed on the sides of the headplate itself and the third is fastened to the headplate in continuation of the line of the bow. It is also required that the headplate should stand at a right angle to the plane of the bow. (Lund Hansen 1970, note 173)

These definitions distinguish cruciform brooches from Nydam brooches, which are described as follows:

> Nydam brooches are of crossbow type. A Nydam brooch may have a returned foot but this is not essential. The Nydam brooch must have a knob placed at the top of the bow and often also has a knob on each end of the spiral axis. The central knob is not placed directly on to the bow but is separated from it by a small, vertical prong, to the centre of which the spiral axis is fastened. The spiral axis can be seen, uncovered, between the bow and the knob. This small area is eventually filled with a small triangular or rhomboidal plate, which is, however, always so narrow that the spiral axis can still be seen. From the Nydam brooch developed the cruciform brooch. (*ibid.*)

True Nydam brooches are rarely found combined with cruciform brooches in Scandinavia (Lund Hansen 1970, 96 and 98; Ethelberg 1986, 20). Moreover two different ranges of jewellery are associated with these two brooch-types. With Nydam brooches are found swastika brooches, metal-sheet brooches lacking profile animal heads, brooches with pressed foil and inlaid stones, and berloque-shaped amber beads. With cruciform brooches are found silver-sheet brooches with profile animal heads and complicated punched ornament,[3] and relief brooches with Nydam-style 'chip-carving' (Lund Hansen 1970, 96–7; *ibid.* 1988, 25ff.; Straume 1987; see also Slomann 1977, 63–4).

Not all scholars have adopted Slomann and Lund Hansen's definition of cruciform brooches. Reichstein, for instance (1975; 1977), allowed the cruciform brooches to include Nydam brooches and Dorchester brooches, which leads to a number of finds with clearly late Roman-period forms being assigned to the Migration Period (Lund Hansen 1988, 25–7).[4]

MIGRATION PERIOD/MEROVINGIAN PERIOD

The transition between the Migration Period and the Merovingian Period appears to have been clear and sharp. In the discussion of this period boundary, importance has been attached to the fact that a set of artefacts, identified as the Migration-period complex, disappeared, and was superseded by new forms (Gjessing 1934; Slomann 1956, 68 and 74–80). Amongst the many changes in material expression it is above all the changes in style which are treated as definitive. A critical point in the debate has been whether Style II replaced Style I at the same time all over the Germanic area, and the date of this stylistic change.

Through their presence in Italy from 568, the Lombards have been the focus of attention as the medium of the impulses which, it has been argued, led to the development of Style II, namely Byzantine ribbon interlace. In this form of interlace, indeed, the rhythm which is a fundamental definitive feature is found. Essential to the change from Style I to Style II is the fact that the animal motifs were subordinated to the compositional principle of ribbon interlace (Haseloff 1981, 607–8, 612–13 and 709, with refs.). This transition has been discussed by Haseloff (1981, 596–614 and 709–10). Byzantine ribbon interlace came to Italy in the second quarter of the 6th century and is supposed to have spread north of the Alps rapidly. This produced not only a synthesis of animal style and ribbon interlace, but also surface-covering interlace or plaitwork composed of multistranded ribbons. These impulses must in turn have reached Scandinavia quickly, if that was the course they followed. The fusion of styles *may* have taken place on the basis of the animal style (Style B) which the Lombards carried into Italy, but is more likely to have taken place north of the Alps, amongst the Alamanni, who had a style characterized by interlaced, ribbon-shaped animals, albeit interlaced in a rhythm different from that of Style II (like our stadium-6 brooches in Style I). The earliest examples of Style II are identified by Haseloff on the Klepsau brooches (grave 4), which were probably made around 565 (around the middle or later in the 6th century according to Koch 1990, 234–5 and 155). Parts of the animal ornament on these brooches are identified as Style I, and other parts as Style II. From the dating of these brooches Haseloff concluded that Style II was developed during the final third of the 6th century (see also Koch 1990, 154–5). The role of the Lombards in this has been discussed and disputed (e.g. Werner 1962, 94–104 with refs.; Roth 1973, 32–3 and refs.) and is possibly of less significance than the occurrence of both styles on the Klepsau brooches and their dating. Ribbon interlace as such seems to appear earlier in Scandinavia, as indicated by early Migration Period spearheads in Denmark (Vang Pedersen 1988), as in the wooden handle of the weaving batten from Kvåle (Ringstad 1989, Fig. 6).

Several scholars have noted that Style I may have lived longer in, for example, western Norway, where the so-called North Sea Style (Style I) could have been concurrent with Continental Style (Style II) in southern and eastern Scandinavia (Brøgger 1925, 52–3; Brøndsted 1940, 294; Nissen Meyer 1935, 102). Such a possibility has also been noted by Solberg (1981, 166–7) for the second half of the 6th century. The discussion of the stylistic change is a complicated matter. In the chronological scheme of this work, however, it is not of such great significance, as in the area under investigation the entire artefact-set was replaced. The period boundary thus defined seems, consequently, to be unproblematic.

PHASE DIVISIONS WITHIN THE MIGRATION PERIOD

The Migration Period has been divided into unequal phases/ sub-periods, partly on the basis of stylistic criteria, and partly on the basis of artefact typology. Shetelig, building on works by Montelius and Salin (see references in Shetelig 1906), regarded, as has been noted, brooches from the Nydam find as the prototype of the early cruciform brooches (1906, 117), and, like Montelius, dated the Nydam find to ca. 350 A.D. Shetelig divided the cruciform brooches into four phases, dated by combinations with silver-sheet brooches and relief brooches (I: 350–400; II: 400–450; III: 450–500; IV: 500–550. Shetelig 1906, 152–3; Tab. 6.1). He considered that cruciform brooches of periods I and II were in use alongside silver-sheet brooches (*ibid.*, 132), while cruciform brooches of period III were contemporary with early relief brooches, and the latest cruciform brooches of period IV were contemporary with later relief brooches, down to the middle of the 6th century (the latest examples are Gyland and Ågedal). Cruciform brooches disappeared around the middle of the 6th century (*ibid.*, 139 and 150) and are not found together with the latest relief brooches (Shetelig 1900, 62).

Nissen Meyer, who dated the beginning and end of the Migration Period in the same way as Shetelig, divided the period into six stadia on the basis of the silver-sheet and relief brooches (1935; Tab. 6.1). She relied primarily on formal features of the brooches (their composition), but also on their decoration (on the basis of Salin's work). Find-combinations were also assessed. Stadium 1 is characterized by silver-sheet brooches, while the cast brooches (with the Nydam Style) appear in stadium 2 (Nissen Meyer 1935, 99). She attached particular importance to the division between the earlier stadia, stadia 2–4 (from the first half of the 5th to the beginning of the 6th century), and the later ones, stadia 5–6 (the first and second halves of the 6th century) (*ibid.*, 34–5 and 103; 1937, 107–8 and 112–26). She regarded the division as clear even though the course of development was gradual. The later stadia are distinguished in that the animal style has reached its fully developed form, and the footplates are cruciform in both the ridged and plane-foot series.[5] Some of the relief brooches of stadium 5, such as those from Gyland (F29) and Fristad (F52) have

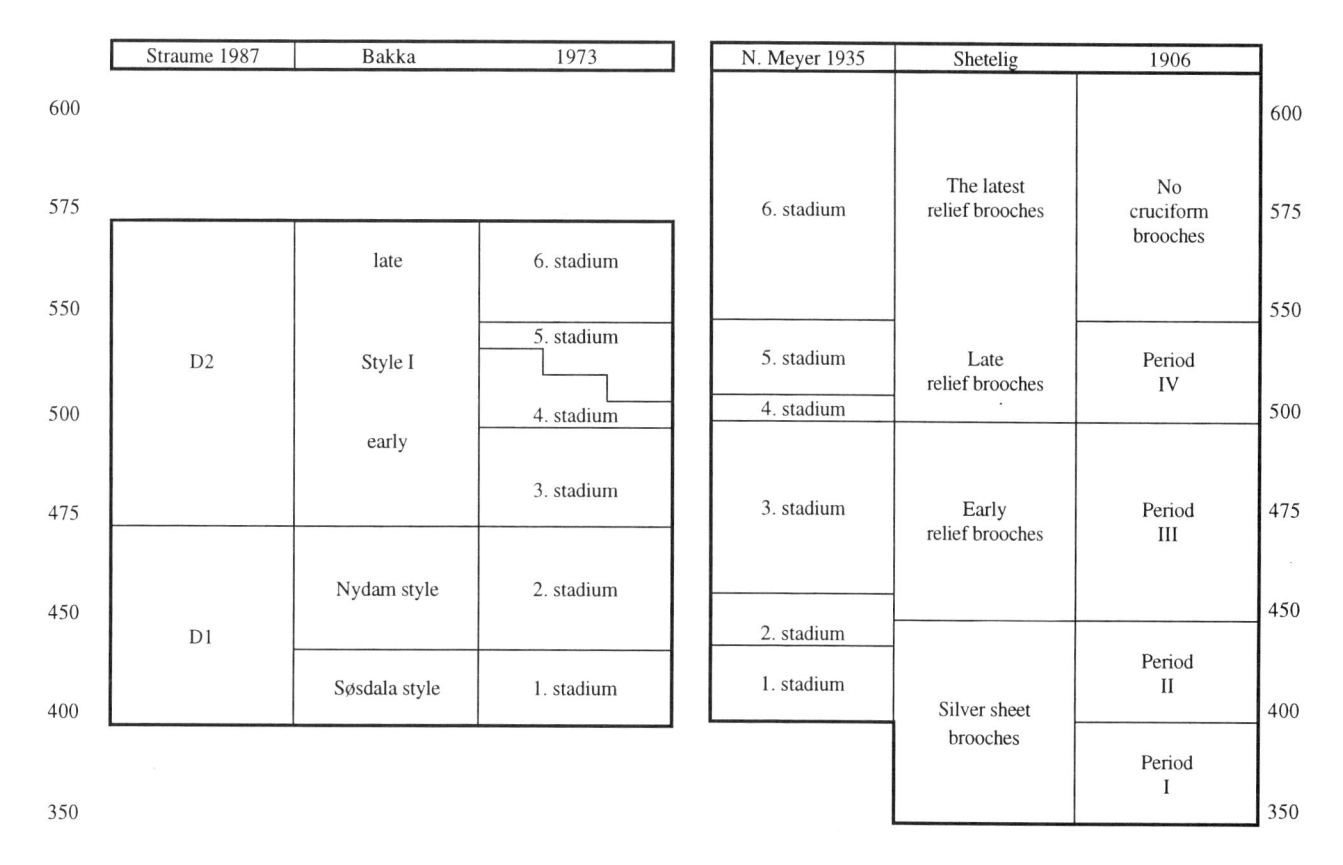

Tab. 6.1 The inter-relationship of various chronological systems.
Die Verknüpfung verschiedener Chronologiesysteme.

Notes on Shetelig's terms 'early', 'late' and 'latest' relief brooches:– Shetelig reckoned brooches such as Lunde (F25) and Tu (F49) as early relief brooches. The brooches of Nissen Meyer's stadim 4 such as Langlo (F1) and Tveitane (F8) he regarded as late relief brooches, together with a number of brooches from stadium 5 such as Ågedal (F17) and Gyland (F29). Other brooches of stadium 5, such as Hauge (F48), he treated as the latest releif brooches along with brooches of stadium 6.

features which link them to brooches of the earlier stadia. Nevertheless Nissen Meyer assigns them to stadium 5 as they also carry features which show that they must have been produced in close association with the relief brooch from Ågedal (F17), which unambiguously belongs with the later brooches (*ibid.*, 63, see also p. 43).

Several scholars have discussed styles in relation to the phase-divisions. The Nydam Style has been of crucial importance in the division of the early Migration Period (D1) (Bakka 1973; Straume 1987). The typical features of the Nydam Style, according to Voss, are spiral ornament in composite patterns executed in geometric chip-carving, beautifully formed niello strips, and punched decoration – the latter being of relatively minor frequency (1954, 176–9). The following figural motifs may be present: animals with their hindparts rolled up *above* the body; and creeping animals with four feet. The animals' bodies may be decorated with incised lines and rows of punched dots. Most characteristic, however, is chip-carved spiral decoration, elegantly formed into complex patterns. In his description of the style, Haseloff (1981, 8–17 and 706) stressed the deep relief of the chip-carving and the presence of late-Roman motifs such as running spirals, palmettes and astragalus patterns. The meander is common, alongside derived forms such as the swastika, step patterns and zigzags. He also emphasized the relationship between chip-carving and niello, and the colour contrast that is produced by this means, as well as, finally, punch-decoration. As in late-Roman metalwork, the animals were originally limited to the outer borders but gradually start to spread in across the surface. They have the character of sea beasts, again showing their connexion to Roman art. Quadrupeds also occur. These are often crouching.

When Forsander (1937) distinguished the Sösdala and Sjörup Styles from Style I, he included both the Nydam Style and the Sjörup Style as it is now understood (cf. Voss 1954) in the Sjörup Style.[6] The Nydam Style itself was distinguished from the Sösdala Style in 1954 (Voss 1954). This style was named after the decorated silver scabbard mounts from the second Nydam bog find, a weapon deposit which was discovered in 1888 (Kjær 1902). It is primarily the relationship between the Sösdala Style (complex punched decoration) and the Nydam Style (relief chip-carving) which has dominated the discussion of relative chronology in the first half of the Migration Period.

According to Voss, there is reason to regard the Sösdala and Nydam Styles as partly contemporary. It was suggested that both were in use around the year 400, when the Nydam Style came into being, a style which came to dominate as the 5th century progressed (Voss 1954, 177). Haseloff placed the Sösdala Style 'immediately before the Nydam Style' (1981, 10). Bakka considered that the Sösdala Style (which belongs in his Stufe I) began around 400 and that the transition to the Nydam Style (which belongs in his Stufe II) took place some time before 450 (1973, 85). He was followed by Hines (1984). Bakka's division between Stufe I and Stufe II did not, however, take account of Lund Hansen's discussion of the Kvarmløse find, in which she discussed the relationship between the Sösdala and Nydam Styles on the basis of punched decoration and its relationship to chip-carving. She considered that both styles arose in central Europe in the interface between late Roman and Germanic cultures:

> That both styles came to Scandinavia is clear, but unfortunately we do not know exactly how they stand in relation to one another chronologically. One thing, however, is certain: that at some stage the two styles were contemporary. They are not only found combined on one and the same object, they are also found on different items in a single grave. Whether or not the fully developed Sösdala Style began before the Nydam Style in Scandinavia we cannot yet determine. (Lund Hansen 1970, 33)

The Nydam Style appears to have spread over large areas in a short time at the beginning of the 5th century (*ibid.*). Haseloff dated the beginning of the Nydam Style thus, and emphasized the same points (1981, 16–17). In his view it is difficult to determine the life-span of the Nydam Style. There are no clear historical dates. The issue cannot, moreover, be clearly defined by Style I.

The transition from the Nydam Style to Style I is more evident in the transformation of Roman ideas and forms in accordance with Germanic preconceptions than in external stylistic features. The change does, however, eventually crystalize into new forms and new means of expression (Haseloff 1981, 172). The sea creatures fade into the background and quadrupeds become more common. These differ from their predecessors by one new feature, the *contour line*. Every part of the body – the head, neck, trunk, forelimb and hindlimb – is surrounded by a contour line. This feature is the most prominent characteristic of Style I (Haseloff 1981, 706). In the Jutlandic brooch group with early Style I, Haseloff found that much of the range of Nydam-style ornament continued, especially Roman motifs such as tongue-patterns, astragalus, running spirals and palmettes (*ibid.*, 18–173). There are, however, differences in the execution of these elements between the two styles. They gradually come to be more freely produced in relation to the tight principals of composition of Roman art, as a result of which surface-covering decoration eventually emerges as characteristic of Style I. In Haseloff's view, however, it is not in these decorative elements that the distinction between the two styles lies; rather in the animals and their handling, primarily the use of the contour line referred to above. The animals' bodies become flat and often covered with decoration. The dividing lines here, however, are not always equally clear. The Jutlandic brooch group stands out because of a series of motifs which partly recall the Nydam Style and partly contrast with it (*ibid.*, 81–139). Human features are more frequent here than otherwise with Style I, especially human heads, in profile or full-face, hanging human heads and anthropomorphic animals. There is also animal ornament in its true form.

On the basis of art-historical arguments, especially Continental finds of Scandinavian animal ornament in Style I, Haseloff nevertheless proposed that Style I had developed in Scandinavia by the last quarter of the 5th century. The boundary is assigned to 475 A.D. Haseloff focussed especially on the Jutlandic brooch group in respect of chronology, allowing the earliest of these (group A) to define the transition from the Nydam Style to Style I which took place in the second half of the 5th century, presumably in the later decades of the century, around 475; he also assigned the latest specimens of this group (group C) to the first decades of the 6th century (1981, 172–3). He further (*ibid.*, 166–7 and 707) divided Style I (with animal decoration in the Jutlandic group) into 4 style phases, A–D. These are not attributed with definite chronological significance, although they do reflect developmental tendencies. Style phase A appears to be earlier than B–D. The characteristic features of style phases B–D may be largely concurrent and partly a matter of regional differences. Haseloff's style phases are derived from the material from a large area and are based upon very general features. The material is not uniform, however, and consists of a range of more or less local traditions, in which the more specialized features may be of just as much value in dating as the general features; certainly when the goal is not a super-regional chronology.

Bakka's aim was to construct a super-regional chronology in which Norwegian finds could be linked to Continental Germanic and Anglo-Saxon material (1958; 1973; 1981). In addition to the decoration, he paid particular attention to objects which could also be found in Continental and English burial contexts, relief brooches and gold bracteates, especially the latter (Bakka 1973, 54, 60 and 73). He divided the Norwegian Migration Period into four 'Stufen' and correlated silver-sheet and relief brooches – in respect of which he based himself upon Nissen Meyer's scheme – with various forms of cruciform brooch and stylistic phases (Tab. 6.1). Stufe III was defined by the occurrence of relief brooches with a plane foot and quadrangular headplate, which begin with the emergence of Style I. Stufe IV begins with the D-bracteates. Bakka's late dating of the D-bracteates and his division between bracteate periods 2 and 3 (1973, 85) was later rejected by Lund Hansen (1988, 27), and in his last article Bakka himself was amenable to an earlier dating of D-bracteates (1981, 27). Bakka has, however, been supported by Hines (1984, 22), who found that the 4-Stufe system worked, with certain adjustments. Bakka's scheme of four Stufen has been a topic of debate since then, and his stylistic phasing is less relevant, at least at a super-regional level, if one accepts Haseloff's view, although it may be more pertinent in the local context.

In contrast to Nissen Meyer herself, Bakka stressed the division between stadia 5 and 6 more than that between stadia 4 and 5, a point which is also related to his desire to produce a super-regional chronology. He also, however, saw some significance in the division between stadia 4 and 5 (1973, 73). This could be correlated with the transition from bracteate period 2 to 3. This dividing line, according to Bakka, may have been more marked in Norway than in southern Scandinavia. Stadium 4 was limited to southern Scandinavia and south-eastern Norway (Vestfold-Telemark), while stadia 5 and 6 were rarely if ever represented here. It is thus possible that the transition to stadium 5 with animal ornament fully dominant and excluding other motifs was a more rapid process in Norway. In Bakka's view, the division between stadia 5 and 6 could be observed over a large area, and thus it served him better as a major chronological watershed in the super-regional perspective.

In 1987, Straume argued that the chronology of the Migration Period relied primarily on the typological development of the cruciform brooch and stylistic analyses of the large silver-sheet and relief brooches (Straume 1987, 14). On the basis of studies by Voss (1954), Bakka (1959), Lund Hansen (1970) and Magnus (1975) she divided the Migration Period into two 'Stufen', D1 and D2 (Tab. 6.1). D1 is represented by the Sösdala and Nydam Styles, D2 by the Sjörup Style and Style I. She placed the boundary between D1 and D2 around 475 (with reference to Hougen 1935; Voss 1954; Bakka 1958; 1977). This division into phases D1 and D2 appears to work well.

CHRONOLOGICAL PHASING IN THIS STUDY

The following study aims to develop a phasing of the period which is capable of grouping a large quantity of the finds in a manner relevant to the questions which have been taken up in the mentioned thesis on relief ornaments (Kristoffersen 1997). Geographically it refers to southern and western Norway, and finds from the regions (*fylker*) along the coast: Vestfold, Telemark, Aust-Agder, Vest-Agder, Rogaland, Hordaland and Sogn og Fjordane are discussed. The relief brooches occupy a major place in the finds under investigation in the thesis and will be of critical importance to this phasing. Nissen Meyer's grouping of the brooches from 1935 seems to stand up extremely well, even though perhaps not all of the stadia should be attributed with equally strong chronological significance, and it will be one of the foundations of what follows.[7] This will be combined with Straume's division into D1 and D2 from 1987. Nissen Meyer's stadia and Straume's phasing can be reconciled via Bakka's study of 1973, in which stadium 2 is associated with the Nydam Style, which defines D1 for Straume. Stadia 3–6 are associated with Style I, which defines D2 for Straume (Tab. 6.2). The boundary between D1 and D2 is thus one of the main divisions in the series of phases. The division between the earlier and later stadia of the relief brooches will form the second main boundary line and thus represent a division of D2 into two parts.

In her dating of the phases, Nissen Meyer attached special importance to combinations with artefact-types such as cruciform brooches, small equal-armed brooches, and bucket-shaped pots with beaded or knotwork decoration: the group around Bøe 1931, figures 324–7 (*ibid.*, 42–3 and

Nissen Meyer 1935	Bakka 1973	Straume 1987
Stadia 3-6	Style I (VWZ III-IV)	D2
Stadia 1-2	Nydam style Søsdala style (VWZ I-II)	D1

Tab. 6.2 The stadia of relief brooches correlated with phases D1 and D2 on stylistic grounds.
Verknüpfung der Stadien der Relieffibeln mit den Phasen D1 und D2 nach stilistischen Argumenten.

99–103).[8] Relatively few relief brooches have been found in good contexts. This is especially the case with stadium-5 brooches. Of the total of 41 relief brooches which are included in this study, 9 were loose finds and 9 from finds which are manifestly problematical. 23 brooches remain whose associated finds can be assessed (Tab. 6.3). Few of the finds are perfect, however, as several of them are incomplete and were not professionally excavated. A relatively unambiguous picture emerges nonetheless, indicating that relief brooches of the earlier stadia are found in combination with cruciform brooches. Cruciform brooches are only once found in combination with a brooch of the later stadia (F84, Nornes). The relief brooches of stadium 6 are found combined with small equal-armed brooches. These brooches are never found in combination with relief brooches of the earlier stadia. Also found combined with relief brooches of stadium 6 are bucket-shaped pots with surface-covering bead- or knotwork decoration.[9]

Chronologically significant artefact-types

Cruciform brooches

Cruciform brooches have been thoroughly discussed in earlier works (Shetelig 1906; Reichstein 1975) and further considered in various other contexts (e.g. Sjøvold 1962; Bakka 1973; Magnus 1975). The typological development of the cruciform brooch has been less emphasized in this study. What is regarded as more important is that these brooches apparently occur up to a certain point within the Migration Period, and then are absent in the final phase of this period, presumably being no longer produced (Shetelig 1900, 62; 1906, 150–1; 1911, 86; Bøe 1931, 200–1; Nissen Meyer 1935, 102–4). Cruciform brooches are richly represented in the whole of the area under consideration. Their absence in the final phase can therefore be treated as highly significant.

In the one case of a relief brooch of the later stadia occurring in a secure combination with a cruciform brooch (F84, Nornes), the grave also contained a definitely late pot, indicating that the cruciform brooch may have been old when it was buried.

Several works have treated the occurrence of cruciform brooches in two crucial finds including relief brooches of stadium 5 as important: Gyland (F29) and Ågedal (F17).

The cruciform brooches in these finds are regarded as the latest cruciform brooches in the Norwegian corpus (Shetelig 1906, 151; Reichstein 1975, 72). The circumstances of the Gyland find are, however, so obscure that it is difficult to place any weight upon it (see catalogue in Kristoffersen 1997). In respect of the Ågedal brooches, meanwhile, it is debatable whether these should be reckoned as cruciform brooches. While the boundary of the brooch-type earlier in time, against the Nydam brooches, has been widely discussed, its later boundary has been neglected. The two identical, nominally cruciform, brooches from Ågedal, only one of which survives, do not have true knobs (Fig. 6.1). The upper part of the brooches, with the 'knobs', is completely flat, and the junction between the headplate and the 'knobs' is blurred. These do not look like knobs attached to the headplate, not even rudimentary knobs as they sit on small projections from the headplate. Nor is the headplate at an angle to the bow, an important criterion in Slomann's definition.[10] These brooches ought probably rather to be considered in the context of the small brooches which developed after the cruciform brooches had fallen out of use.

Small equal-armed brooches

These equal-armed brooches form a small group within a wider range of small brooches which came into use after the cruciform brooch had been abandoned (Fig. 6.2). The equal-armed type has been selected here because it is more distinctive than other small brooches, which need more

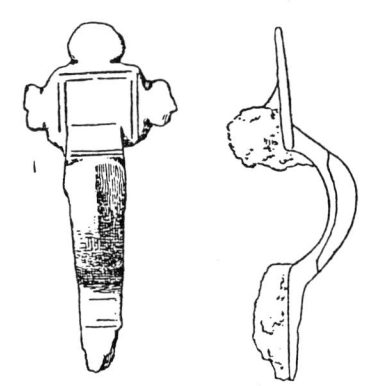

Fig. 6.1 Small brooch from Ågedal (F17) similar to cruciform brooches. From Shetelig 1906, Fig. 190.
Kleine Fibeln aus Ågedal (F17) ähnlich den kreuzförmigen Fibeln.

RELIEF BROOCHES	FIND PLACE	CRUCIFORM BROOCHES	EQUAL ARMED BROOCHES	BUCKET SHAPED POTS
Stadium 2	Ommundrød (F4)	●		
	Lunde (F25)	●		
	Erga (F46)	●		
	Hauge (F47)	●		
	Tu (F49)	●		
Stadia 3-4	Langlo (F1)	●		
	Tveitane (F8)	●		
	Tveitane (F9)			
	Falkum (F11)	●		
	Søtvet (F12)			
	Skjervum (F82)	●		
	Kvåle (F86)	●		
Stadia 5-6	Ågedal (F17)	○		
	Kvassheim (F35)		●	
	Eikeland (F45)		●	●
	Nord-Braut (F51)		○	
	Østbø (F62)			●
	Indre Ålvik (F69)			
	Døsen (F74)		●	●
	Indre Arna (F77)		●	●
	Holum (F83)			●
	Nornes (F84)	●		●
	Kvåle F85			●

Tab. 6.3 Combinations involving relief brooches in the various stadia with cruciform brooches, equal-armed brooches and late bucket-shaped pots.
Fundkombinationen von Relieffibeln verschiedener Stadien mit kreuzförmigen Fibeln, gleicharmigen Fibeln und späten eimerförmigen Töpfen.

careful analysis to separate them from earlier types. The equal-armed brooches have not been discussed in detail before, although they have been considered in various contexts (Shetelig 1911, 84–6; Åberg 1924, 53–4, Figs. 28–131; 1953, 61–7). Shetelig called this group 'small equal-armed bronze brooches ("beak shaped")' and illustrated examples of this type in his figures 69–72 (1911, 85). According to Shetelig, this type of brooch came into use in Norway for a short time towards the end of the Migration Period. They are not known in any finds from before the middle of the 6th century and they disappear at the end of the period (*ibid.*, 86).

There are 18 small equal-armed brooches in the finds included in the collection analysed in the study of relief

100   *Siv Kristoffersen*

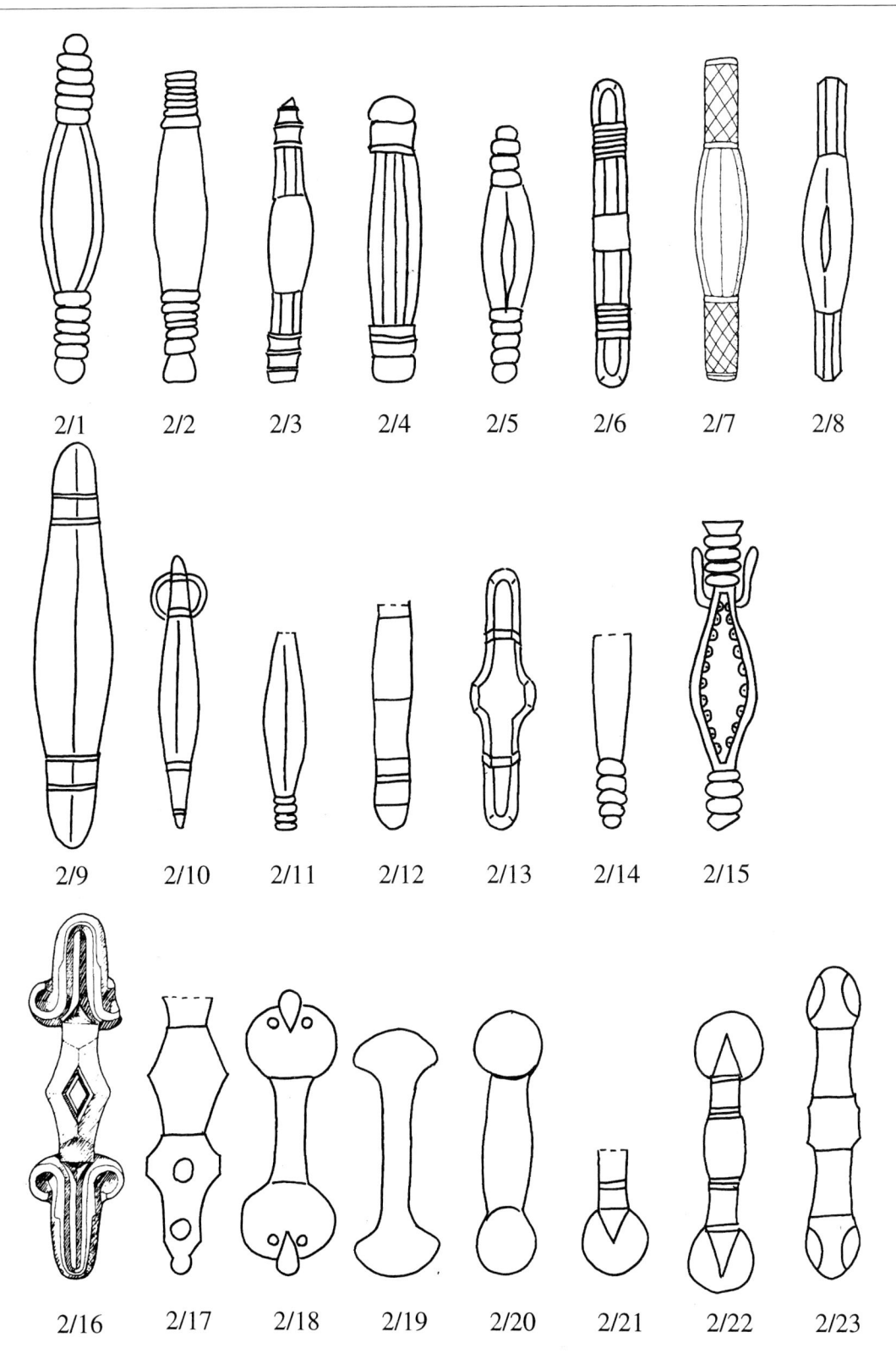

Fig. 6.2 Small equal-armed brooches. Figures 2/1–2/15 represent the most common type with moulded arms. Figures 2/16–2/23 represent a smaller group with different types of plate on the arms. Figure 2/1 (F64/B10202IIp) (earlier no. S2617), 2/2 (B5607), 2/3 (F31/S308), 2/4 (B14186/192), 2/5 (B6375) and 2/6 (B6658) sketches by the author; 2/7 (B11694II) from the accession list. Fig. 13, 2/8 (F74/B6090I), 2/9 (S2417a), 2/10–11 (F39/S3741), 2/12 (F77/B565), 2/13 (F85/B6516d–e), 2/14 (F71/B6474) and 2/15 (F61/S4116) sketches by the author; 2/16 (B11694II), from the accession list, Fig. 14; 2/17 (B8989), 2/18–19 (F81/BB6691), 2/20 (S1928), 2/21 (F54/C1638), 2/22 (B5839) and 2/23 (F35/5362) sketches by the author. All at approximately 1:1.

Kleine gleicharmige Fibeln. Nr. 1–15 vertreten den häufigsten Typ. Nr. 16–23 vertreten die kleinere Gruppe mit verschiedenen Typen von Scheiben auf dem Arm.

ornaments (Kristoffersen 1997).[11] Fifteen of these belong to or are very similar to the most common type, as in Shetelig 1911 figures 69–71 (Fig. 6.2/1–15). It was really only this type which Shetelig discussed. Three equal-armed brooches have a different form with plates on the arms. Both of the brooches from Hove (F81) are of this type: one has round plates with a point (Fig. 6.2/18), the other semi-circular plates (Fig. 6.2/19) (Shetelig 1917, Figs. 12–13). The surviving brooch from Kvassheim (F35) has plates shaped as animal heads (Fig. 6.2/23). The two lost brooches from this find were apparently of similar form. The brooch from Lunde (F54) had round plates (only one is preserved) with three-sided points running down from the bow (Fig. 6.2/21).

From finds within the area (Vestfold, Telemark, Aust-Agder, Vest-Agder, Rogaland, Hordaland and Sogn og Fjordane) which are not included in the study of relief ornament, we have a group of 17 small equal-armed brooches.[12] This overview, which is based upon an examination of accession lists and the inspection of all the brooches in Bergen Museum, comprises all the brooches of this type from Hordaland and Sogn og Fjordane. The corpus is not believed to be complete for the remainder of the area, although the group is unlikely to be very much larger. This brooch-type is rare east of Rogaland. Ten of these brooches are of the type of Shetelig 1911 figures 69–71. Four brooches have round plates on the arms. One of these, from Rygg, Etne (Fig. 6.2/22), is like the brooch from Lunde with three-sided points running down on to the plates. Three brooches belong to neither of these two main types: the two identical brooches from Skaim (Fig. 6.2/16) and the brooch from Eikenes (Fig. 6.2/17). The brooches from Skaim have elongated plates formed from a ribbon rolled up at the sides. The brooch from Eikenes, which is much more simple, has a plate of a form comparable to those on the brooches from Skaim.

Equal-armed brooches of the type in Shetelig 1911 figures 69–71 or with plates on the arms are also found in Sweden (Åberg 1924, 53–4), the latter in a somewhat different form from the Norwegian ones (Åberg 1953, 61–2, Figs. 65–8 and 77–80). The Norwegian brooches, especially those from Hove, are more like Continental types of equal-armed brooches, types which are believed to originate in Lombard territory in Italy (Böhner 1958, 89). They occur at Krefeld-Gellep, for example (with round plates: Pirling 1966, Fig. 16, 15) and in Frankish graves in the Trier region (with round and semi-circular plates: Böhner 1958, Pl. 12, 1–5). Brooches of this type (but not other small equal-armed brooches) are supposed not to occur on the Continent before the 7th century (Böhner 1958, 89–91; Pirling 1966, 171; Wamers 1986, 50 and refs.).

Within the collection analysed in the study of relief ornaments (Kristoffersen 1997) there are two small brooches which are not entirely symmetrical but which nevertheless have a form which is very much like that of the equal-armed brooches with plates on the arms: S2451c Nord-Braut (F51) and B6474 Hæve (F71) (Fig. 6.3). There are another four brooches of this type in finds within the area of investigation which is not included in the study of relief ornaments.[13] This group may well be larger.

Of this total of 41 brooches of equal-armed or nearly equal-armed form, two were found together with cruciform brooches, Vestly and Birkeland. Vestly (B2535) is clearly a mixed find. In the double burial

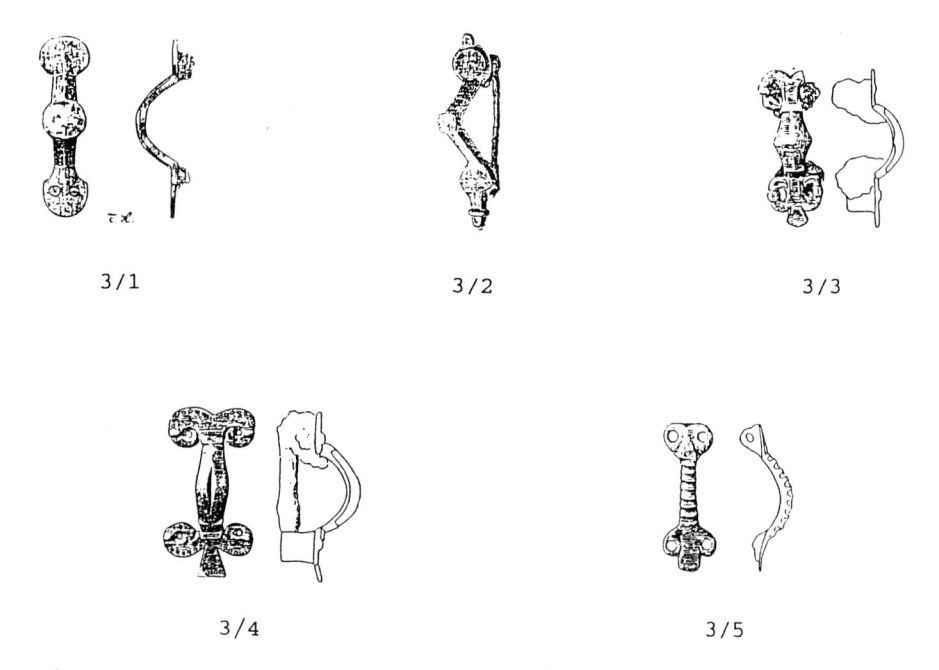

Fig. 6.3 Small brooches similar to equal-armed brooches. 3/1 (S2417b) from the accession list, Fig. 2; 3/2 (F51/S2451c) from the accession list, Fig. 6; 3/3 (B7145a) from the accession list, Fig. 1; 3/4 (F71/B6474a) from Shetelig 1911, Fig. 64; 3/5 (B6309) from Shetelig 1911, Fig. 65. All at approximately 1:2.
Kleine Fibeln ähnlich den gleicharmigen Fibeln.

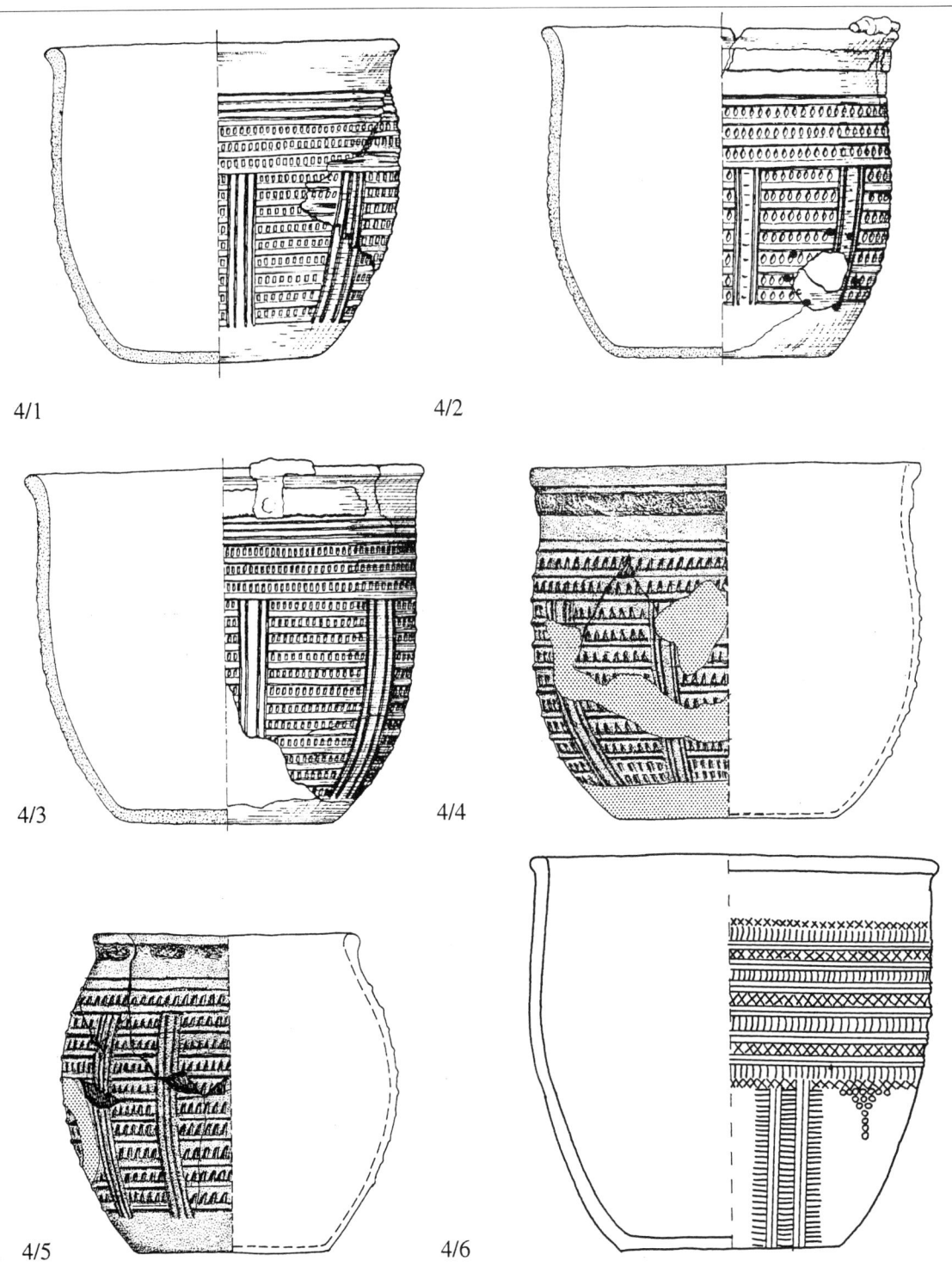

Fig. 6.4 Bucket-shaped pots with surface-covering beading and knotwork. 4/1–3 (B9614IIIx, B9614IIy, B9614IIIj), from Magnus 1965; 4/4–5 (F83/B8045k,j) drawn by E. Hoff, Bergen Museum; 4/6 (B6611) sketch by the author; 4/7 (F79/B5705d) and 4/8 (F64/B10202IIn), drawn by E. Hoff, Bergen museum; 4/9 (B11694II), from Straume 1987, plate 60:4; 4/10 (F70/B6809i), drawn by E. Hoff, Bergen museum; 4/11 (B5114a) and 4/12 (B5769g) sketches by the author. All at approximately 1:2.
Eimerförmige Töpfe mit flächendeckenden Buckeln und Perlen.

from Birkeland there were 2 bucket-shaped pots, both of the late type (Shetelig 1917, 30–8, Pl. IV; Bøe 1931, 200–1), and the cruciform brooch was probably old when buried. In Reichstein's catalogue of cruciform brooches (1975, 114–28), Birkeland stands as the only find from the area under consideration in which a cruciform brooch is associated with an equal-armed one.

This indicates that the small equal-armed brooches, and brooches very similar to them, form a group which belongs late in the Migration Period and which became

4/7 4/8
4/9 4/10
4/11 4/12

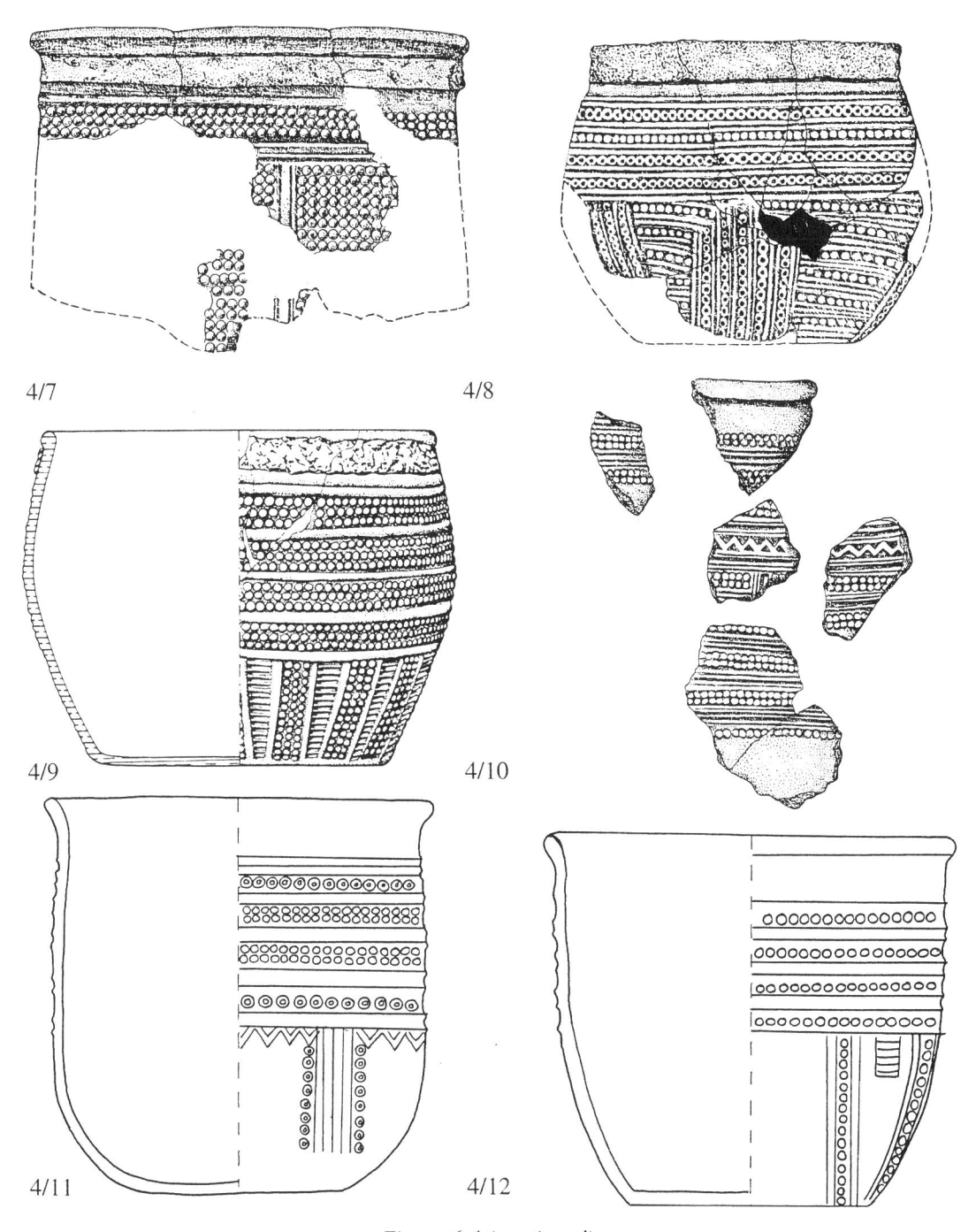

Figure 6.4 (continued).

part of the artefact inventory after cruciform brooches had gone out of use.

Bucket-shaped pots
The decoration of bucket-shaped pots can be a useful form of chronological evidence, even though some forms of decoration appear to have remained in use for long periods and therefore cannot be dated very precisely. There are, for instance, a number of distinctive types of decoration which form clear groups that appear, in turn, to belong to specific periods. The pots related to Bøe 1931 figures 324–7 form one such group. The decoration of these pots varies somewhat, but is characterized by stamped beading and knotwork which covers the surface of the wall of the vessel (Bøe 1931, 198–201) (Fig. 6.4). According to Bøe, these vessels belong to the period following the demise of cruciform brooches, even though they have been found associated with such brooches in a few cases such as the find from Birkeland noted above (*ibid.*, 201). These vessels are also seldom associated with handled pots (R361) (*ibid.*). Bucket-shaped pots with this form of decoration have been correlated with spears too, occurring in combinations with spears of Solberg's Types I.3 and II of the late Migration Period

and early Merovingian Period (Solberg 1981, 163–4).

There are 15 bucket-shaped pots with some variety of surface-covering beading or knotwork in the finds included in the collection analysed in the study of relief ornaments (Kristoffersen 1997).[14] In finds from the area not included in this study there is a collection of 56 vessels of this type.[15] This corpus, which is based upon examination of accession lists and inspection of all the vessels in Bergen Museum, comprises all of the vessels of this type from Hordaland and Sogn og Fjordane. The list is not believed to be complete for the remainder of the area, although the group is unlikely to be very much larger as Hordaland and Sogn og Fjordane are the principal domain of this vessel-type. These vessels are especially numerous in Hordaland, in the Voss area above all, with a total of 19 specimens. Typically, the three specimens from Telemark are from high up in the west of the region, in and close to the mountain ranges connected to the west. The same applies to the find from Aust-Agder, which is from Bykle at the upper end of the valley of Setesdal.

Elements of bead and knotwork decoration are also found combined with other motifs. Most common are surface-covering ribbon interlace of the type in Bøe 1931 figures 316 and 318–19 (Fig. 6.5), and hanging arches of the type in Bøe 1931 figures 309–12 (Fig. 6.6), often in complicated patterns. Only beading is found together with ribbon interlace, while hanging arches are found with both beading and knotwork.

Both surface-covering ribbon interlace and hanging arches have themselves been identified as late decorative styles (Shetelig 1904; Bøe 1931; Magnus 1984; Solberg 1981). Surface-covering ribbon interlace have been linked to the Byzantine ribbon interlace which is believed to be introduced late in the period (see above). Shetelig assigned ribbon interlace in a simple form to the beginning of the 6th century (1904, 84–6). It gradually developed into a surface-covering style spreading over most of the outer surface of the vessel and often combined with stamped beading. Shetelig dated this form of decoration to 550–600. Bøe dated vessels with surface-covering ribbon interlace, represented by his figures 316, 318 and 319, to the late 6th century (1931, 196). Earlier in the century there is a narrower form of ribbon interlace. Bøe admittedly recognized the possibility that surface-covering ribbon interlace could have developed earlier, but only in respect of the basketwork pattern in his figure 317 which he sees as developed interlace pattern. This, however, must be regarded as highly dubious. There is no clear connexion between ribbon interlace and the decoration in Bøe's figure 317. Likewise the view that basketwork may have developed from ribbon interlace cannot be used as an argument against the late dating of the latter (Bøe 1931, 198). Solberg picked out surface-covering ribbon interlace as one of the forms of decoration which is found in association with the late spears of her Type I.3 (Solberg 1981, 163, Fig. 19). Pots with surface-covering ribbon interlace are placed in Magnus's group of vessels with four or five strands of ribbon interlace (Magnus 1984, 150–3). In Magnus's view these were produced from around 500 up to the end of the Migration Period. Her

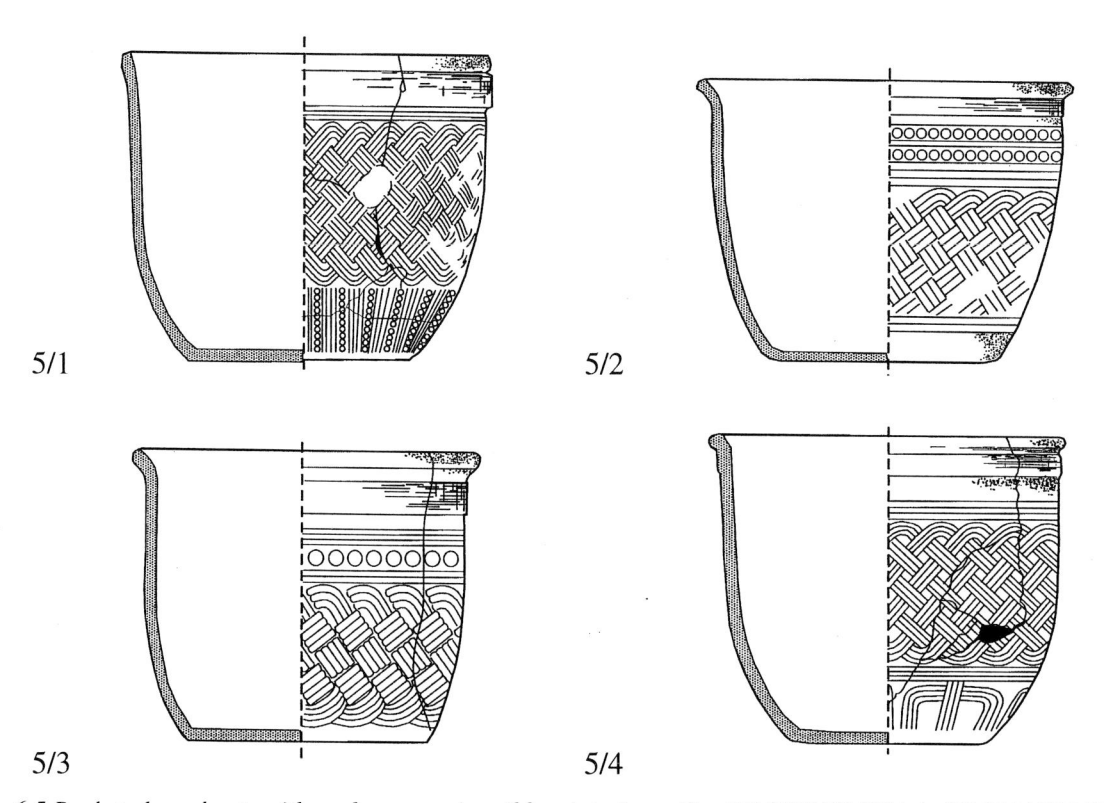

Fig. 6.5 Bucket-shaped pots with surface-covering ribbon interlace. Fig. 5/1 (F19/C26001y), 5/2 (F44/S8635o) and 5/3 (F64/B10202IIm) and 5/4 (B9614IIIz) drawn by E. Hoff, Bergen museum. All at approximately 1:3.
Eimerförmige Töpfe mit flächendeckenden verflochtenen Bändern.

Migration Period chronology in Norway

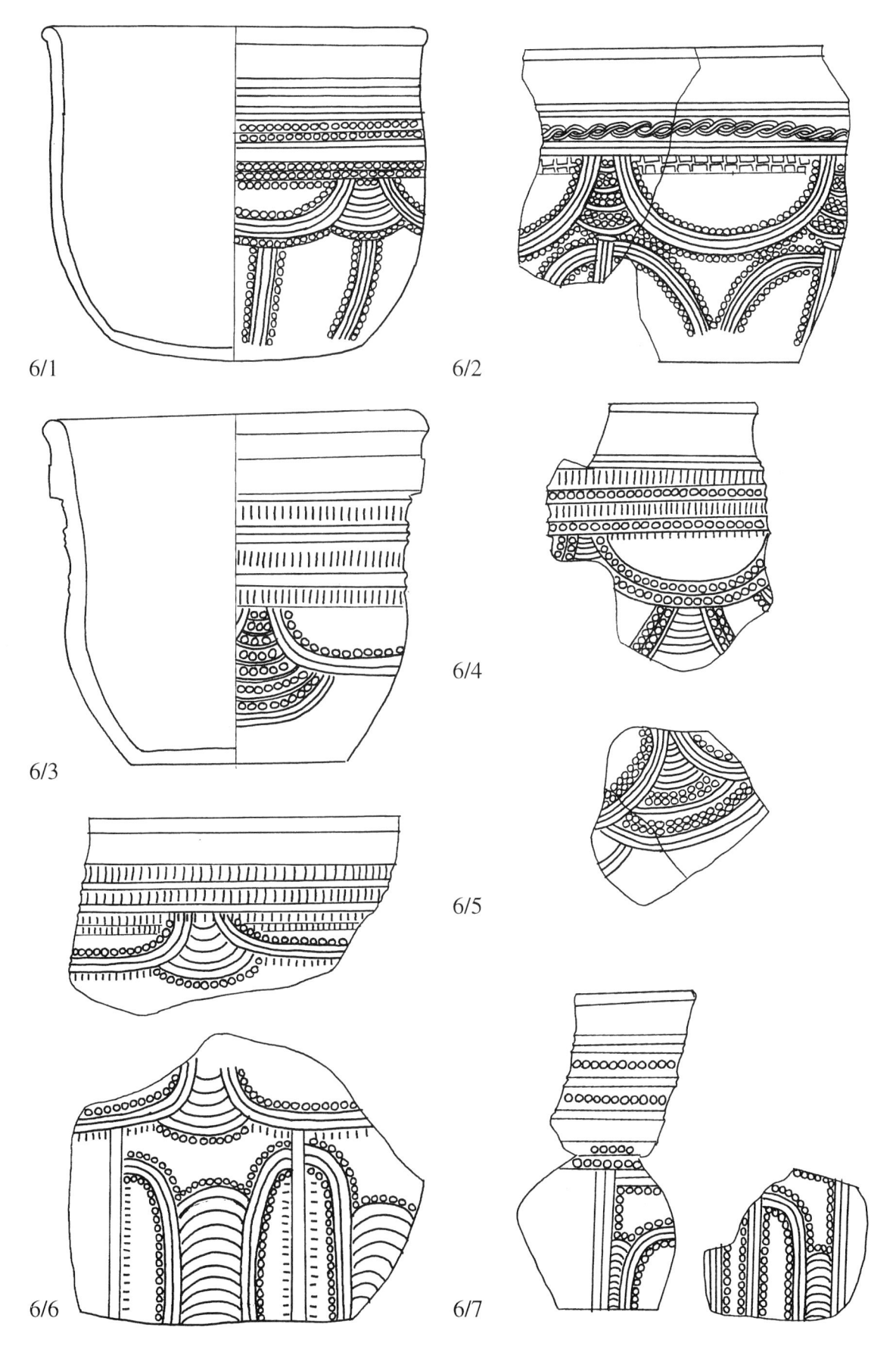

Fig. 6.6 Bucket-shaped pots with hanging arches. 6/1 (B6091Ie), 6/2 (F90/B6656m), 6/3 (B9506Ib), 6/4 (B10299d), 6/5 (F18/C28026q), 6/6 (B5587f) and 6/7 (B4923o) sketches by the author. All at approximately 1:2.
Eimerförmige Töpfe mit hängenden Bögen.

early dating of the group is, however, possible due to the fact that vessels with narrower ribbons are also included in this group, such as a vessel from Krågedal, which is combined with a cruciform brooch (*ibid.*, 144 and 150).

Hanging-arch decoration has attracted less attention. Bøe dated it to a relatively short phase around the middle of the 6th century (1931, 194). Shetelig, who drew attention to its association with beading, assigned the hanging arches to the second half of the 6th century (1904, 83–5). Their disagreement is due to a different view of the Ågedal grave (F17) which contains a vessel of this type.

Hanging arches are found on five vessels within the collection included in the study of relief ornaments (Kristoffersen 1997).[16] Two of these, Ågedal and Anda, represent a more simple form, while Snartemo (Fig. 6.6/5), Sørheim and Sandal (Fig. 6.6/2) are of a form of more complex composition. From finds within the area of investigation which are not in the collection included in the mentioned study, there are 25 pots with hanging arches.[17] The representativity of this corpus is the same as that of the vessels with beading and knotwork. Amongst this group there are vessels of the simpler form, such as Presthus, Kvalbein and Røysland, from Vest-Agder and parts of Rogaland close to Vest-Agder. Most of the vessels, however, have more complex decoration and are of very high quality. On the vessels from Lyngdal and Hægebostad the decoration is of a somewhat different type, as the hanging arches are combined with deep points instead of

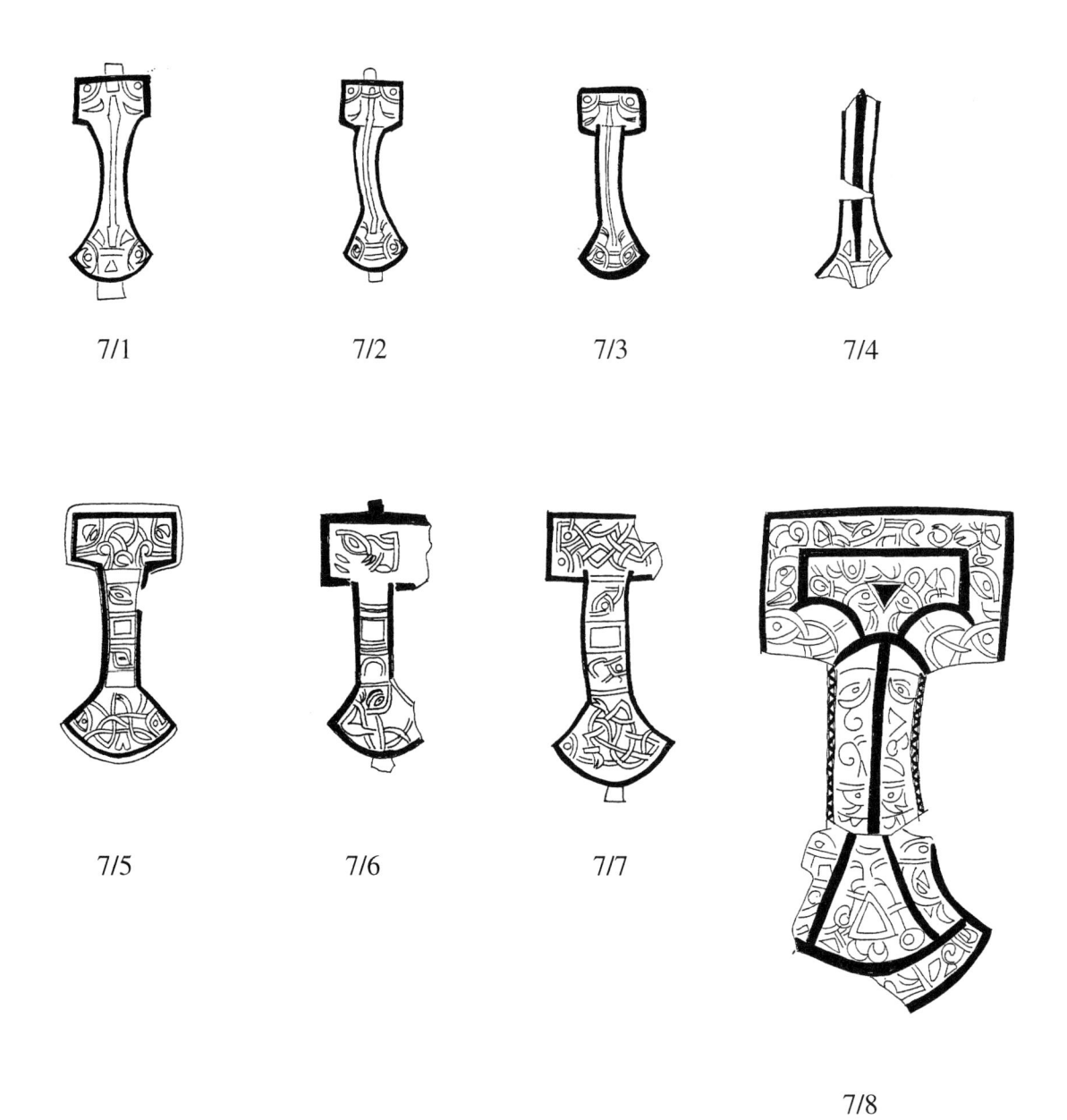

7/1 7/2 7/3 7/4

7/5 7/6 7/7

7/8

Fig. 6.7 Relief brooches with a spatulate foot. The group with no border outside the frame on headplate and foot. 7/1–2 (F34/B5994a), 7/3 (F81/B6691a), 7/4 (F39/S3741d), 7/5 (F31/S307), 7/6 (F78/B4846a), 7/7 (F69/B6899b), 7/8 (F40/S440). Drawn by the author. All at approximately 2:3.
Relieffibeln mit schaufelförmigem Fuß. Gruppe der Fibeln ohne Bord außerhalb des Rahmens von Kopf und Fuß.

beading and a surface-covering beaded chequer pattern respectively.

This form of decoration is most common in Hordaland and the Sogn area, with nine and eight vessels respectively. In all assessable cases, the hanging arches are of the complex form and the vessels of high quality. Only the specimens from Noreim and Skrøppa are of somewhat coarser form. There are seven vessels from Vest-Agder. Both decoration and quality vary here. As in the case of vessels with beading and knotwork, the pot from Mogen is associable with the vessels in the west.

There is a typical decorative motif on several of the pots with beading and knotwork or with hanging arches which is principally composed of narrow and wider horizontal ribbons.[18] This motif appears as a dense fringe of chevrons or chequers surrounding small pits. That pattern is found on 3 vessels from the finds in the collection included in the study of relief ornaments (Kristoffersen 1997).[19] On the vessel from Rivjaland it is placed in the zone beneath the rim and in narrow vertical panels on the side of the vessel.

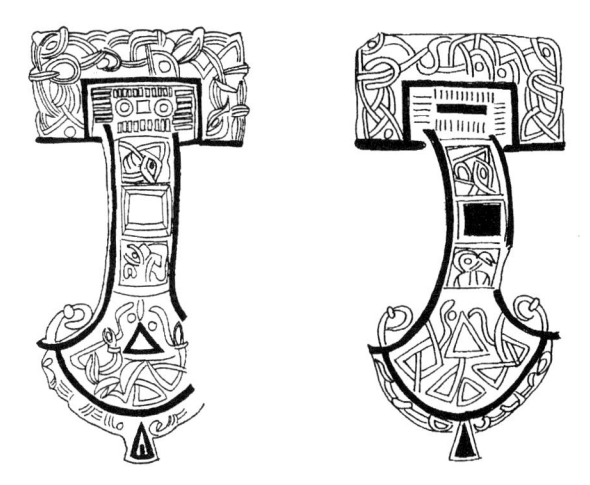

Fig. 6.8 Near identical relief brooches with spatulate foot from Ågedal (F17/B3410i) and Vatland (F59/S2772c). Drawn by the author. Approximately 2:3.
Eng verwandte Relieffibeln mit schaufelförmigem Fuß aus Ågedal und Vatland.

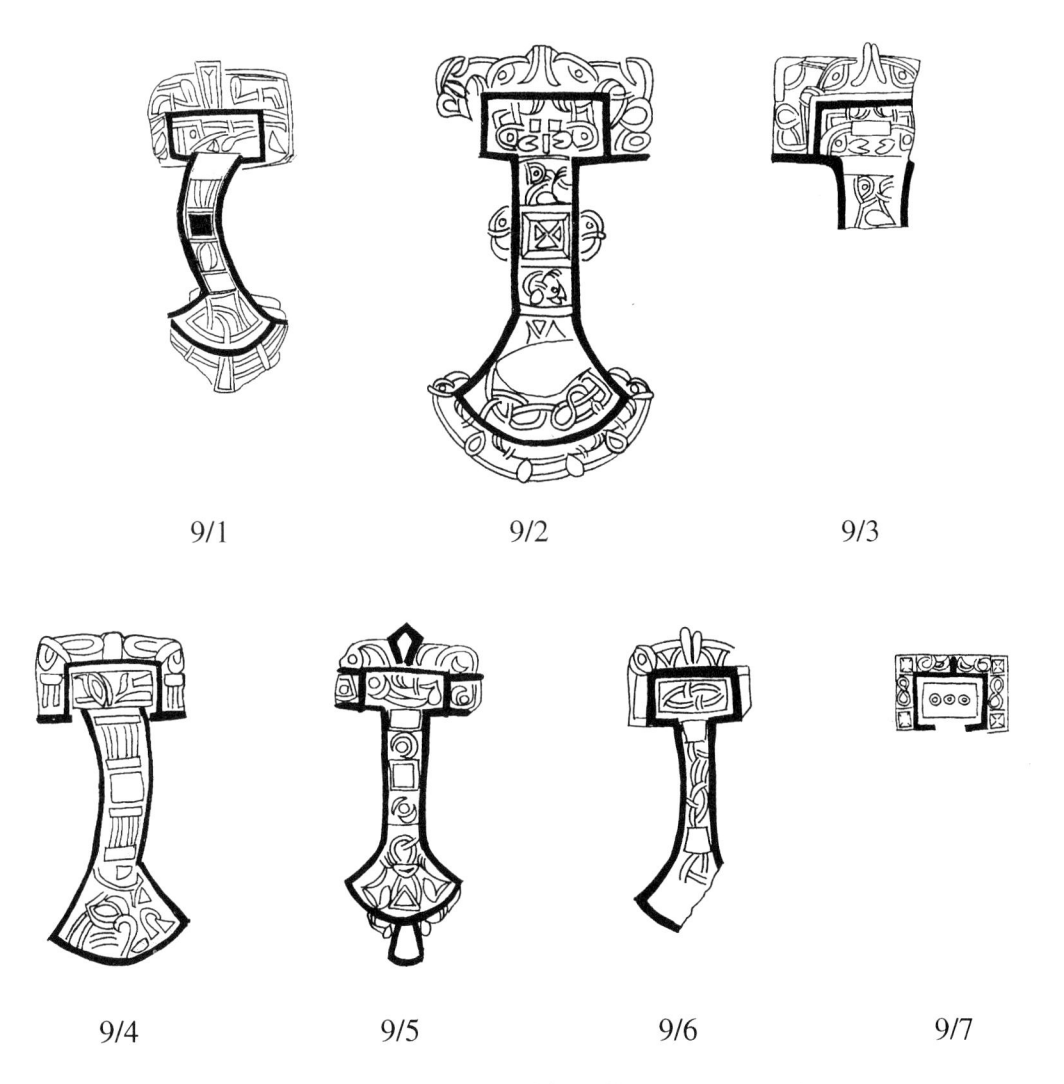

9/1   9/2   9/3

9/4   9/5   9/6   9/7

Fig. 6.9 Other relief brooches with spatulate foot. 9/1 (F36/S2062), 9/2 (F43/B1784), 9/3 (F54/C1638), 9/4 (F17/B3410k), 9/5–6 (F85/B6516b–c), 9/7 (F65/B10205d). Drawn by the author. Approximately 2:3.
Weitere Relieffibeln mit schaufelförmigem Fuß.

The above discussion has considered four decorative forms which are found in combination with one another. Of all of the vessels which have been referred to, pots with beading and knotwork are found in combination with cruciform brooches only in the graves from Birkeland, Nornes and Haugland. Birkeland and Nornes, however, contain other late elements, an equal-armed brooch and a relief brooch of the later stadia respectively. It is likely that the cruciform brooches in these graves were old when they were buried. Haugland is manifestly a mixed find. A vessel with hanging arches has been found associated with a cruciform brooch in one find, Skrøppa, where the cruciform brooch has a semi-circular foot. This vessel, however, is not absolutely typical of the group, and is difficult to assess, being very coarse.

It is, in conclusion, reasonable to infer that these forms of decoration belong to the phase which follows the general demise of cruciform brooches. A large number of the finds are, admittedly, weapon graves, in which one would not expect cruciform brooches to occur. It has been shown, however, that some of these finds contain late weapon-forms. The late spearheads of Solberg's types are found together with vessels with beading and knotwork, surface-covering ribbon interlace and hanging arches (Solberg 1981). A number of weapon finds also contain three-edged arrowheads (Fett 1940, 39).

All four forms of decoration occur in combinations including relief brooches of the later stadia and equal-armed brooches. This is shown in respect of vessels with beading and knotwork by Table 6.3. In addition, the vessels from Skjerpe (F39), Skaim and Nedre Aure were found together with equal-armed brooches. A vessel with surface-covering ribbon interlace was found together with a relief brooch of the later stadia and equal-armed brooches in the Eikeland find (F45). A vessel with hanging arches was found together with a relief brooch of the later stadia at Ågedal. There is also a vessel of this kind in the grave from Sandal (F90) which also included a late relief brooch, and in the grave from Sørheim (F65) from which there is also an equal-armed brooch, although in both these cases there are problems with the find context.

The combinations are not numerous, but those we have indicate that these forms of decoration are contemporary with relief brooches of the later stadia and with equal-armed brooches.

The form of the vessels has not been discussed here. The majority of the vessels with these forms of decoration belong to Category E (Magnus 1984, 141). There are, however, vessels which undoubtedly belong to the late types with the form of Category C, such as, for instance, the pot from Kvåle (F85). There are also bowl-shaped vessels.[20] It is possible that these vessels represent the very latest bucket-shaped pots. It was noted above that the find-circumstances of the Gjukastein find are unclear, but it is perhaps no mere coincidence that we have a bowl-shaped pot as well as a small disc-on-bow brooch from this find.

Relief brooches with a spatulate foot

Relief brooches with a spatulate foot have not previously been discussed as a group, though reference has been made to them in various contexts. Shetelig dated them to the 6th century without going into any detail (1911, 66). Nissen Meyer picked out the brooches from Garpestad and Lunde, which she linked to her Rogaland group and to stadium 6 (1935, 45–6). Bakka referred to this brooch-type in connexion with the bracteates, and regarded the brooches from Ågedal as associable with stadium-5 relief brooches on the strength of their decoration, which he described as 'advanced Style I', and put them late in Stufe III (1973, 67 and 69). With reference to the relief brooches with spatulate feet from Kvåle and Hove, he concluded that the type continued into Stufe IV. He dated the Garpestad brooch on the basis of the Ågedal find. Sjøvold has included a small group of these brooches in his book (1993, Pl. 32).

Relief brooches with a spatulate foot have a distribution which coincides closely with the area under investigation. There is a total of 19 brooches from 15 finds within an area from Vest-Agder to Sogn. The majority of finds are from Rogaland and Hordaland. Just one brooch has been found outside of the area from Vest-Agder to Sogn, the small specimen B3264 from Storeneset on Giske in Sunnmøre, a brooch which is similar to the examples from Kvassheim and Hove. This brooch was found on its own.

Relief brooches with a spatulate foot vary in size from the large brooch from Torland, which is 10 cm long, to the smallest, from Hove and Kvassheim, which are 3–4 cm long. Table 6.4 shows the typological groups of this brooch-type defined by formal features such as the division of the bow and the presence or absence of borders outside the frame on the headplate or foot. There is a group of relatively complex brooches, such as Ågedal, Garpestad, Vatland, Lunde, Anisdal and Kvåle. This group contains the larger brooches of high quality such as Ågedal and Garpestad. Vatland and Lunde are parallels to these two specimens respectively, although Lunde is undoubtedly of much simpler quality. The brooches from Kvassheim, Skjerpe and Hove are the simplest examples. The large brooch from Torland belongs with these, although otherwise it stands on its own, as an individualistic piece which is similar to the larger relief brooches. There are several intermediate forms between these two groups, amongst which Indre Ålvik, Hodneland and Abeland (Figs. 6.7/5–7) form a fairly consistent group.

Several of the brooches, such as those from Garpestad (Fig. 6.9/2), Lunde (Fig. 6.9/3) and Torland (Fig. 6.7/8), can, for typological and stylistic reasons, be associated with the relief brooches of the Rogaland group, especially brooches such as Hovland (F32) and Veierland (F2) with a virtually identical division and decoration of the bow, but also with the relief brooch from Hauge (F48) on the basis of the division of the headplate. Table 6.5 shows that three brooches were found in association with relief brooches, all of them of the later stadia. They can also be

	Border outside the frame on foot	Border outside the frame on the head plate	Tripple division of the bow	No border outside the frame on foot	No border outside the frame on the head plate	Central ridge on the bow
Ågedal (F17)	●	●	●			
Garpestad (F43)	●	●	●			
Vatland (F59)	●	●	●			
Lunde (F54)		●	●			
Anisdal (F36)	●	●	●			
Kvåle (F85)	●	●	●			
Sørheim (F65)		●				
Kvåle (F85)		●		●		
Ågedal (F17)		●	●	●		
Ågedal (F17)		●	●	●		
Indre Ålvik (F69)			●	●	●	
Hodneland (F78)			●	●	●	
Abeland (F31)			●	●	●	
Kvassheim (F34)				●	●	●
Kvassheim (F34)				●	●	●
Hove (F81)				●	●	●
Torland (F40)				●	●	●
Skjerpe (F39)						●

Tab. 6.4 Typological features on relief brooches with a spatulate foot.
Typologische Elemente an Relieffibeln mit schaufelförmigem Fuß.

linked to the relief brooches of these stadia by way of combinations with equal-armed brooches and late pottery, while at the same time they are never found in association with cruciform brooches.

PHASING

On the basis of the above discussion, a three-phase division of the Migration Period, into phases D1, D2a and D2b, is proposed. The phases are defined by the presence and combination of the following styles and artefact-types:-

D1: The Nydam Style. Relief brooches of stadium 2. Cruciform brooches.
D2a: Style I. Relief brooches of stadia 3 and 4. Cruciform brooches.
D2b: Style I. Relief brooches of stadia 5 and 6. Equal-armed brooches. Bucket-shaped pots with beading and knotwork, surface-covering ribbon interlace and hanging arches. Relief brooches with a spatulate foot.

The relief brooches of stadium 5 are assigned to the same major phase as those of stadium 6 on the strength of the primary division which Nissen Meyer identified between the brooches of stadia 4 and 5. There are few brooches of stadium 5, within the research area, and only one of these was found in a satisfactory context. This is a general problem affecting the relief brooches of stadium 5, and it leads to a degree of uncertainty as to their chronological place.

As far as combinations are concerned, brooches from stadium 5 have no definite associations with cruciform brooches, although it is possible that we have such a combination in the Gyland find. They have been found

	Stadia 3-4	Cruciform brooches	Stadia 5-6	Equal armed brooches	Bucket shaped pots
Ågedal (F17)			●		
Abeland (F31)				●	
Kvassheim (F34)					
Anisdal (F36)					
Skjerpe (F39)				●	●
Torland (F40)					
Garpestad (F43)					
Lunde (F54)					
Vatland (F59)					
Sørheim (F65)				●	●
Indre Ålvik (F69)			●		
Hæve (F71)				●	●
Hodneland (F78)					●
Hove (F80)					●
Kvåle (F85)			●	●	●

Tab. 6.5 Combinations involving relief brooches with a spatulate foot, relief brooches of various stadia, cruciform brooches, equal-armed brooches and late bucket-shaped pots.
Fundkombinationen von Relieffibeln mit schaufelförmigem Fuß, Relieffibeln der verschiedenen Stadien nach Nissen Meyer, kreuzförmigen Fibeln, gleicharmigen Fibeln und späten eimerförmigen Töpfen.

together with pottery with hanging arches and brooches with a spatulate foot but never with equal-armed brooches. The connexion between stadia 5 and 6 seems, however, to be supported by typological features of the brooches with a spatulate foot dicussed above.

DEUTSCHE ZUSAMMENFASSUNG

Der Beitrag gibt eine Übersicht über die Forschungsgeschichte zur völkerwanderungszeitlichen Chronologie in Norwegen (ca. 400–550/575 n.Chr.). Vorgeschlagen wird eine chronologische Gliederung dieser Periode in drei Phasen: D1, D2a und D2b. Sie beruht auf den älteren Studien norwegischer Forscher, die Gliederung kann aber hier auf der Grundlage eines erheblich umfangreicheren Materials diskutiert werden. Die Phase D1 ist durch den Nydamstil, Relieffibeln des Stadiums 2 nach Nissen Meyer und kreuzförmige Fibeln charakterisiert. Die Phase D2a wird umrissen durch den Stil I, Relieffibeln der Stadien 3 und 4 sowie kreuzförmige Fibeln. Phase D2b wird charakterisiert durch den Stil I, Refieffibeln der Stadien 5 und 6 oder solche mit schaufelförmigem Fuß, dazu kleine, gleicharmige Fibeln sowie eimerförmige Töpfe mit dichten Buckelchen, flächendeckendem Flechtband oder hängenden Bögen.

APPENDIX

Finds with an F-number which are incorporated in the thesis referred to (Kristoffersen 1997) – not all are mentioned in the present paper.
B = Accession number, Bergen Museum
C = Accession number, Universitetets Oldsaksamling, Oslo
S = Accession number, Arkeologisk museum, Stavanger

All these finds are published in the accession lists of the museums. A catalogue of lists published up to 1950 can be found in *Universitetets Oldsaksamlings Årbok* 1949/50: Gjessing, H. and Fett, P: 'Register over trykte tilvekster av norske oldsaker'. Hines has a catalogue which has been extended up to 1993 (Hines 1993, 103–6).

pgd. = parish; k. = administrative district

F1. LANGLO, STOKKE pgd. and k. VESTFOLD C5947–5962
F2. VEIERLAND, STOKKE pgd. NØTTERØY k. VESTFOLD C18714–15
F3. NORDHEIM, HEDRUM pgd. LARVIK k. VESTFOLD C19858
F4. OMMUNDRØD, HEDRUM pgd. LARVIK k. VESTFOLD C29300
F5. ROLIGHETEN, HEDRUM pgd. LARVIK k. VESTFOLD C14338–50, 14534, 89–90, 711
F6. SKÅRA, TJØLLING pgd. LARVIK k. VESTFOLD C18892–904, 18917–18, 19095
F7. BERG, BRUNLANES pgd. LARVIK k. VESTFOLD C19227
F8. TVEITANE, BRUNLANES pgd. LARVIK k. VESTFOLD C11220–36
F9. TVEITANE, BRUNLANES pgd. LARVIK k. VESTFOLD C11237
F10. BRATSBERG, GJERPEN pgd. SKIEN k. TELEMARK C26566
F11. FALKUM, GJERPEN pgd. SKIEN k. TELEMARK C21856
F12. SØTVET, SOLUM pgd. SKIEN k. TELEMARK C9440–49, 9811
F13. STENSTAD, HOLLA pgd. NOME k. TELEMARK Nat. mus.Kbh.8031,8306–08,8411,8420
F14. NORDGÅRDEN, SELJORD pgd. and k. TELEMARK C19269–19280, 19615–16
F15. VIK, FJÆRE pgd. GRIMSTAD k. AUST-AGDER C7072–7082
F16. TRYGSLAND, BJELLAND pgd. MARNARDAL k. VEST-AGDER K DCCXI–III, DCCXV, DCCXXII–IVb, DCCCXXXII–VII
F17. ÅGEDAL, BJELLAND pgd. AUDNEDAL k. VEST-AGDER B3410a–t, B4132
F18. SNARTEMO II, HÆGEBOSTAD pgd. and k. VEST-AGDER C28026, C8897
F19. SNARTEMO V, HÆGEBOSTAD pgd. and k. VEST-AGDER C26001
F20. HÆGEBOSTAD-ØDEGÅRDEN, N.-AUDNEDALEN pgd. LINDESNES k. VEST-AGDER C13697
F21. LØLAND, NORD-AUDNEDALEN pgd. LINDESNES k. VEST-AGDER C18301–309
F22. GITLEVÅG, SØR-AUDNEDALEN pgd. LYNGDAL k. VEST-AGDER B5060
F23. BERGSAKER, LYNGDAL pgd. and k. VEST-AGDER C25813
F24. HØYLAND, VANSE pgd. FARSUND k. VEST-AGDER B5037
F25. LUNDE, VANSE pgd. FARSUND k. VEST-AGDER B3543
F26. SLETTEN, VANSE pgd. FARSUND k. VEST-AGDER B4234
F27. SPANSKSLOTTET, VANSE pgd. FARSUND k. VEST-AGDER B4286
F28. ÅMDAL, LISTA pgd. FARSUND k. VEST-AGDER C25077
F29. GYLAND, BAKKE pgd. FLEKKEFJORD k. VEST-AGDER C7453–63, 7539–40, 7563
F30. ÅDLAND, BAKKE pgd. FLEKKEFJORD k. VEST-AGDER C8713–21
F31. ABELAND, HELLELAND pgd. BJERKREIM k. ROGALAND S306–311
F32. HOVLAND, HELLELAND pgd. EGERSUND k. ROGALAND S2276
F33. KVASSHEIM, EGERSUND pgd. HÅ k. ROGALAND B5343
F34. KVASSHEIM, EGERSUND pgd. HÅ k. ROGALAND B5994
F35. KVASSHEIM, EGERSUND pgd. HÅ k. ROGALAND B5362

F36. ANISDAL, HÅ pgd. and k. ROGALAND S2062–2066
F37. VOLL, HÅ pgd. and k. ROGALAND S927–938
F38. JORENKJØL AV SKRETTING, HÅ pgd. and k. ROGALAND S6970
F39. SKJERPE, HÅ pgd. and k. ROGALAND S3741
F40. TORLAND, HÅ pgd. and k. ROGALAND S440
F41. RIMESTAD, HÅ pgd. and k. ROGALAND S4268
F42. FOSSE, TIME pgd. and k. ROGALAND S6697
F43. GARPESTAD, TIME pgd. and k. ROGALAND B1781–1784, 1877
F44. VESTLY, TIME pgd. and k. ROGALAND S8635
F45. EIKELAND, TIME pgd. and k. ROGALAND S9181
F46. ERGA, KLEPP pgd. and k. ROGALAND S7131
F47. HAUGE, KLEPP pgd. and k. ROGALAND B2269–82, 88–92, 94–99
F48. HAUGE, KLEPP pgd. and k. ROGALAND B4000
F49. TU, KLEPP pgd. and k. ROGALAND C21407
F50. ANDA, KLEPP pgd. and k. ROGALAND B2973–74
F51. NORD-BRAUT, KLEPP pgd. and k. ROGALAND S2451
F52. FRISTAD, KLEPP pgd. and k. ROGALAND S1961
F53. VATSHUS, KLEPP pgd. and k. ROGALAND C3300–3313
F54. LUNDE, HØYLAND pgd. SANDNES k. ROGALAND C1638
F55. HOGSTAD, HETLAND pgd. SANDNES k. ROGALAND S1520–26
F56. SYRE, SKUDESNES pgd. KARMØY k. ROGALAND S9269
F57. MELBERG, STRAND pgd. and k. ROGALAND S7577, S7858
F58. RIVJALAND, HJELMELAND pgd. and k. ROGALAND S2547
F59. VATLAND, JELSE pgd. SULDAL k. ROGALAND S2772
F60. NÆRHEIM, SULDAL pgd. and k. ROGALAND S2848
F61. ÅM, SKJOLD pgd. VINDAFJORD k. ROGALAND S4116
F62. ØSTBØ, VIKEDAL pgd. VINDAFJORD k. ROGALAND S2695
F63. ETNE pgd. HORDALAND B2049
F64. GRINDHEIM, ETNE pgd. and k. HORDALAND B10202 II (S2617)
F65. SØRHEIM ETNE pgd. and k. HORDALAND B10205 (S2850)
F66. SÆBØ, FJELLBERG pgd. KVINNHERAD k. HORDALAND B3358
F67. ØVSTHUS, FJELLBERG pgd. KVINNHERAD k. HORDALAND B3731
F68. NORDHUS, FJELLBERG pgd. KVINNHERAD k. HORDALAND B4096
F69. INDRE ÅLVIK, KVAM pgd. and k. HORDALAND B6899
F70. LØINING UNDER ØYSTESE, KVAM pgd. and k. HORDALAND B6809
F71. HÆVE, VOSS pgd. and k. HORDALAND B6474
F72. MITTUN, VOSS pgd. and k. HORDALAND B7190
F73. GJERMO, VOSS pgd. and k. HORDALAND B7607
F74. DØSEN, OS pgd. and k. HORDALAND B6090 I
F75. HAUGLAND, FANA pgd. BERGEN k. HORDALAND B5541
F76. HARTVEIT, HAUS pgd. OSTERØY k. HORDALAND B4291, B5208
F77. INDRE ARNA, HAUS pgd. BERGEN k. HORDALAND B564–569
F78. HODNELAND, LINDÅS pgd. and k. HORDALAND B4704, B4846
F79. HODNELAND, LINDÅS pgd. and k. HORDALAND B4950, B5705
F80. HOVE, VIK pgd. and k. SOGN og FJORDANE B319
F81. HOVE, VIK pgd. and k. SOGN og FJORDANE B6691
F82. SKJERVUM, VIK pgd. and k. SOGN og FJORDANE B8830
F83. HOLUM, LEIKANGER pgd. and k. SOGN og FJORDANE B8045

F84. NORNES, SOGNDAL pgd. and k. SOGN og FJORDANE B9688
F85. KVÅLE, SOGNDAL pgd. and k. SOGN og FJORDANE B6516
F86. KVÅLE, SOGNDAL pgd. and k. SOGN og FJORDANE B13954
F87. UGULEN, HAFSLO pgd. LUSTER k. SOGN og FJORDANE B6071, B6092I–II
F88. SØRHEIM, LUSTER pgd. and k. SOGN og FJORDANE B3720
F89. BOLSTAD, LUSTER pgd. and k. SOGN og FJORDANE B3724
F90. SANDAL, JØLSTER pgd. and k. SOGN og FJORDANE B6656
F91. GJEMMESTAD, GLOPPEN pgd. and k. SOGN og FJORDANE B12549
F92. EVEBØ, GLOPPEN pgd. and k. SOGN og FJORDANE B4590
F93. INDRE BØ, STRYN pgd. and k. SOGN og FJORDANE B4842

NOTES

1 Although it might cause some confusion, I have chosen to maintain a division between the finds which belong to the primary collection of finds in the thesis (finds which contain artefacts with relief ornament and which are given an F-no.) and the finds which have been introduced secondarily into the chronological discussion (finds from the research area which do not contain artefacts with relief ornament). The finds in the primary collection are systematically treated and presented in a catalogue. A list of these finds is given below.

2 In several places (Slomann 1977, 63; Lund Hansen 1988, 25) reference is made to Shetelig (1906) in respect of the definition of a cruciform brooch. It is not entirely clear where Shetelig stood on this matter. He described Nydam brooches as cruciform (1906, 18) and discussed them in the chapter on the 'Origin of the type'. In the next chapter, 'The early cruciform brooches', he referred to them as prototypes (his figs. 17–19). His figure 20, which has a trapezoid plate which is narrower than the bow, was clearly regarded as a cruciform brooch and was also referred to in the conclusion (pp. 152–3) as a typical example of period I. The development of wider plates occurs only some time into this period (ibid. p. 24, from fig. 26). In 1911 he described brooches whose 'plate is of the same width as the bow, and narrow higher up so as to serve as the base for the end knob' as the 'very first stage in the development of this form [i.e. the cruciform brooch]' (Shetelig 1911, 53–4).

3 The Sösdala Style, according to Forsander's definition (see below). The extent to which punched decoration can be used as a criterion was later discussed by S. Jensen (1978). See also Ethelberg 1976, 20.

4 See also the classification of these brooches by German scholars cited by Lund Hansen (1970, note 173).

5 This is explained further as follows. For the late ridge-foot brooches 'the side lobes are raised above the middle and the contours are rounded, so that they turn into proper cross-arms. Animal heads are placed as terminals on all three lobes (except on nos. 53, 59 and 61–2). On the lower borders there are often animal figures, but these always follow the outline of the edge-frame closely. The bow is either parallel-sided or expanded in the middle. The headplate is composed in various ways. Its border is covered in animal ornament in place of the regularly repeated knobs or en-face masks.
The triangles on the foot disappear or are reduced to a small motif in the middle. Heads as marginal decoration disappear. Geometrical decoration and spirals give way to animal ornament. The bodies of the animals grow long and ribbon-like.' (Nissen Meyer 1935, 36. On late plane-foot brooches, ibid. p. 62.)

6 The Sjörup Style is now thought to be later than both the Sösdala and the Nydam Styles.

7 A new classification of the relief brooches (Sjøvold 1993) appeared so late that it did not appear appropriate or necessary to allow it to influence this study. Sjøvold's study is, however, taken into consideration in relation to those brooches which were not considered by Nissen Meyer (different types and more recent finds). Similarly the high quality of the photographic illustrations in Sjøvold's book has been a great help.

8 Nissen Meyer identified several such artefact assemblages and their association with stadium-6 brooches. *'Diese Fibeln sind nur einmal mit einem Brakteaten (aus Rivjaland) zusammen gefunden worden. Dagegen werden sie oft mit den kleinen gleicharmigen Fibeln, die fast innerhalb dieses Zeitraumes beschränkt scheinen, zusammen angetroffen. Andere kleine, einfache Fibeltypen gibt es auch, diese ersetzen, wie erwähnt, die Armbrustfibeln* (here = cruciform brooches). *Typisch für die Gräber sind endlich eimerförmige Tongefässe mit Perlenreihen, hängenden Bögen und Bandgeflechten.'* (Nissen Meyer 1935, 103).

9 The relief brooches of stadia 5 and 6 in Rogaland are seldom found in the late combinations presented in Tab. 6.3. The picture would be filled out if other late, small brooches were included. Several of these are discussed in the catalogue in the thesis (Kristoffersen 1997), and some of them can be linked together through combinations with other relief brooches. See also the discussion of the late bucket-shaped pots in Rogaland.

10 This is, however, a feature which serves to distinguish cruciform brooches from Nydam brooches.

11 S308 Abeland, Bjerkreim k., Rogaland (F31) (1 brooch); B5362b-d Kvassheim, Hå k., Rogaland (F35) (1 brooch of an original 3); S3741c Skjerpe, Hå k., Rogaland (F39) (2 brooches); S9181f, k and n Eikeland, Time k., Rogaland (F45) (4 brooches); S4116a Åm, Vindafjord k., Rogaland (F61) (1 brooch); B10202IIp Grindheim, Etne k., Hordaland (F64) (1 brooch); B10205c Sørheim, Etne k., Hordaland (F65) (1 brooch); B6474b Hæve, Voss k., Hordaland (F71) (1 brooch); B6090Ic Døsen, Os, Hordaland (F74) (1 brooch); B565 Indre Arna, Bergen k., Hordaland (F77) (1 brooch); B6691c-d Hove, Vik k., Sogn og Fjordane (F81) (2 brooches); B6516d-e Kvåle, Sogndal k., Sogn og Fjordane (F85) (2 brooches). In addition, there is an equal-armed brooch in the find C1638 Lunde, Sandnes k., Rogaland (F54), in uncertain association with the small relief brooch with a spatulate foot.

12 C28758e Øvre Moi, Kvinesdal, Vest-Agder (1 brooch); S1928 Hundvåg, Stavanger, Rogaland (1 brooch); S2372c Ombo, Jelsa, Rogaland (1 brooch); S2417a Meling, Håland, Rogaland (1 brooch); S9327 Unknown provenance, Rogaland (1 brooch); S9335 Unknown provenance, Rogaland (1 brooch); B2535 Vestly, Time, Rogaland (1 brooch); B5607a Bryne, Time, Rogaland (1 brooch); S2748a Eide, Ølen, Hordaland (1 brooch); B5839 Rygg, Etne, Hordaland (1 brooch); B6658d Birkeland, Kvam, Hordaland (1 brooch); B11694h,i Skaim, Aurland, Sogn og Fjordane (3 brooches); B6375 IIb Stedje, Sogndal, Sogn og Fjordane (1 brooch); B14186/192 Skrivarhelleren, Årdal, Sogn og Fjordane (1 brooch); B8989g Eikenes, Stryn, Sogn og Fjordane (1 brooch). C12323c Hafsø, Egersund, Vest-Agder may also be an equal-armed brooch. According the accession list, this find also included a cruciform brooch,

although that is not included in Reichstein's catalogue. These finds were identified by going through the accession lists. Only the brooches in Bergen museum have been checked. There are, however, illustrations of brooches S1928 and S2417a.

13 S1626 Bø, Hå, Rogaland (1 brooch); S2417b Meling, Håland, Rogaland (1 brooch); B7145a Nedre Aure, Voss, Hordaland (1 brooch); B6309 Ervik, Selje, Sogn og Fjordane (1 brooch). These finds were identified by examining accession lists. Only the brooches in Bergen museum have been checked. There are, however, illustrations of S1625 and S2417b.

14 S3741b Skjerpe, Hå k., Rogaland (F39); S9181s Eikeland, Time k., Rogaland (F45); S2695f Østbø, Vindafjord k., Rogaland (F62); B10202IIn Grindheim, Etne k., Hordaland (F64); B6809i Løining, Kvam k., Hordaland (F70); B6090II Døsen, Os k., Hordaland (F74); B569 Indre Arna, Bergen k., Hordaland (F77); B4846b Hodneland, Lindås k., Hordaland (F78); B5705d Hodneland, Lindås k., Hordaland (F79); B6691t Hove, Vik k., Sogn og Fjordane (F81); B8045j Holum, Leikanger k., Sogn og Fjordane (F83); B9688n Nornes, Sogndal k., Sogn og Fjordane (F84); B6516s Kvåle, Sogndal k., Sogn og Fjordane (F85); B6092IIg Ugulen, Luster k., Sogn og Fjordane (F87).

15 C30088c Mogen, Rauland, Telemark; C30088e Mogen, Rauland, Telemark; C32325c Koren, Nedigård Håtveit, Mo, Telemark; C33984b Skarg under Tveiten, Bykle, Aust-Agder; K DXLVIII Audne, Audnedal, Vest-Agder; S4022a Årstad, Egersund, Rogaland; S4384a Hæstad, Helleland, Rogaland; S7086e Forsand, Høgsfjord, Rogaland; S? Gudmestad, Hå, Rogaland (found in 1984); B352b Rekve, Voss, Hordaland (2 pots); B653 Gjukastein, Voss, Hordaland; B2957 Hauge, Os, Hordaland; B5407d Skjelde, Voss, Hordaland; B5637e Hauge, Voss, Hordaland; B5769g Sævareide nedre, Etne, Hordaland; B6138 Sørheim, Etne, Hordaland; B6187 Solheim, Manger, Hordaland; B6227I Byrkje, Voss, Hordaland; B6227II Byrkje, Voss, Hordaland; B6227III Byrkje, Voss, Hordaland; B6658n Birkeland, Kvam, Hordaland (2 pots); B6727n Løn, Voss, Hordaland; B7145g Nedre Aure, Voss, Hordaland; B7935a Østebø, Etne, Hordaland; B8019a Østebø, Etne, Hordaland; B8203b Huse, Ullensvang, Hordaland; B8579b Østre Mjelde, Osterøy, Hordaland; B8588a Rekve, Voss, Hordaland; B8588b Rekve, Voss, Hordaland; B8588d Rekve, Voss, Hordaland; B9015i Rongve, Osterøy, Hordaland; B9608a Nordre Kvåle, Voss, Hordaland; B9614IIIj Bolstad, Voss, Hordaland; B9614IIIx Bolstad, Voss, Hordaland; B9614IIIy Bolstad, Voss, Hordaland; B9995o Kvåle, Voss, Hordaland; B14491 Lydvo, Voss, Hordaland; B14954/22 Nerhus, Kvinnherad, Hordaland; B4506 Mindresunde, Stryn, Sogn og Fjordane; B6037IIb Sande, Gloppen, Sogn og Fjordane; B8567a Indre Opedal, Lavik, Sogn og Fjordane; B9197i Nybø, Leikanger,

Sogn og Fjordane; B12924 Vambeset, Gloppen, Sogn og Fjordane; B14061/1 Kvamme, Lærdal, Sogn og Fjordane; B14078/7 Hellingbøen under Hovland, Årdal, Sogn og Fjordane; B14124/1-4, 6-10 Lægereid, Årdal, Sogn og Fjordane. Four vessels which do not strictly belong to Bøe's types but which nevertheless should be counted as part of the group are B5114a Mæle, Osterøy, Hordaland; B6611g Indre Haugland, Os, Hordaland; B10018o Grimstad, Lindås, Hordaland; and B6654 Norddalen indre, Fjaler, Sogn og Fjordane. There are additional decorative motifs on these vessels, but beading and knotwork are so predominant that they ought to be counted in with the main group.

16 B3410a Ågedal, Audnedal k., Vest-Agder (F17); C28026q Snartemo II, Hægebostad k., Vest-Agder (F18); B2974 Anda, Klepp k., Rogaland (F50); B10205b Sørheim, Etne k., Hordaland (F65); B6656m Sandal, Jølster k., Sogn og Fjordane (F90).

17 C30088a Mogen, Rauland, Telemark; C9000(b) Øvre Vemestad, Lyngdal, Vest-Agder; C9396 Presthus, Lindesnes, Vest-Agder; C9400h Presthus, Lindesnes, Vest-Agder; S2462c Rægje, Håland, Rogaland; C26307b Ytre Vatne, Hægebostad, Vest-Agder; C16290a,b Kvalbein, Egersund, Rogaland (2 pots); S4200g Røisland, Bjerkreim, Rogaland; B4881b Skjerpe, Hå, Rogaland; B4522 Nernes, Etne, Hordaland; B4857f Hjellestad, Fana, Hordaland; B4923o Varberg, Eidfjord, Hordaland; B5693a Nedre Ullestad, Voss, Hordaland; B5976b Nordheim, Etne, Hordaland; B6091Ie Døsen, Os, Hordaland; B10299d Øvre Ullestad, Voss, Hordaland; B14492 Lydvo, Voss, Hordaland (2 pots); B3871a Hamre, Leikanger, Sogn og Fjordane; B4593b Evebø, Gloppen, Sogn og Fjordane; B6588c Skrøppa, Gloppen, Sogn og Fjordane; B5587f Bortnheim, Dale, Ytre Holmedal, Sogn og Fjordane; B9506Ib Stauri, Stryn, Sogn og Fjordane; B11472uu Modvo under Setre, Luster, Sogn og Fjordane.

18 B6611 Indre Haugland, Os, Hordaland; B10018o Grimstad, Lindås, Hordaland; B4857 Hjellestad, Fana, Hordaland. This decoration is also found on other vessels such as C27939IIIj Sostelid, Åseral, Vest-Agder; B4214 Valand, Audneland, Vest-Agder; B2528 Vestly, Time, Rogaland; and B5992 Kvassheim, Hå, Rogaland.

19 S2547 Rivjaland, Hjelmeland parish and k., Rogaland (F58); B10202IIo Grindheim, Etne p. and k., Hordaland (F64); B10205a Sørheim, Etne p. and k., Hordaland (F65).

20 Grindheim, Etne, Hordaland (F64), and Holum, Leikanger, Sogn og Fjordane (F83), both within the set under examination. This form is also found at Gjukastein and at Skjelde, Voss, and Skaim, Sogn, all of which specimens have surface-covering beading and knotwork. The form is also, however, found in vessels with hanging arches, such as Skjerpe, Rogaland, and perhaps also Varberg, Hordaland.

BIBLIOGRAPHY

Åberg, N. 1924. *Den nordiska folkevandringstidens kronologi*. Stockholm.

Åberg, N. 1953. *Den historiska relationen mellan folkevandringstid och vendeltid*. Stockholm.

Annaler for Nordisk Oldkyndighet 1844-1845. Copenhagen.

Bakka, E. 1958. 'On the beginning of Salin's Style I in England'. *Univ Bergen Årb* 1958 Hist-ant rekke 3.

Bakka, E. 1973. 'Goldbrakteaten in norwegischen Grabfunden: Datierungsfragen'. *Frühmittelalterliche Stud* 7, 53-87.

Bakka, E. 1977. 'Stufengliederung der nordischen Völkerwanderungszeit und Anknüpfungen an die kontinentale Chronologie'. In G. Kossack & J. Reichstein (eds), *Archäologische Beiträge zur Chronologie der Völkerwanderungszeit*, 57-60. Bonn.

Bakka, E. 1981. 'Scandinavian-type gold bracteates in Kentish and continental grave finds'. In V. I. Evison (ed.), *Angles, Saxons and Jutes. Essays presented to J. N. L. Myres*, 11-38. Oxford.

Bøe, J. 1931. *Jernalderens keramikk i Norge*. Bergens Mus Skrifter 14. Bergen.

Böhner, K. 1958. *Die fränkischen Altertümer des Trierer Landes*. Germanische Denkmäler der Völkerwanderungszeit B 1. Berlin.

Brøgger, A.W. 1925. *Folkevandringstidens og Vikingetidens kunst*. Norges Kunsthistorie 1. Oslo.

Brøndsted, J. 1940. *Danmarks Oldtid. III: Jernalderen*. Copenhagen

Ethelberg, P. 1986. *Hjemsted - en gravplads fra 4. og 5. årh. e.Kr.* Skrifter fra Museumsrådet for Sønderjyllands Amt 2. Haderslev.

Fett, P. 1940. 'Arms in Norway between A.D. 400-600. Part I'.

Bergens Mus Årb 1938 Hist-ant rekke 2. 'Part II'. *Bergens Mus Årb* 1939–40 Hist-ant rekke 1, 1–45.

Forssander, J. E. 1937. 'Provinzialrömisches und germanisches. Stilstudien zu den schonischen Funden von Sösdala und Sjörup'. *Meddelanden från Lunds universitets historiska museum* 1937.

Gjessing, G. 1934. *Studier i norsk merovingertid. Kronologi og oldsakformer*. Oslo.

Haseloff, G. 1981. *Die germanische Tierornamentik der Völkerwanderungszeit*. 3 vols. Berlin.

Hines, J. 1984. *The Scandinavian Character of Anglian England in the pre-Viking Period*. BAR British Series 124. Oxford.

Hines, J. 1993. *Clasps–Hektespenner–Agraffen. Anglo-Scandinavian Clasps of Classes A–C of the 3rd to 6th centuries A.D. Typology, Diffusion and Function*. Stockholm.

Hougen, B. 1935. *Snartemofunnene: Studier i folkevandringstidens ornamentikk og tekstilhistorie*. Norske Oldfunn VII. Oslo.

Jensen, S. 1978. 'Overgangen fra romersk til germansk jernalder i Danmark'. *Hikuin* 4, 101–16.

Kjær, H. 1890–1903. 'Et nyt fund fra Nydam Mose'. *Nordiske Fortidsminder* 1, 180–94. Copenhagen.

Koch, U. 1990. *Das fränkische Gräberfeld von Klepsau im Hohenlohekreis*. Forschungen und Berichte zur Vor- und Frühgeschichte in Baden-Württemberg 38. Stuttgart.

Kristoffersen, S. 1997. *Dyreornamentikkens sosiale sammenheng og maktpolitiske tilhørighet. Nydamstil og Stil I i Sør- og Sørvestnorge*. University of Bergen doctoral thesis. Unpublished.

Lund Hansen, U. 1970. 'Kvarmløsefundet – en analyse af Sösdalastilen og dens forudsætninger'. *Aarb nordisk Oldkyndighed og Historie*, 63–102.

Lund Hansen, U. 1988. 'Hovedproblemer i romersk og germansk jernalders kronologi i Skandinavien og på Kontinentet'. In P. Mortensen & B. M. Rasmussen (eds), *Fra Stamme til Stat i Danmark. 1: Jernalderens stammesamfund.*, 21–35. Århus.

Magnus, B. 1965. Bolstad. In *Inventaria Archaeologica: Norway*. Bonn.

Magnus. B. 1975. *Krosshaugfunnet*. Stavanger Mus Skrifter 9. Stavanger.

Magnus, B. 1984. 'The interlace motif on the bucket-shaped pottery of the Migration period'. In *Festskrift til Thorleif Sjøvold på 70-årsdagen*, 139–57. Universitetets Oldsaksamlings Skrifter, Ny rekke 5. Oslo.

Montelius, O. 1896–1900. 'Den nordiska jernålderns kronologi'. *Svenska Fornminnesföreningens Tidsskrift* IX–X.

Montelius, O. 1906. *Kulturgeschichte Schwedens*. Leipzig.

Nerman, B. 1935. *Die Völkerwanderungszeit Gotlands*. Stockholm.

Nielsen, J. N., Bender Jørgensen, L., Fabech, E. and Munksgaard, E. 1985. 'En rig germanertidsgrav fra Sejlflod, Nordjylland'. *Aarb nordisk Oldkyndighed og Historie*, 1983, 66–112.

Nissen Meyer, E. 1935. 'Relieffspenner i Norden'. *Bergens Mus Årb* 1934 Hist-ant rekke 4.

Nissen Fett, E. 1937. 'Nordische Relieffibeln der Völkerwanderungszeit'. *IPEK (Jahrbuch für prähistorische und ethnographische Kunst)* 11, 106–16.

Pirling, R. 1966. *Das römisch-fränkische Gräberfeld von Krefeld-Gellep*. Germanische Denkmäler der Völkerwanderungszeit B 2. Berlin.

Reichstein, J. 1975. *Die kreuzförmige Fibel*. Neumünster.

Reichstein, J. 1977. 'Stufengliederung der späten Kaiserzeit und der Völkerwanderungszeit anhand von Grabfunden mit Kreuzförmigen Fibeln'. In Kossack, G. and Reichstein, J. (eds) *Archäologische Beiträge zur Chronologie der Völkerwanderungszeit*, 53–6. Bonn.

Ringstad, B. 1989. 'Ein Webschwert der Völkerwanderungszeit mit Entrelacs-Ornamentik aus Kvåle, Sogndal, Westnorwegen'. *Offa* 1988, 145–58.

Roth, H. 1973. *Die Ornamentik der Langobarden in Italien. Eine Untersuchung zur Stilentwicklung anhand der Grabfunde*. Berlin.

Rygh, O. 1885. *Norske Oldsager*. Christiania.

Salin, B. 1935. *Die altgermanische Thierornamentik*. 2nd ed. Stockholm.

Shetelig, H. 1900. 'Vaabengrave fra Norges ældre jernalder'. *Foreningen til norske fortidsminnesmerkers bevaring Årb* 1900, 46–67.

Shetelig, H. 1904. 'Spandformede lerkar fra folkevandringstiden'. *Foreningen til norske fortidsminnesmerkers bevaring Årb* 1904, 42–91.

Shetelig, H. 1906. 'The cruciform brooches of Norway'. *Bergen Museums Årb* 1906 No. 8.

Shetelig, H. 1911. 'Smaa bronsespænder fra folkevandringstiden'. *Oldtiden* 1, 51–99.

Shetelig, H. 1912. *Vestlandske graver fra jernalderen*. Bergen.

Shetelig 1916–17. 'Nye jernaldersfund paa Vestlandet'. *Bergens Mus Årb* 1916–17 Hist-ant rekke 2.

Sjøvold, Th. 1962. *The Iron Age Settlement of the Arctic Norway I: Early Iron Age*. Tromsø Museums Skrifter X, 1. Oslo.

Sjøvold, Th. 1993. *The Scandinavian Relief Brooches of the Migration Period: An Attempt at a New Classification*. Norske Oldfunn XV. Oslo.

Slomann, W. 1956. 'Folkevandringstiden i Norge. Spredte trekk og enkelte problemer'. *Stavanger Mus Årb* 1955, 63–82.

Slomann, W. 1977. 'Der Übergang zwischen der späten Kaiserzeit und der frühen Völkerwanderungszeit in Norwegen'. In G. Kossack & J. Reichstein (eds), *Archäologische Beiträge zur Chronologie der Völkerwanderungszeit*, 61–4. Bonn.

Solberg, B. 1981. 'Spearheads in the transition period between the early and the late Iron Age in Norway'. *Acta Archaeol* 51, 153–172.

Stjernqvist, B. 1961. 'Über die Kulturbeziehungen der Völkerwanderungszeit'. *Die Kunde*, Neue Folge 12, 16–43.

Straume, E. 1987. *Gläser mit Facettenschliff aus skandinavischen Gräbern des 4. und 5. Jahrhunderts n. Chr*. Oslo.

Vang Pedersen, P. 1988. 'Nydam III – et våbenoffer fra ældre germansk jernalder', *Aarb nordisk Oldkyndighed og Historie* 1987, 105–37.

Voss, O. 1954. The Høstentorp silver hoard and its period. *Acta Archaeol* 25, 169–219.

Wamers, E. 1986. 'Schmuck des frühen Mittelalters'. *Archäologische Reihe* 7, 51–68. Museum für Vor- und Frühgeschichte, Frankfurt am Main.

Werner, J. 1962. *Die Langobarden in Pannonien*. Munich.

7. The assemblage from Hade in Gästrikland and its relevance for the chronology of the late Migration Period in eastern Sweden

Bente Magnus

INTRODUCTION

In July 1845 a charcoal burner found an assemblage of copper-alloy and iron objects while digging in a meadow at the farm of Hade, Hedesunda parish in Gästrikland, Sweden. According to his own recorded statement, the objects were found together at the depth of a quarter of a Swedish ell, i.e. 15–16 cm below the surface, close to a small cairn. The find was delivered to the local authorities who forwarded it to the Royal Swedish Academy of Science and Letters. Here it was entered in the collections in September and given an inventory number, No. 1209. Due to the unusual character of the find the charcoal burner, Mr. Andersson, was paid a reward, but no inquiry was held into the circumstances of the discovery.

The objects were kept well wrapped up for more than 20 years, until the archaeologist Hans Hildebrand went through the collections in connection with planned fieldwork in Gästrikland in the early 1870's. He was fascinated by the assemblage and published the find (Hildebrand 1873, 304ff.; 1874, 180–4; 1875, 67ff.; Montelius 1872, 122, fig. 406). Hildebrand travelled to Hade (Mr. Andersson must have died by then) and was shown the find spot, which was said to be on the outskirts of an Iron-age cemetery. Consequently Hildebrand excavated several of the low burial cairns there, hoping to solve the problem of where the find was made. But the cairns contained cremation graves of the Vendel Period, and the fragments of metal finds were in a very poor state. In these circumstances Hildebrand's interest in the early find from Hade evidently faded. But he believed that Mr. Andersson had lied about the circumstances of the find and that the charcoal burner had assembled the find through grave plundering (Bellander 1938, 57–8).

Since Hildebrand's articles appeared in the 1870's, single items from the Hade find have appeared in numerous publications (Åberg 1920; Lindqvist 1926; Nissen Meyer 1934; Gjessing 1934; Holmqvist 1972; Arrhenius 1973; Stjernquist 1977–78; Andrzejowski 1991; Sjøvold 1994 etc.). But the only publication of the whole assemblage appeared in 1938, by the archaeologist Erik Bellander. He wrote a settlement history of the Iron Age in Gästrikland, in which he also tried to solve the problem of the find from Hade (Bellander 1939, 53–60). He made an inventory of all ancient monuments and finds and interviewed local people. In respect of the find spot for the assemblage from Hade he was shown to an enclosed pasture between the farm and the local school. But there were no burial cairns, only irregular cairns from clearing the field of stones. So he got no further than Hildebrand had 65 years previously, and believed that the assemblage came from the Iron-age cemetery north of the farm where Hildebrand had conducted his excavations. Bellander also published all of Hildebrand's finds from this excavation (SHM [= Statens Historiska Museum, Stockholm] 5412; 1938, 54–9). The modern farm of Hadeholm was created by the Swedish king Gustav Vasa, who, around 1550, turned the four farms of the village of Hade, which was the biggest in the parish, into royal property through exchange. The king intended to open a silver mine in a promising hill in the vicinity (Hedblom 1958, 99).

My interest in the assemblage from Hade was aroused by my studies of the relief brooches of Sweden (forthcoming). Both its chronology and chorology are of importance for the understanding of the transition from the Migration Period to the Vendel Period in eastern Sweden.

The assemblage from Hade (SHM 1209) comprises 16 objects, whereof three are in more than one part. Four objects are made of iron, twelve of copper alloy. Thirteen items are in a very good state of preservation, and three in a more fragmentary state. The objects are as follows (Fig. 7.1):

1. One square-headed relief brooch of gilt copper alloy. Length 10.2 cm.
2. One square-headed relief brooch of copper alloy with traces of gilding on the front side and a coating of a white metal alloy on the back. Length 9.5 cm.
3. One equal-armed relief brooch of copper alloy. Length 20.8 cm.
4. One equal-armed relief brooch of copper alloy very similar to no. 3 but with the end of one arm missing. Length 16.7 cm.

Fig. 7.1 SHM 1209. The assemblage from Hade, Hedesunda, Gästrikland. 1:4. Photo: G. Hildebrand 1996.
Der Fundkomplex von Hade, Kirchspiel Hedesunda (Gästrikland, Schweden).

5. One disc-on-bow brooch of copper alloy inlaid with garnets. Length 6.4 cm.
6–10. Remains of copper-alloy mounts for at least three large drinking horns, about 9 cm in diameter at the rim:
6. Circular rim binding of copper-alloy sheet which curves over and secures the wall of copper alloy on which there are the remains of two different silver bands, one with traces of gilding with a frieze of embossed running animals, the other with embossed bovine heads en face. Diameter between 9.2 and 8.3 cm; somewhat out of shape.
7. Two circular bindings of copper-alloy sheet from the wall below the rim with two ridges.
8. Two identical copper-alloy ferrules with holes for nails. Length 6.0 cm, width 2.3 cm with one loose nail of copper alloy and two remaining *in situ*. Length 0.8 cm.
9. Copper-alloy ferrule consisting of two discs with a tube of copper-alloy sheet in between. Length 1.9 cm. Diameter of the lower disc 3.7 cm.
10. Copper-alloy spiral. Length 1.3 cm.
11. Iron spearhead with a very short socket and a prominent midrib on the blade, which has a curved outline and star-shaped cross-section. Length 22.3 cm.
12. Iron lancehead with a long socket and a long blade with a midrib and wavy outline. The socket is square in cross-section. Length 36.8 cm.
13. Iron knife with a prominent step between the tang and the blade. Length 26.4 cm, width 2.4 cm at the base of the blade.
14. Claw-like instrument of iron with two dents and the hollow shaft bent at an angle. Length 10.2 cm.

FIND CATEGORY

Before entering on a detailed discription of each object it is of importance to ascertain what kind of assemblage this is. Hans Hildebrand was convinced that the find represented grave robbing but he was unable to show that this had been the case. During his excavations at Hade he found nothing but cremations with a few surviving objects from the Vendel Period in a poor state of preservation. No finds from earlier periods are known from Hade.

Studying the assemblage to-day, one is struck by its good state of preservation and by the fact that both the copper alloy and the iron have a special patina. The thin copper-alloy sheet mounts have kept its shape even though they are torn and parts are missing. The objects show no signs of having been on a funeral pyre, although some of the objects show signs of deliberate destruction, e.g. the case with the drinking horns. In addition, both the equal-armed brooches lack fastening pins and catches on the back and the end of one of the arms is broken off, a feature observed on many relief brooches (see below). The square-headed relief brooches and the disc-on-bow brooch, on the other hand, are very well preserved, with traces of gilding and whitemetal coating, and garnets in their frames.

The short spearhead has had its tip hammered out of line and its socket torn; the longer lancehead also has part of the lower end of its socket missing and the outline 'broken off' on both sides. In my opinion there is nothing which indicates that this assemblage came from one or more graves. Cremation was the general burial custom in Sweden in the Iron Age and the assemblage most probably, therefore, represents a hoard. The question remains whether it was a cache of metal or a ritual hoard.

There is no reason to disbelieve the official statement from 1845 that Olof Andersson made his discovery when digging in a meadow. Andersson was probably one of the crofters of the Hade manor (to-day named Hadeholm). On a modern map of Hedesunda parish a charcoal burner's croft is marked to the south-west of the farm. According to older maps in 1845 the farm was surrounded by a lot of wet land due to its proximity to the mighty Dala river (Fig. 7.2). Work to gain more arable land by drainage was in progress in Sweden at the time when the find was made. It is therefore possible that Olof Andersson was engaged in digging ditches for the drainage of the wetland when he found the objects. The state of preservation of the find and the patina of the iron and copper alloy indicates that they had survived under favourable conditions like a marsh or flooded meadow. The meadow that was shown to Bellander may be the area where the find was located. But it is a fact that when wet land is drained and made into arable land it undergoes a profound transformation and changes shape so that the original find spot would be difficult to remember. The 'small cairn' that Andersson reported he had noticed close to the find spot may either have had nothing to do with the find, or may have been a cairn from field clearing, a small burial cairn from the Iron-age cemetery nearby, or may have been remains of a structure raised to mark the place. We will return to the question of whether the assemblage represents a metal cache or a hoard later.

The assemblage from Hade divides readily into three groups of finds according to function, one of brooches, one of weapons and one of 'feasting gear'.

THE BROOCHES (FIG. 7.3)

1. The small relief brooch with the ridged foot has traces of gilding on the front side and whitemetal alloy on the back where there are also remains of an iron pin. The decoration of the headplate consists mainly of a broad frame with small square fields at the upper corners and a rudimentary string of knobs between

Fig. 7.2 View from the Iron-age cemetery at Hade towards the Dala river. Photo: E. Bellander 1931.
Blick vom eisenzeitlichen Gräberfeld von Hade auf den Dalälv.

Fig. 7.3 SHM 1209. The brooches from Hade, Hedesunda, Gästrikland, 1:2. Photo: G. Hildebrand 1996.
Die Fibeln von Hade.

them. Otherwise the pictorial field is divided into three parts by way of two curved ridges emanating from the the upper end of the bow. The bow is decorated with longitudinal mouldings with a few curved lines in between. The most prominent feature of the footplate are the lobes which terminate in circular discs with a striated triangular projection. This is a dominant feature amongst the fragments of clay moulds for relief brooches found at Helgö (Holmqvist 1972, 249, Fig. 103; Lundström 1972, 142). The side fields of the brooch just below the lower end of the bow show a bird's head with one round eye and a curved beak. The broad dorsal ridge of the footplate leaves little room for decoration; such as there is consists mostly of curved lines. Apart from the side fields, with their bird's heads, there is no clear pattern discernible either on the headplate or the footplate of the brooch. Viewed from some distance, however, the different decorative lines make up a mask with the birds' beaks being the eyes and some curved lines forming a wide mouth. The form of brooch, the shape of the lobe terminations, the rudimentary ornamentation and the *trompe l'oeil* figure of a mask indicate that the brooch was made in a Style-I milieu but probably very late. There are good reasons to believe that the brooch is a product of one of the workshops at Helgö.

2. The square-headed brooch with plane footplate. The front is gilded. This brooch has three parallels, all found in northern Sweden, which seem to have been cast after a common model. The decoration is well executed and shows two doubled-up human figures being squeezed in between the inner and outer frame of the headplate (Arrhenius 1973, Pl. IX). Four human heads in profile look out from the bow divided into four sections by ridges. When viewed from the side they make up two masks. The upper arms of the footplate have been reduced to two eyes with elongated eyebrows; the lower arms end in triangular fields with transverse ribs; and the foot terminates in a medallion with a circular field in the middle giving the impression of a large, staring eye. From this medallion extends a triangular field bordered by a high ridge where an eye and a human leg with foot can be seen encased. The

pictorial field of the foot shows two eyes in profile and two human limbs, possibly arms. From the frame of the lower part of the foot, two tongue-like figures protrude on both sides.

There are two other brooches which have obviously been cast in nearly identical moulds, presumably made after one of them (Holmqvist 1972, 233–7). These were found at Bjällsta, Indal parish (SHM 4046) and Sörfors, Attmar parish (SHM 12220). An equal-armed relief brooch from Rallsta, Svedvi parish in Västmanland (VLM [= Västmanlands läns museum] 14547:22) is constructed by joining two footplates of the same kind of brooch together by way of a bow. Only the latter brooch was professionally excavated (in 1946) but the report was lost so the information is scanty. The brooch was found in a cremation grave together with fragments of a comb, a simple iron bow brooch, a penannular brooch of iron and a trapezoid bone object of unknown function (Lamm 1979, 130–1). Both the bow brooch and the pennanular brooch belong early in the fifth century, while the comb is of a type which was in use from the late fifth to the late sixth century (E. Brynja 1998, 133). The trapezoid object is also of early Migration-period date. The cremated bone fragments from this grave amounted to 4 litres, and it is possible that the burial cairn covered more than one cremation.

A Finnish equal-armed relief brooch from Gulldynt, Vöyri (Erä-Esko 1965, Pl. VI) was evidently produced after a related model.

The three almost identical square-headed brooches are discussed in numerous publications, most recently by Thorleif Sjøvold (1994, 45–8, with map and references). Nissen Meyer groups the brooches with her Bothnian group (1934, 80–4) and dates them, in agreement with Åberg, to the second half of the sixth century. Arrhenius assigns this type of brooch to the late 6th century as well (1986, 151). She considers that this type of brooch with ornaments in East Scandinavian Style I was probably contemporaneous with early Style II.

3. Equal-armed relief brooch made of copper alloy with no traces of gilding or white metal. Length 20.8 cm (Fig. 7.3). On the back is a catchplate and a fastener for a pin. The ornamentation is structured by way of an outer and an inner ridge which give an narrow outer and an broad inner pictorial field. Both parts of the brooch terminate in a mask. The bow has straight parallel sides and is divided into four fields by way of a median and a transverse ridge. Apart from the masks the motif(s) are very unclear, but the design of the other brooch from Hade of the same type gives the general idea. The broad pictorial field in the middle of each arm has a circular knob, reminiscent of a mounted garnet or a piece of coloured glass as seen on the equal-armed relief brooch from Gillberga, Svennevad, Närke (Åberg 1953, 72; Arrhenius 1973, Pl. X; Magnus 1995). On both sides of the brooch close to the bow can be discerned a long beak and an eye. The terminal masks are zoomorphic, with distinct eyebrows joining over the bridge of the nose and ending in a triangular field. A single, hook-beaked bird's head is visible on one side of one of the masks. Below the triangular field and outside the outer frame at both ends of the brooch a rhombic figure juts out. The lack of distinct ornamental details makes this brooch either very late or badly executed.

4. Equal-armed relief broooch of copper alloy of a similar type to the preceding, with no trace of gilding but of better craftmanship. Length 16.7 cm (Fig. 7.4). The bow is of the expanded type, being divided into four panels by a transverse ridge. One end of the brooch has been broken off, making the original length close to 20 cm. The design is structured by way of a double outer and inner ridge with a groove in the middle for a niello inlay. The most prominent ornamental features apart from the terminal masks are the long beaked birds on either side of the inner pictorial field. Otherwise mostly loose limbs and eyes are discernible. The knobs are decorated with a cross. The terminal masks are zoomorphic and stand out in higher relief than the rest of the ornaments. The masks seem to represent some sort of monster with slanting eyes under an arched forehead continuing in the ridge of the nose which runs over an open mouth with broad lips and ending in a triangular field jutting out over the outer contour of the brooch. Inside the triangle a couple of limbs and what are probably two eyes are to be seen. Above the forehead the head terminates in two raised, semicircular ears and two semicircular ridges, one above the other. The cheeks are indicated by a double transverse ridge between the eyes and the mouth.

The semicircular ears are common on equal-armed relief brooches. They probably developed from the arched eyebrows which are such a prominent feature of the terminal mask on some late square-headed relief brooches found in Norway, like the one from Dalum, Sparbu, Nord-Trøndelag (Sjøvold 1993, Pl. 21).

The triangular extension at both ends of the Hade brooch is a common feature of late square-headed relief brooches as well as of equal-armed brooches. On well-designed brooches it emerges from the mouth of the mask and obviously represents the breath, if not the 'soul'. Since many of the triangular fields contain figures of limbs, eyes and heads, like the smaller of the square-headed Hade brooches, it does not seem too far fetched to relate them to the Germanic perception of the dual soul.

5. Disc-on-bow brooch of copper alloy with a coating of whitemetal alloy. Length 6.4 cm (Fig. 7.5). Both headplate and footplate are decorated with circular garnets in round settings with a collar of twisted silver thread. Two of the garnets make up the eyes of two

bird's heads with inward curving beaks. The surface of the brooch is otherwise covered with half-moon and Y-shaped punchmarks. The disc on the bow is set with 5 rectangular garnets, the setting of the middle one showing faint traces of gilding. The colour of the stones is not the deep red but a lighter, more brownish, red.

This small brooch is the only one not belonging to the traditional jewellery of the late Migration Period in eastern Sweden. With its total lack of animal ornamentation apart from the two bird's heads, its many garnets and the punched decoration signal changes. The Y-shaped punchmarks with small dots have parallells on the upper part of the large buckle and on the squarish mount from Åker, Vang, Hedmark, Norway (Cleve 1942, 12–23) and on mounts from Gotland. One belt mount from Grötlingbo, Gotland, is covered with Y-shaped punchmarks, and its four silver bosses have the same collar of twisted silver thread as the disc-on-bow brooch from Hade (Arwidsson 1942). This gives an indication of its chronological position.

Gjessing (1934), Cleve (1942, 13–23) and Nissen Meyer (reluctantly; 1934, 83–4) all date the assemblage from Åker to the middle of the sixth century. Arrhenius discusses the Åker buckle and its Y-shaped punchmarks and rejects Cleve's assumption that such stamps indicate a Nordic provenance for the objects in question (1986, 142 and refs.). The Y-shaped punchmark may, on the other hand, have developed from related punches which are typical of the artistic tradition of Gotland in the Roman Period (Andersson 1993, 13, Fig. 6 nos. 58–60; Andersson 1995, 88), a fact which supports Cleve's conclusion and points to eastern Scandinavia for the origin of this punch-type.

In conclusion: Of the five brooches from Hade four have a common stylistic character already noted by Nissen Meyer who maintained that they were contemporary (1934, 83–4). This stylistic variant was later named East Scandinavian Style I and is characterized by the frequent use of small ornamental fields bordered by high ridges (Erä-Esko 1965; Arrhenius 1973, 26). The two square-headed brooches may have been manufactured at Helgö, where most of the moulds for buttons and relief brooches show late stylistic features (Holmqvist 1972, 230ff.). The equal-armed relief brooches all have an eastern bias, but the two from Hade were not, as far as we can judge, made at Helgö. The small disc-on-bow brooch has no direct parallells but belongs in a group of small brooches found in Scandinavia heralding the Gotlandic ones of the Vendel Period (Gjessing 1934, 135–8). Its punched ornamentation with Y-shaped and half-moon punchmarks is restricted in Gotland to Nerman's Vendel Period I. The brooches from Hade indicate deposition during a time when the old brooch-types and animal Style I was still in use alongside

Fig. 7.4 SHM 1209. The smaller of the two equal-armed brooches from Hade, Hedesunda, Gästrikland. 1:1. Photo: G. Hildebrand 1996.
Die kleinere der beiden gleicharmigen Fibeln von Hade.

objects with a new language of form and decoration. This must have been during a short time in the middle of the sixth century.

As far as I am aware, brooches from this period have very seldom if ever been found together in the same grave. But it seems reasonable to believe that conditions in real life were different from the burial situation, and that the women kept and wore both older and newer pieces of jewellery. How long this lasted is uncertain; maybe a generation?

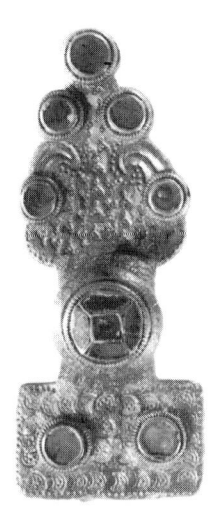

Fig. 7.5 SHM 1209. The disc-on-bow brooch. 1:1. Photo: G. Hildebrand 1996.
Die Bügelscheibenfibel von Hade.

THE WEAPONS

1209:6. Lancehead of iron with a long, narrow socket and a long blade with a midrib and two sets of wing-like extensions to the blade above the shoulder. Length 36.8 cm. (Fig. 7.3). The transition between socket and blade is marked by four transverse grooves. The socket is torn at the base.

This type of spearhead is seldom found, but has a few interesting parallels. One is from the well known weapon grave, grave V, from Snartemo, Vest-Agder, Norway (Hougen 1935, Pl. III:2). This lancehead has only one set of extensions on the blade. Hougen based his dating of grave V on the well-decorated sword whose different decorated elements he found to be related to the ornamentation of relief brooches of Nissen Meyer's stage 5 and consequently to belong to the middle of the sixth century (Hougen 1935, 52). In a recently published study of the Norwegian weapon graves, this type of lancehead has been called the Snartemo type (Bemmann and Hahne 1994, 432, Fig. 64.5, map 81). The only difference between the lanceheads from Snartemo and Hade is that the socket of the Hade lancehead is circular in section while the one from Snartemo is square in section. According to the map published by Bemmann and Hahne the Snartemo type occurs mainly in Norway. The only place in eastern Scandinavia where this type occurs is in the sacrificial bog find of Skedemosse, Öland, where one lancehead of this type with a socket of circular section was found (Hagberg 1967, 68 and 898). Without being able to go more deeply into this question here, there is a possibility that this type of lancehead differs slightly between the east and the west.

1209:7. Spearhead of iron with a short socket and prominent midrib on the blade which has a curved outline and a star-shaped section. Length 22.3 cm (Fig. 7.3). The spearhead is torn at the socket and the tip has been bent (or hammered) out of line. It belongs to the Moldestad type (Ilkjær 1990, 40 and 79–85) and a variant of this type labelled the Vestly type (Bemmann and Hahne 1994, 428 and 585, Fig. 64,3, map 78). According to Bemmann and Hahne there are two more spearheads of this sub-type from northern Sweden, both loose finds. As with the lancehead from Hade, the Norwegian specimens have a socket with a square section, while the northern Swedish finds have a round-sectioned socket. The well-furnished male grave from Vestly, Time, Rogaland, Norway, which contained such a spearhead is dated to the latter part of the Migration Period (Straume 1987, 15 and 84).

1209: 8. Knife of iron with a distinct step between the tang and the dorsal side of the blade, and its edge in line with one side of the tang. Length 26.4 cm (Fig. 7.3). The blade measures 15.7 cm and it is debatable whether it should be classified as an ordinary knife or a combat knife. According to both Gjessing (1934, 70ff.) and Gudesen (1980) this is an ordinary knife (the blade being < 50 cm) which displays features which are seen both in the late Migration Period and the early Vendel Period.

Conclusion: The weapons of the assemblage have parallels in eastern Sweden and in late Migration-period graves in western Norway, where they occur in traditional weapon-sets consisting of a long double edged sword, a javelin, a lance, a shield and an axe. The combat knife does not occur in these grave finds.

THE 'FEASTING GEAR'

1209: 9. Iron fork with two dents and the socket bent at angle. Length 10.2 cm (Fig. 7.1).

Similar forks with three dents are known from the cemetery of Valsgärde, Uppland, for instance grave 6 (Arwidsson 1942, Fig. 61) where the item is interpreted as a meat fork. As far I know this type of tool is not known from Migration-period burial contexts, although fishing spears of iron are known from mound no. 3 at Högom, Medelpad, Sweden, dated to the Migration Period, and from sixth-century mountain graves in Norway (Ramqvist 1992, 159).

1209: 10–12. The drinking horn mounts are made of copper alloy and sheet copper-alloy and stem from three horns. They consist of one rim mount, two different sections from below the rim and three finials (Fig. 7.1).

10: The only rim mount carries decoration in thin silver foil with an embossed frieze of running goats or chamois just below the rim. They are running with lifted head looking forward and with their tails straight out. Underneath is a small fragment of a similar silver-sheet band with a frieze of embossed stylized bovine heads with horns.

This type of decoration is well known from Hemmoor-type buckets of provincial Roman origin and silver drinking cups with a gilt band with embossed decoration just below the rim as known from the Danish grave finds of Himlingøje and Nordrup grave J, Sjælland. The goats or chamois of the Himlingøje cups are also running, but with their heads bent and their tails hanging. These drinking cups are believed to be of eastern Danish manufacture and belong in phase C1b of the Roman Iron Age, the early 3rd century A.D. (Lund Hansen 1995, 237, 377 and 473 Taf.2). In the early 4th-century male grave from Lilla Jored, Kville, Bohuslän, Sweden (phase C2) there are the remains of two silver drinking cups with a frieze of goats or chamois with heads turned backwards. The idea of adorning a drinking cup with an ornamented frieze continues in the Migration Period and is known on glass vessels in the form of a strip of gilt silver sheet below the rim with impressed decoration in Style I from Vestly, Snartemo, Kvassheim, Evebø and Rimestad in western Norway (Straume 1987, Tafn. 42, 58, 61, 73 and 86).

A set of drinking-horn mounts from Karleby in Uppland, Sweden, have rectangular panels of gilt copper alloy with embossed decoration in advanced Style I with distinct Style-II elements as well (Holmqvist 1951, 36–38 and 60–62). An ornamental lower border of pearled silver thread in the shape of rosettes and embossed pearled frames of the decorated panels are decorative features pertaining to Vendel Style A (Arwidsson 1942), but may be a revival of embossed and gilt decorative sheet silver from the 4th century (Holmqvist 1951, 42).

This is worth mentioning when studying the frieze of chamois and bovine heads on the Hade mount in more detail. Frontal bovine heads as a symbolic decorative element are known from gold objects of the early Roman Period: the gold neckring from Havor, Gotland, dated to the time of the birth of Christ or somewhat earlier has bovine heads with large horns on the collars below the knobs. But two of the biconical golden beads from the Swedish hoard from Hede, Möklinta parish, Västmanland also have bovine heads in filigree and are considered to be Gotlandic pieces (Andersson 1995, 85–88 and 156 ff.). The youngest of them belongs to phase C1. A Gotlandic golden neckring from Vestringe, Ethelhem parish, has a stamp which seems to be a stylized bovine head (Andersson 1993, 207, Fig. 86). This is obviously a Gotlandic motif with a long tradition (Andersson 1995, 88). The chronological discrepancy between the decoration of the rim mount from Hade and the objects mentioned above has troubled many scholars (Bellander 1938, 57; Stenberger 1971, 377) and has led to the conclusion that the Hade objects must represent plundered goods from several graves of different age.

But there are parallells. Mounts for drinking horns of the type from Hade are well known from the southern Baltic region (Latvia, Lithuania and Prussia). In Lithuania they occur in male inhumation graves from the 4th century to the 7th and parallels to the fragments from Hade decorated with embossed motives in silver or copper alloy have been found in graves of the cemetery at Plinkaigalis, Lithuania, excavated during the years 1978–1984 (Kazakevičius 1987,

Fig. 4). The rim mounts carrying friezes of silver embossed with animal and other non-figurative decorative motifs of the best quality are considered to be of Scandinavian origin and are dated to the second half of the fifth century, and are believed to have been manufactured on the eastern side of the Baltic Sea or under influence from Scandinavia (Kazakevičius 1987, 62). The Iron-age chronology of the Baltic region is far from clear, however, and Marianne Schaumann-Lönnqvist has drawn my attention to a Finnish find from Kaakuri in Uleåborg. This is a piece of silver sheet which, according to its shape, must have a binding for a sword or knife sheath. It has four parallel horizontal rows of frontal bovine heads divided by lines. It belongs to a Migration-period grave find, but the find circumstances are unclear and there is no excavation report for the cemetery (Schauman-Lönnqvist 1995, 103). Andrzejowski groups both the mount from grave 61 at Plinkaigalis and the mount from Hade with one from Havor, Gotland grave 222 (1991, 50, Fig. 11).

The three knobs from the tips of the drinking horns are of two types, one ending in a round knob between two discs and one with two discs separated by a concave piece. The latter have a parallel in grave 332 at Plinkaigalis, Lithuania (Kazakevičius 1987, Fig. 9), while the others belong to Andrzejowski's group D (Andrzejowski 1991, 28 and map 11). In his thorough article on the chronology of drinking horns (which needs to be published in full in a translated version) Andrzejowski places the drinking-horn mounts from Hade in his group D2c, a numerous and highly varied group, but gives no dating other than to the Migration Period (Andrzejowski 1991, 119).

Conclusion: The mounts of the three drinking horns are of a type rarely found in Scandinavia but have their closest parallels in Lithuania, where they occur in the late 5th century, according to local chronology.

DISCUSSION

The weapons from Hade belong to the late Migration Period, when warrior leaders carried a double-edged sword, a shield, a lance, a javelin and an axe. The Vendel-period weapon set consisted of a lance, a throwing axe, a sax and a shield; in other words these weapon sets represented two different fighting techniques. In a transitional phase, when new families were acquiring powerful central positions, they must have been able to muster small armies of warriors with different equipment. Their leaders were probably mounted and carried a lance, shield and sword, while others were foot soldiers and fought with traditional weapons. This is supported by the figurative foils from the helmets of Vendel and Valsgärde. Arwidsson recorded one of eight motifs where the warriors carry one lance and one javelin. Two other foils display javelins as well, although these are unknown in Scandinavian Vendel-period finds (Arwidsson 1983, 88).

According to Bergljot Solberg the transitional period for changes in the weaponry was of longer duration in western and northern Norway than in eastern Norway, Sweden, the island of Gotland and on the Continent. The latest phase of Style I in the west was thus contemporary with the so called Åker phase in eastern Norway, the early Vendel Period and Vendel Style A in the east, i.e. the second half of the 6th century (Solberg 1980). This view, which has been held by several eminent Scandinavian scholars such as Nissen Meyer and Åberg, is also shared by Arrhenius for Scandinavian material in general (1986, 151). She maintains that the transition between the Migration Period and the Vendel Period took place around 520–530 A.D. Solberg's conclusions have lately been challenged by Bemmann and Hahne (1994, 334ff.). Their conclusion is that the lanceheads of western Norwegian origin and those from Gotland which she contrasts them with belong to the same type. They consider the Norwegian forms as ones that signal the Vendel Period rather than seeing them as indicators of a retarded Migration Period.

I have problems in seeing the distinction, as both Bemmann and Hahne and Solberg agree that there was a transitional period before the full Style II and the new weapon set and jewellery were accepted all over this large and – compared to the Frankish Continent – peripheral part of Europe. Before the new dynasties had consolidated their power whether in Uppland, Sweden, at Åker in eastern Norway or at Karmøy in Rogaland, a varying length of time elapsed during which rival groups represented by Style I, relief brooches and traditional weapons were fighting to keep their power. Style I was used in Langobardic Italy as a sort of demonstration against Romanization and to underline Scandinavian ancestry (Høilund Nielsen 1997). Similar phenomena are well known from more recent history.

Judging from the numerous finds of relief brooches and buttons decorated in East Scandinavian Style I and the enormous material of fragmented moulds from the production of brooches and buttons with the same stylistic features at Helgö, this late stylistic variant was still in use when the earliest graves at Vendel and Valsgärde were built and Ottar's Mound near old Uppsala was erected. Some Vendel-period jewellery was produced at Helgö, and the settlement, with its cemeteries, continued to exist in the same locality (Hyenstrand 1988, 66). In Scandinavia as a whole there were obviously marked regional differences with very different political and economic development in the late Migration Period (Widgren 1988, 273ff.). The south of Norway, as well as the north of Sweden and the islands Gotland and Öland, were subject to a critical development which led to the desertion of farms and villages and consequently to a reduced population (Näsman 1988, 227ff.). To my mind the assemblage from Hade fits well into the picture as a non-funerary sacrificial deposition in a border zone of emerging polities (Hines 1989, 193ff.).

Hade is a farm with a significant strategic position. It lies on the south side of an important natural bridge over the mighty Dala river. The 'bridge' is formed by a boulder ridge left by a retreating glacier. This part of the Sweden to-day belongs to the *landskap* of Gästrikland, but was in the medieval period part of Uppland with its centre at Uppsala (Hedblom 1958, 27). This is an iron-producing district, and in addition to numerous fairly recent mining shafts, there are an equal number of hollows with traces of Iron-age exploitation of bog-iron resources in and around Hade and across the 'bridge' to Hedesunda on the north side (Jensen 1983, 76ff.). The burial cairns excavated by Hildebrand at Hade contained iron slag in the filling. Enormous coniferous forests north of the Dala river provided room for grazing cattle and hunting grounds, as well as wood for charcoal.

As has been pointed out by several scholars, the medieval topographical name Jernbäraland, which means the iron-carriers' land, very probably referred to an area including Gästrikland, on account of its wealth in both bog iron and rock iron (Kumlien 1963). The broad Dala river was easy to traverse by way of a couple of ferry boats and was the natural bridge between Hedesunda and Hade, so that this route must have been of vital importance for transport of iron and other products from the north of Sweden to the south. The topographical names Hade and Hedesunda are obviously connected, although neither can satisfactorily be interpreted due to their age (Hedblom 1958, 91ff.).

A small gold hoard from Rångstad in the north of the Hedesunda parish contained finger rings, bits of gold rod and a solidus of the Byzantine emperor Leo I (457–474) (Bellander 1938, 41). This is a very different find from the hoard from Hade, not only because it comprises only gold objects, but also because it came from a typical agricultural locality. The objects making up the hoard from Hade are connected to both female and male outfits. It may have been a collection belonging to a group of people rather than to one or two individuals. We will probably never know the exact find spot, and the time and circumstances of the deposition remain an open question. But sometime in the second half of the 6th century, when the new elite was well established on the flat and fertile Uppsala plain, this fortune was sacrificed in the hope of future gain.

DEUTSCHE ZUSAMMENFASSUNG

Im Jahre 1845 kam beim Ackerbau in einem nassen Gelände bei Hade im Kirchspiel Hedesunda (Gästrikland, Schweden) ein ungewöhnlicher Fund der späten Völkerwanderungszeit zu Tage. Der Komplex umfaßt 14 Gegenstände, die in drei Gruppen eingeteilt werden können: Schmuck, Waffen und 'Festgeschirr'. Der Schmuck umfaßt zwei Relieffibeln mit rechteckiger Kopfplatte, zwei gleicharmige Relieffibeln und eine kleine Bügelscheibenfibel mit Granateinlagen.

Als Waffen wurden zwei ungewöhnliche Lanzenspitzen und ein großes Messer gefunden. Zu dem 'Festgeschirr' gehören bronzene Beschläge von drei großen Trinkhörnern und eine große eiserne Fleischgabel.

Der Beitrag behandelt drei Problemkreise: die Kategorisierung, die Chorologie und die Chronologie des Komplexes. Nach einer Analyse der einzelnen Gegenstände wird vorgeschlagen, den Komplex als Depot mit Opfercharakter zu deuten. Der Hof von Hade liegt an der südlichen Seite eines von zwei natürlichen, nord-süd-verlaufenden Kiesrücken, die hier den breiten, von West nach Ost fließenden Dalälv queren; diese günstige Lage muß eine entscheidende Bedeutung für die Niederlegung des Fundes gehabt haben. Nördlich des Dalälv grenzen weite Bereiche mit Mooreisenvorkommen an. Die Kontrolle über diese Rohstoffe sowie die Transportwege müssen für die neue Elite, die sich in der Ebene um Uppsala in der späten Völkerwanderungszeit etablierte, von großer Bedeutung gewesen sein. Die Mitte oder der späte Abschnitt des 6. Jahrhunderts, in die die Deponierung des Opfers fällt, gehört zu einer turbulenten Periode der Geschichte Skandinaviens.

BIBLIOGRAPHY

Åberg, N. 1920. *Den Nordiska Folkvandringstidens Kronologi*. Stockholm.

Åberg, N. 1953. *Den Historiske Relationen mellan Folkvandringstid och Vendeltid*. Stockholm.

Andersson, K. 1993. *Romartida Guldsmide i Norden. I: Katalog*. Uppsala.

Andersson, K. 1995. *Romartida Guldsmide i Norden*. Uppsala.

Andrzejowski, J. 1991. 'Mountings for drinking-horns from the late pre-Roman and Roman Periods in central and northern Europe'. *Materialy Starożytne i Wczesnośredniowieczne* 6, 7–120.

Arrhenius, B. 1973. 'East Scandinavian Style I – a review'. *Medieval Archaeol* 17, 26–42.

Arrhenius, B. 1986. 'Einige christliche Paraphrasen aus dem 6. Jahrhundert'. In H. Roth (ed), *Zum Problem der Deutung frühmittelalterlicher Bildinhalte*, 129–51. Sigmaringen.

Arwidsson, G. 1942. *Valsgärde 6*. Uppsala.

Arwidsson, G. 1983. 'Valsgärde'. In J. P. Lamm and H.-Å. Nordström (eds), *Vendel Period Studies*, 71–82. Stockholm.

Bellander, E. 1939. *Gästriklands järnåldersbebyggelse. 1: Fornlämningar och Fynd*. Gävle.

Bemmann, J. and Hahne, G. 1994. 'Waffenführende Grabinventare der jüngeren römischen Kaiserzeit und Völkerwanderungszeit in Skandinavien. Studie zur zeitliche Ordnung anhand der norwegischen Funde.' *Berichte Römisch-Germanische Komm* 75, 283–640.

Brynja, E. 1998. *Kammar från Mälardalen, AD 350–600*. Licentiatavhandling. Arkeologiska Forskningslaboratoriet, Stockholms Universitet.

Cleve, N. 1942. 'Det stämpelornerade remgarnityret i fyndet från Åker i Norge'. *Finskt Mus* 49, 13–23.

Erä-Esko, A. 1965. *Germanic Animal Art of Salin's Style I in Finland*. Helsinki.

Fett, E. Nissen 1945. 'Åkerfunnet'. *Bergens Mus Årb* Hist-ant rekke 7, 1–18.

Gjessing, G. 1934. *Studier i Norsk Merovingertid*. Oslo.

Gudesen, H. G. 1980. *Merovingertiden i Øst-Norge: Kronologi, kulturmønstre og tradisjonsforløp*. Oslo.

Hagberg, U. E. 1967. *The Archaeology of Skedemosse. I: The Excavations and the Finds of an Öland Fen, Sweden*. Stockholm.

Hedblom, F. 1958. *Gästriklands Äldre Bebyggelsesnamn: En Förberedande Undersökning*. Gävle.

Hildebrand, H. 1869. 'Den äldre järnåldern i Norrland'. *Antik Tidskrift för Sverige* 2, 304–15.

Hildebrand, H. 1874. 'Fornlämningar vid Dalelfven'. *Månadsblad*, 181–2.

Hildebrand, H. 1875. 'Ett spänne från Öje socken i Södermanland'. *Månadsblad*, 65–8.

Hildebrand, H. 1880. 'Studier i jämförande fornforskning. Bidrag till spännets historia'. *Antik Tidskrift för Sverige* 4, 1872–80.

Hines, J. 1989. 'Ritual hoarding in Migration-period Scandinavia: a review of recent interpretations'. *Proc Prehistoric Soc* 55, 193–205.

Høilund Nielsen, K. 1997. 'Retainers of the Scandinavian kings: an alternative interpretation of Salin's Style II (sixth-seventh centuries AD)'. *J European Archaeol* 5, 151–69.

Holmqvist, W. 1951. 'Dryckeshornen från Söderby-Karl'. *Fornvännen* 46, 33–65.

Holmqvist, W. 1972. 'Relief brooches: comparative analyses of the A-B-C elements at the Helgö workshop'. In W. Holmqvist (ed), *Excavations at Helgö IV: Workshop Part I*, 230–62. Stockholm.

Hougen, B. 1935. *Snartemofunnene: Studier i Folkevandringstidens Ornamentik og Tekstilhistorie*. Norske Oldfunn VII. Oslo.

Hyenstrand, Å. 1981. *Excavations at Helgö VI: The Mälaren Area*. Stockholm.

Hyenstrand, Å. 1988. 'Helgö, Birka and the church of St Gautbert'. In H.-Å. Nordström and A. Lundström (eds), *Thirteen Studies on Helgö*, 65–71. Stockholm.

Ilkjær, J. 1990. *Illerup Ådal. 1: Die Lanzen und Speere*. Århus.

Jensen, R. 1983. 'Bebyggelse och lågtekniska järnframställningsplatser i Gävleborgs län – en rumslig analys'. *Arkäologiska järnforskning 1980–83*, 61–112. Stockholm.

Kazakevičius, V. 1987. 'Motifs of animal decorative pattern on bindings of the 5th-6th century drinking horns from Plinkaigalis burial ground (Lithuania)'. *Finskt Mus* 94, 45–63.

Kumlien, K. 1963. 'Järnbäraland'. *Kulturhistorisk Leksikon for Nordisk Middelalder* VIII, 52–3. Copenhagen.

Lamm, J. P. 1979. 'De folkvandringstida reliefspännena från Hamre och Rallsta'. *Västmanlands Fornminnesförening Årsskrift* 57, 126–34.

Lindqvist, S. 1926. *Vendelkulturens Ålder och Ursprung*. Stockholm.

Lund Hansen, U. 1987. *Römischer Import im Norden*. Copenhagen.

Lundström, A. 'Relief brooches: introduction to form-element and variation'. In W. Holmqvist (ed.), *Excavations at Helgö IV*, 132–230. Stockholm.

Magnus, B. 1995. 'Praktspennen fra Gillberga'. *Från Bergslag och Bondebygd. Årsbok för Örebro läns hembygdsförbund och Stiftelsen Örebro läns museum* 46, 29–40.

Näsman, U. 1988. 'Den folkvandringstida ?krisen i Sydskandinavien'. In U. Näsman and J. Lund (eds), *Folkevandringstiden i Norden: En Krisetid mellem Ældre og Yngre Jernalder*, 227–55. Århus.

Nissen Meyer, E. 1934. *Relieffspenner i Norden*. Bergen Mus Årb Hist-Ant rekke 4.

Pedersen, P. V. 1991. 'Nye fund af metalsager fra yngre germansk jernalder'. In P. Mortensen and B. M. Rasmussen (eds), *Fra Stamme til Stat i Danmark 2: Høvdingesamfund og Kongemagt*, 49–66. Århus.

Ramqvist, P. 1992. *Högom: The Excavations 1949–1984*. Neumünster.

Schaumann-Lönnqvist, M. 1995. 'Lokal finsk smyckeproduktion under yngre romartid och tidig folkvandringstid'. In H. G. Resi (ed), *Produksjon og Samfunn: Om Erverv, Spesialisering og Bosetning i Norden i 1. Årtusen e. Kr.*, 103–16. Oslo.

Sjøvold, Th. 1993. *The Scandinavian Relief Brooches of the Migration Period: An Attempt at a New Classification*. Norske Oldfunn XV. Oslo.

Stenberger, M. 1971. *Det Forntida Sverige*. Uppsala.

Stjernqvist, B. 1978. 'Mountings for drinking-horns from a grave found at Simris, Scania'. *Meddelanden från Lunds Univ Hist Mus* 1977–78, 129–50.

Straume, E. 1987. *Gläser mit Facettenschliff aus skandinavischen Gräbern des 4. und 5. Jahrhunderts n. Chr.*. Oslo.

Ström, F. 1961. *Nordisk Hedendom: Tro och Sed i Förkristen Tid*. Stockholm.

Widgren, M. 1988. 'Om skillnader och likheter mellan regioner'. In U. Näsman and J. Lund (eds), *Folkevandringstiden i Norden: En Krisetid mellem Ældre og Yngre Jernalder*, 273–87. Århus.

8. The chronology of the Scandinavian gold bracteates

Morten Axboe

This paper can only offer a brief preliminary presentation of my bracteate chronology, which will be treated in full in the concluding volume of *Die Goldbrakteaten der Völkerwanderungszeit* (Hauck et al. 1985–. Vol. 4, in prep.). A more detailed discussion of the methods used will be published there, as well as definitions of the variables and detailed descriptions of the groups resulting from the analyses. I shall also abstain from any attempt to summarize previous endeavours on bracteate chronology (see Malmer 1963, 76–105; Axboe 1992).

With Egil Bakka (1973, 54) I have distinguished between the *internal* and the *external* chronology of the bracteates, the first being a relative chronology based on a typological treatment of the variables inherent in the bracteate pictures themselves, the latter being their integration into the general chronology of the Migration Period, and their absolute dates. Most emphasis will be placed on their internal chronology, where I have worked out a relative chronology for the A-, B- and C-bracteates, while the relative dating of the D- (and F-) bracteates can only be tentative. The absolute dating of the beginning of bracteate production is profoundly dependent on the general dating of the beginning of Style I. On the other hand it seems that bracteates themselves can contribute to the discussion of the dating of the end of Style I and the Migration Period.

The pictures on the gold bracteates include a lot of different details: human heads or figures, animals, inscriptions, and decorative or symbolic signs like swastikas, triskeles etc. To be able to treat A-bracteates with a human bust, B-bracteates with one or more full human figures, and C-bracteates with a human head over a quadruped together on an equal basis in this analysis I have concentrated on the variables of their large human heads. Thus neither the D/F-bracteates, which only show highly stylized animals, nor some bracteates with small human heads (like Hauck's *Drei-Götter-Brakteaten* with three standing figures: M 6:11–16; IK 20, 39, 40, 51,1–3, 66 & 165) have been included. The medallion-imitations, which in several graves can be dated to the Late Roman Iron Age, have also been excluded, together with some closely related pieces. The database was closed in 1988; later finds have not been added.

The bracteate motifs were divided into details – eyes, ears, hair-style, diadems etc. – to obtain the types of the seriation; see Appendix A for short descriptions of the types. The units were the bracteate dies. Using sorted matrices and correspondence analysis alternately, I checked and refined my classifications to improve the parabolas in the correspondence analysis (see Axboe 1993a). The programs used were the Bonn Archaeological Statistics Package Ver. 4.5 (Herzog and Scollar 1987), combined with a program for correspondence analysis (Wright 1985) as the latter method had not yet been added to the BASP when I started my work.

For the large human heads of the A-, B- and C-bracteates I ended up with a database with 342 dies (representing c. 500 bracteates), 51 types and 2445 incidences. Figure 8.1 shows the resultant correspondence analysis, figure 8.2 a simplified version of the sorted matrix. To check the seriation and search for possible geographical variations in the material I also made separate seriations. One was for southern Scandinavia (present-day Denmark, Schleswig, Halland, Skåne and Blekinge) and included 150 dies, 47 types (details) and 1159 incidences. The other consisted of the Norwegian finds, with 54 dies, 30 types and 353 incidences. The resulting correspondence analyses are seen in figures 8.3–8.4.

It is evident that the curves of the two local seriations are slimmer and thus better than the general seriation of all large human heads in figure 8.1, but that was to be expected as the objects come from smaller areas and thus can be expected to be more uniform. It is more remarkable that it was actually possible to establish an acceptable parabola like figure 8.1 for bracteates dispersed over an area stretching from northern Norway to Hungary and from England to Poland. From a methodological point of view it might have been more appropriate to start by establishing local chronologies, based on seriations like figures 8.3–8.4 and then to try to build a supra-regional

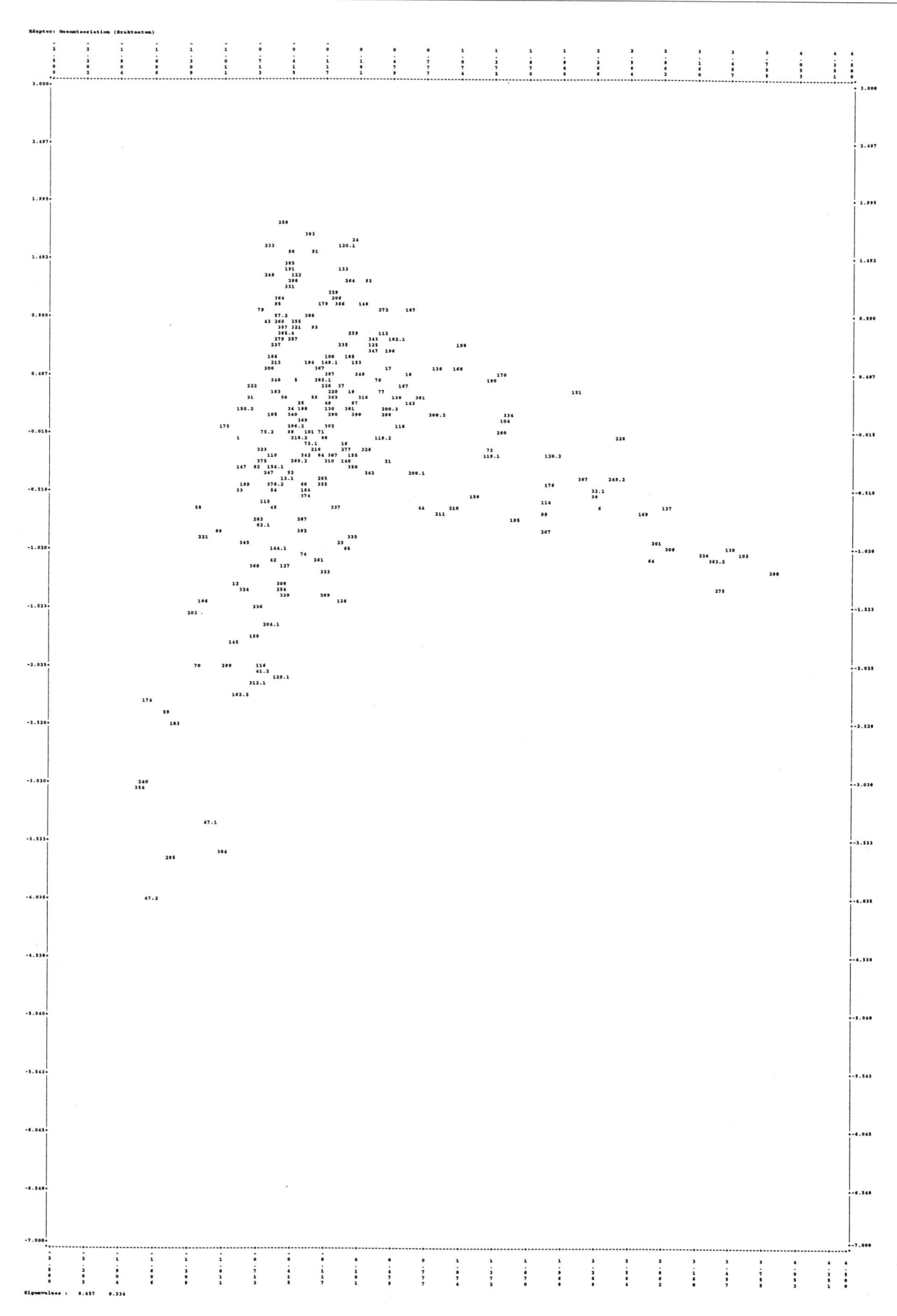

Fig. 8.1 Correspondence analysis of large human heads on 342 A-, B- and C-bracteate dies.
Korrespondenzanalyse der großen menschlichen Häupter. Grundlage sind 342 Model von A-, B- und C-Brakteaten.

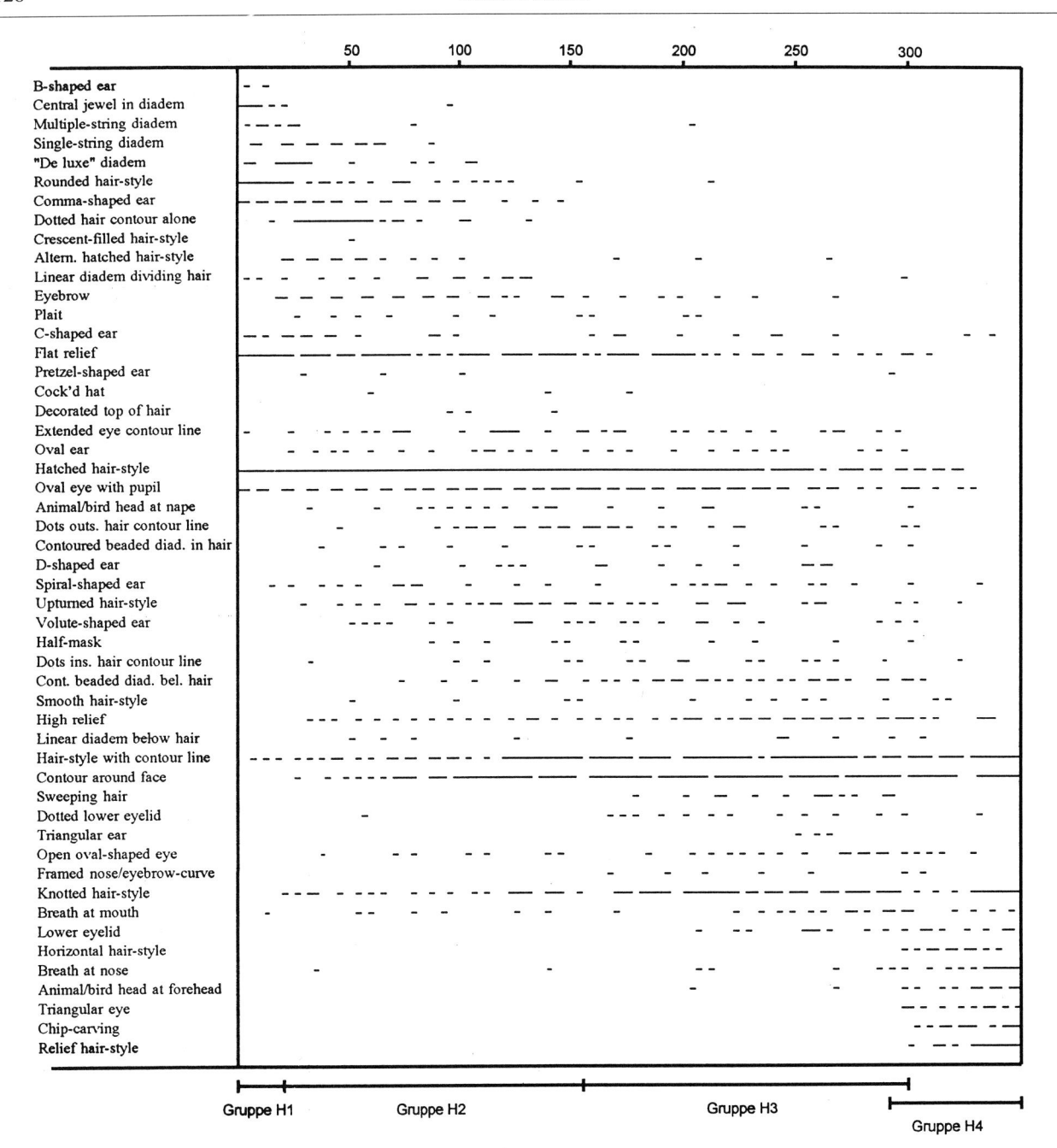

Fig. 8.2 Simplified sorted matrix corresponding to figure 8.1, with groups H1–H4 marked. See Appendix A for descriptions of the types.
Vereinfachtes Kombinationsdiagramm der A–C-Brakteaten (vgl. Abb. 8.1) mit Markierung der Gruppen H1–H4.

chronology on this basis. But to be able to include as many details and bracteates as possible, including the relatively numerous finds of unknown origin, I have chosen to use the total material as presented in figures 8.1–8.2. Attempts to seriate bracteates from other areas (e.g. central mainland Sweden, Skåne or Öland/Gotland) were, in fact, unsuccessful. In any event, the southern Scandinavian finds form the backbone of the whole corpus.

Both the general and the local seriations include several long-lived types, although removing these could possibly have resulted in less crowded matrices and much slimmer curves in the correspondence analyses and thus in 'better' seriations. I have chosen not to exclude these types for several reasons. It can be seen from the sorted matrices that although some types may be present through most of a matrix, the beginning or ending of such types can help to characterize the early and late groups. Also some long-lived types change in frequency, e.g. being scarce in one half of the matrix and dominant in the other, and thus constitute an aspect of general stylistic change. Finally, excluding such types would leave quite a few bracteates

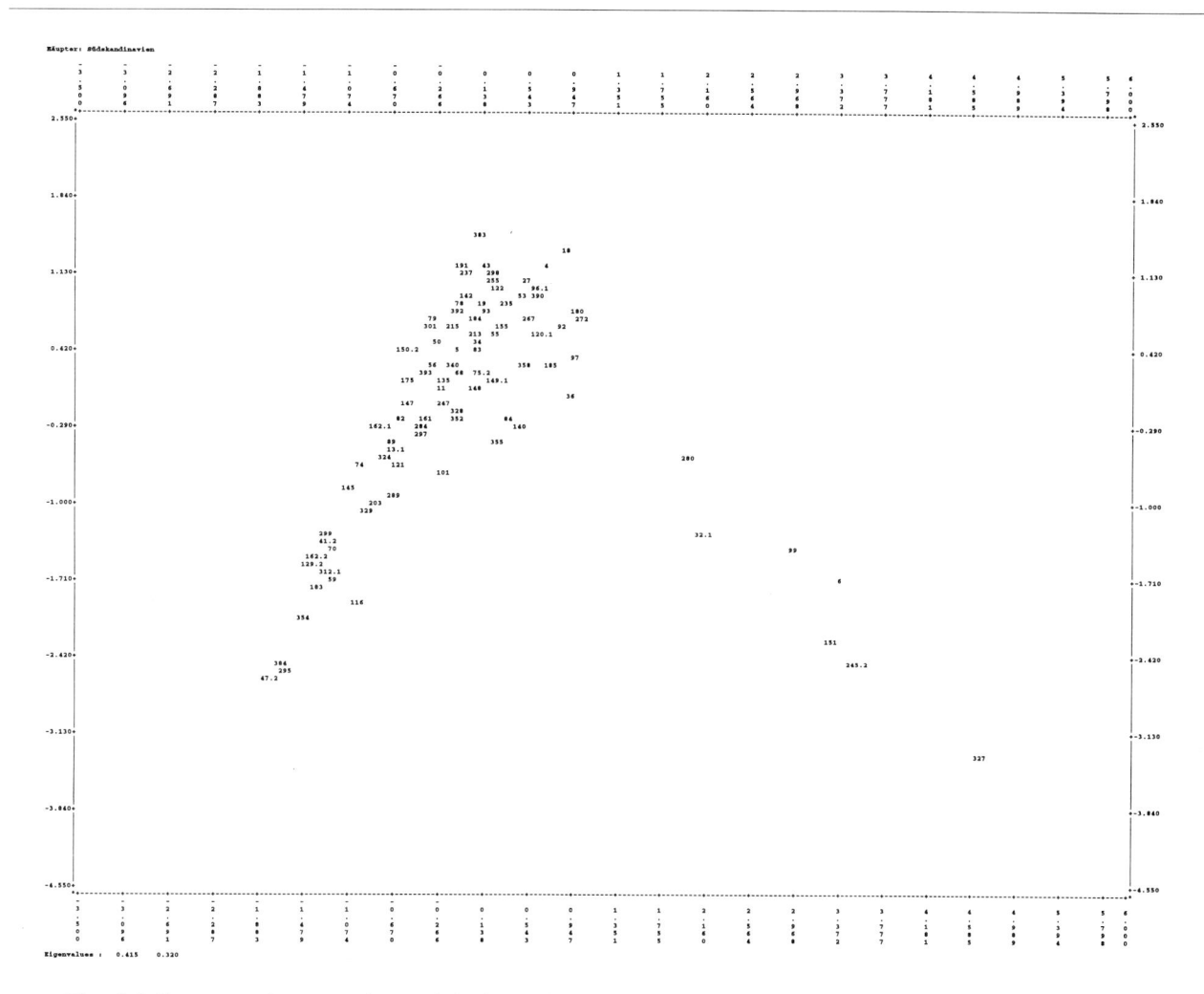

Fig. 8.3 Correspondence analysis of the large human heads from 150 dies found in southern Scandinavia.
Korrespondenzanalyse der großen menschlichen Häupter. Grundlage sind 150 Model aus Südskandinavien.

with so few variables that they would have to be excluded from the seriation too.

The great variation inherent in the bracteate pictures does, however, influence the shape of the correspondence analysis curves, as each die may have from 3 to 12 types of detail recorded and the types may appear on everything from a few up to 276 dies. This results in curves which are broader at the top and may have prolonged tips, as it can be shown that the objects and variables with only few incidences appear to be 'blown' away from the 0/0-point of the diagram compared to those with many incidences (Axboe 1993a, Fig. 37.5).

Correspondence analysis gives an impression of the quality and consistency of the seriation, better than the sorted matrix which was nevertheless superior for the detailed trimming of the database and also for establishing chronological divisions (see Axboe 1993a–b with further references). Basically, the typology-based internal bracteate chronology is a chronology of the production of the pieces while the next step, the external chronology, involves the period of use of objects found together and thus the problem of the time of production compared with the time of deposition (see Steuer forthcoming; for discussions of finds relevant to bracteate chronology Bakka 1981; Hawkes and Pollard 1981, esp. 337–351).

But even an apparently coherent and purely internal seriation, based on typology, may have inherent problems, forcing us to take it with a substantial pinch of salt (Axboe 1993a). This was demonstrated by bracteates from different dies, which can be linked through the occurrence of identical punch-marks in their borders or by other specific details. Within each such set the dies can be astonishingly different and they are placed accordingly in the diagrams (Fig. 8.5; Appendix B).

Another reason not to take the seriations at face value in every detail is the fact that some bracteates – mostly of type A – were copied from Roman coins and medallions. This is evidently one of the premises for the start of the production of medallion-imitations and gold bracteates, but we must remember that Roman coins may have been available as a source of inspiration at any time throughout the entire sequence of bracteate production.

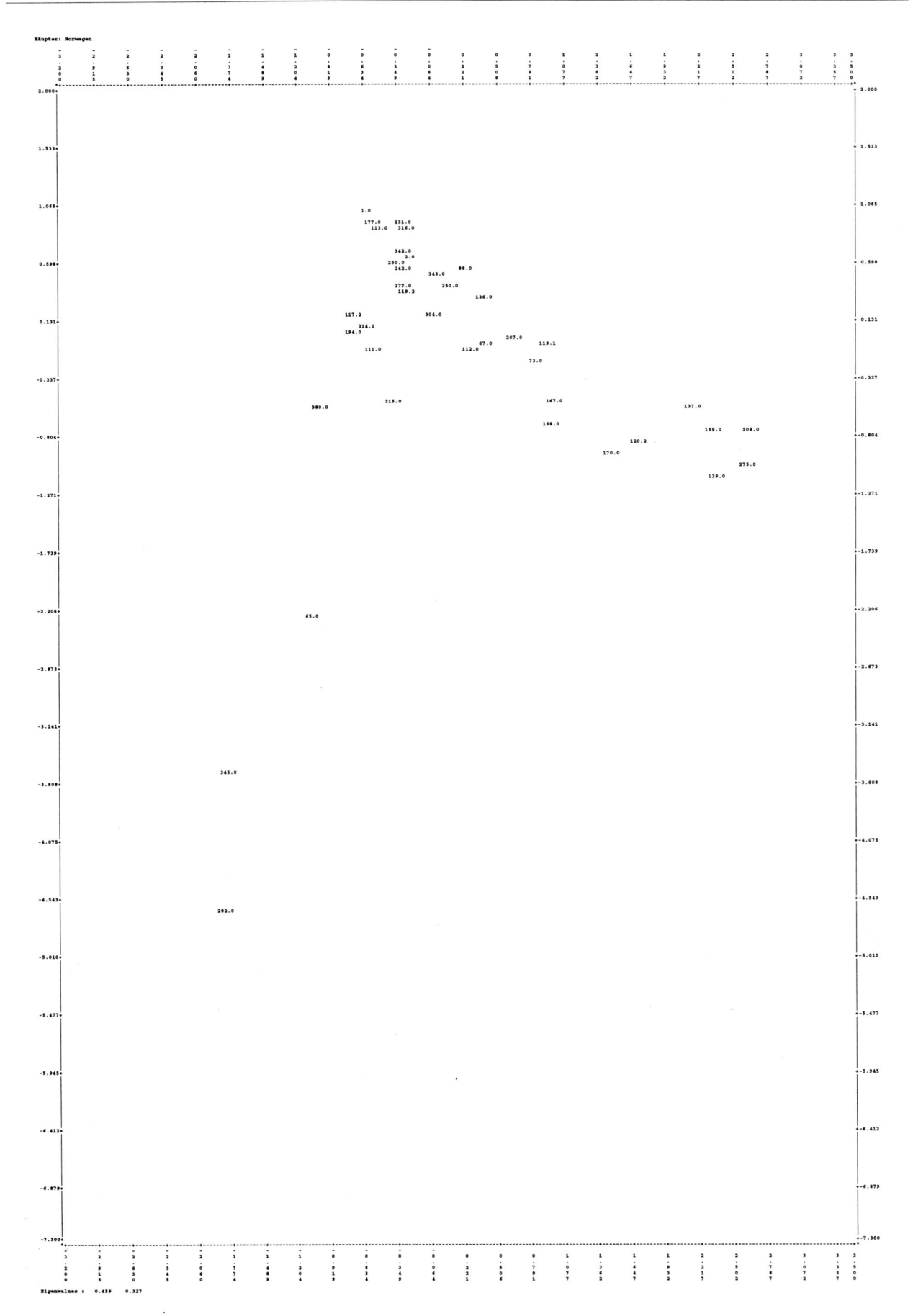

Fig. 8.4 Correspondence analysis of the large human heads from 54 dies found in Norway.
Korrespondenzanalyse der großen menschlichen Häupter auf 54 Modeln aus Norwegen.

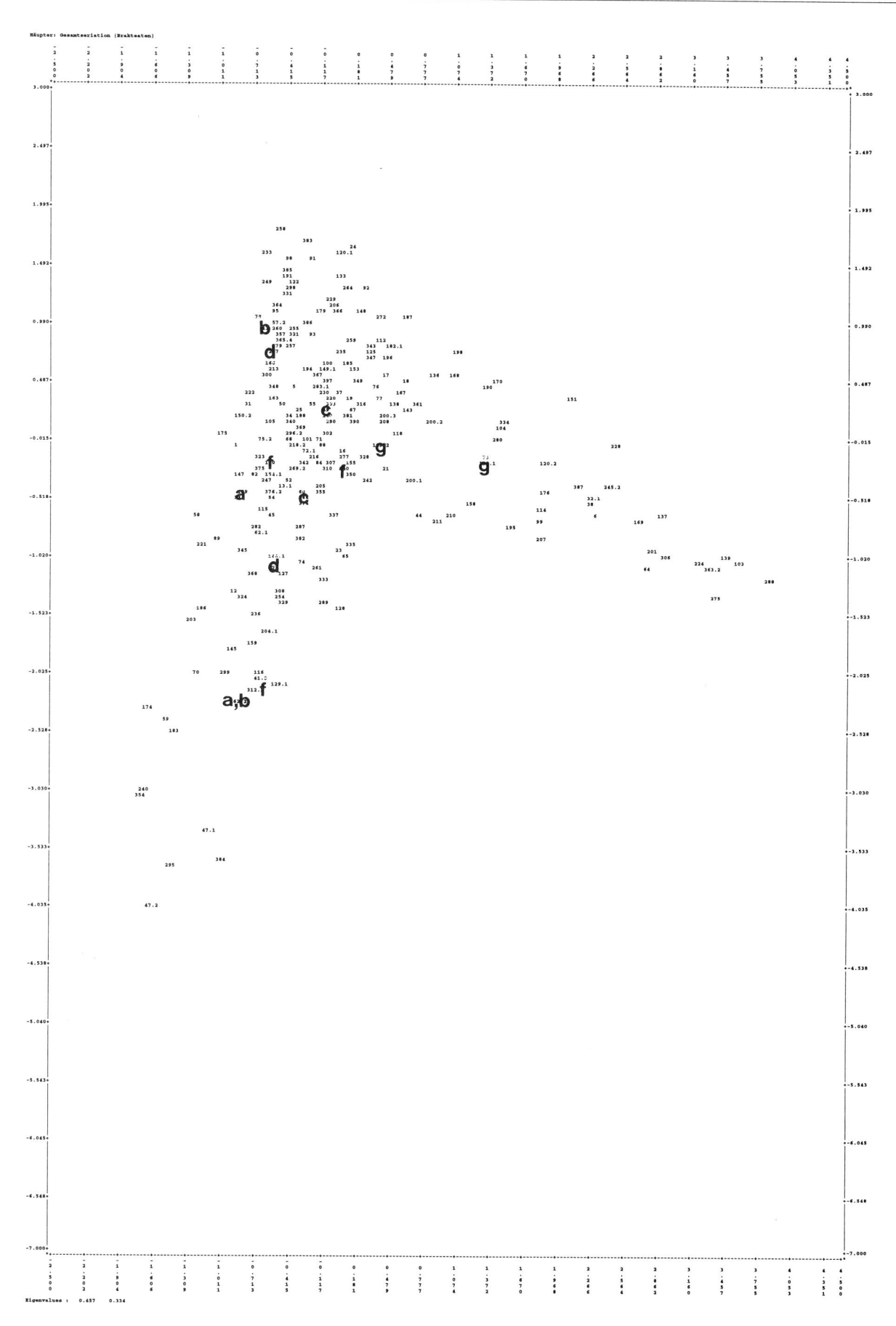

Fig. 8.5 Correspondence analysis of the large human heads with indication of dies linked through border punches or other specific details. For further information, see Appendix B.
Korrespondenzanalyse der großen menschlichen Häupter mit Hervorhebung der Model, die durch Punzen oder andere Details eng verknüpft sind.

If the goldsmith followed his prototype closely enough the result will inevitably be placed early in the seriation, no matter how late it was actually produced.

Thus we must be aware that a typological seriation is a construct which may help us to grasp a general course of development, but which should not be the basis for statements like 'object A appears to be slightly later than object B'. Such detailed relations may in a few cases be established with the help of other criteria, e.g. the identical punch-marks mentioned above or other tool-traces (Axboe 1982, 30, 37f., 52ff.; Axboe and v. Padberg 1977; Benner Larsen 1982–83; 1985; 1987). But while these isolated connections give important information on the pieces involved and glimpses of a reality not revealed by the seriations, they are much too rare to form the basis of a chronological system.

Although archaeological chronology tends, for practical reasons, to establish more or less square-cut phases/periods/*Stufen*, etc., the methods used here favour a picture of more gradual development. No doubt this is closer to reality than the chronological 'boxes', but some sort of classification seems necessary to handle the results.

To this end I have divided the bracteate sequence into a number of *groups*, hoping by this more cautious term to avoid the impression of sharply delimited chronological entities. For the seriation of all large human heads the divisions into the groups H1–H4 is indicated at the bottom of figure 8.2 – 'H' stands for 'Heads/*Häupter*', while the groups in the seriation for southern Scandinavia are labelled S1–S4, the Norwegian groups N1–N3 (Fig. 8.11). Unfortunately only a few types appeared to be characteristic of one group only and thus qualified as '*Leittypen*', and none of these appears on very many dies. So to draw the limits between the groups I had to work out those zones in the sorted matrices where several types started or ended within a short space; that is, where major changes in the repertoire of the bracteate makers seemed to take place. For practical reasons I have then drawn a line within the border-zone, to enable calculations of percentages etc. Only in few cases, like Group H3–H4, are the types characteristic of each group so numerous and diagnostic that it seems sensible to let the groups overlap.

Figures 8.6–8.9 present examples of dies from Group H1–H4. They appear in strict seriation order and are chosen to indicate the development through the groups, as well as demonstrating the variation within each group.

As a general tendency my study shows a dominance of *flat relief* in the first half of the development, although *high relief* becomes increasingly important and is dominant in Group H3. In H4 *chip-carving* appears as a distinctive, late development.

The human heads in **Group H1** (Fig. 8.6) are close to the Roman prototypes, showing a rather naturalistic *rounded hair-style* without any additions at the nape. The diadems of Group H1–H2 are often of rather elaborate forms, and in Group H1 they may display the imperial *central jewel* (Fig. 8.6 nos. 1–2 and indicated on no. 4). The central jewel

Fig. 8.6 Group H1: Selection of dies. 1: IK 47,2 Broholm-A, Fyn (M 3:6). 2: IK 174 Småland(?)-C (M 9:3). 3: IK 384 Vindum Stenhuse-B, North Jutland (M 6:8). 4: IK 183 Tjurkö (III)-A, Blekinge (M 4:3). 5: IK 41,1 Darum (II)-A, West Jutland (M =4:8). 6: IK 129,2 Darum (IV)-B, West Jutland (M 5:16).
Ausgewählte Brakteaten der Gruppe H1.

is a diagnostic type for Group H1, together with the *B-shaped ear* (Fig. 8.6 no. 1 left, 3).

Group H2 (Fig. 8.7) adds new types of hair-style with a *knotted* (e.g. Fig. 8.7 nos. 1, 9, 12) or *upturned* hair-style (Fig. 8.7 nos. 6–8, 11). Both can be found later as well. This applies also to the *animal/bird heads placed at the nape* (Fig. 8.7 nos. 6–8, Fig. 8.8 no. 12), and to the *breath-signatures at the mouth* of the human head (Fig. 8.7 nos. 4, 7, 12; Fig. 8.8 nos. 6, 11; Fig. 8.9 no. 4). *Plaited hair-style* (Fig. 8.7 nos. 5, 10) and *hair-style with alternately hatched sections* (Fig. 8.7 nos. 1, 4, 8) are hardly found outside Group H2 and can be considered diagnostic of this group, as can hair *contoured with a dotted line only* (Fig. 8.6 no. 6; Fig. 8.7 nos. 1, 4, 5) and the more unusual details

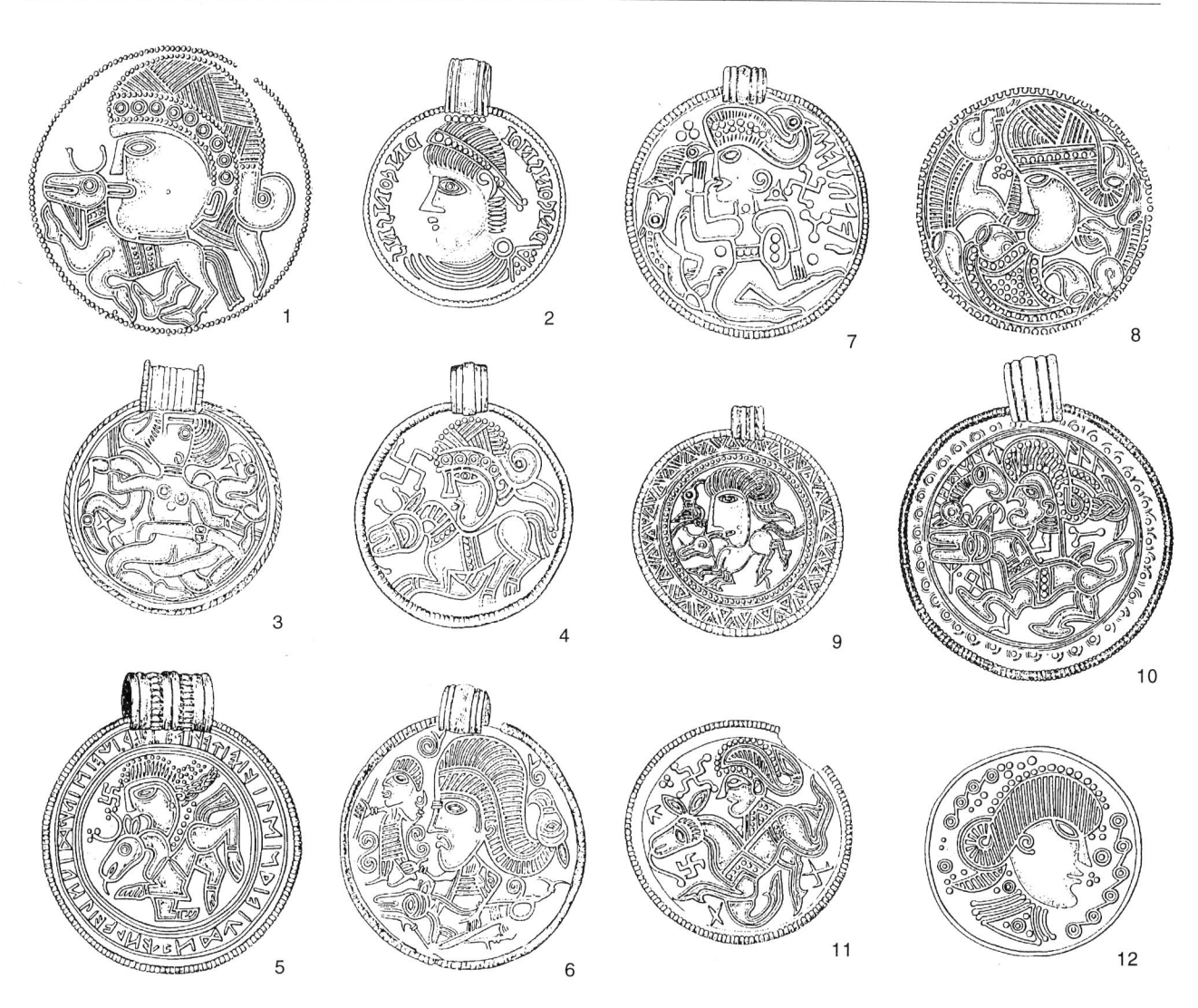

Fig. 8.7 Group H2: Selection of dies. 1: IK 203 Vä-C, Skåne (M 12:2). 2: IK 282 Hov-A, Nord-Trøndelag (M 3:5). 3: IK 308 Nebenstedt-B, Niedersachsen (M 5:9). 4: IK 121 Maen-C, Halland (M 11:1). 5: IK 1: Ågedal-C, Vest-Agder (M 10:5). 6: IK 50 Near Esrom Sø-C, Sjælland (M 8:22). 7: IK 105 Lellinge Kohave-B, Sjælland (M 5:2). 8: IK 279 Holmetorp-A, Öland (M 4:18). 9: IK 81 Near Hjørring-C/Stejlbjerg(?), North Jutland (M 13:19). 10: IK 300 Gummersmark-C, Sjælland (M 6:20). 11: IK 215 Aversi-C, Sjælland (M 7:22). 12: IK 273 Near Hjørring-A/Stejlbjerg(?), North Jutland (M 3:16).
Ausgewählte Brakteaten der Gruppe H2.

crescent-filled hair-style, *decorated top of hair*, and *pretzel-shaped ear* (not illustrated).

The Roman rounded hair-style has almost disappeared in **Group H3** and the diadems appear mostly in stylized forms, like in figure 8.8 nos. 1, 6, 7, 9–12. Four types of motif-detail can be considered diagnostic of this group. The two most numerous are the *sweeping hair-style* where the hair starts vertically from the forehead and continues unbroken to the tip of the knot (Fig. 8.8 nos. 6, 8), and the *dots below the eyelid* (Fig. 8.8 nos. 10, 11). Rarer are the *framed nose/eyebrow-curve* (Fig. 8.8 no. 12) and the *triangular ear* (not illustrated).

Characteristic of **Group H4** is the frequent use of *chip-carving*, which, *inter alia*, may appear as infill in the *relief hair-style* (Fig. 8.9 nos. 2, 5–6). Also characteristic is the *horizontal hair-style* where the hair starts horizontally over the forehead (Fig. 8.9 no. 4), as well as a *triangular eye* (Fig. 8.9 nos. 4–6). Complete diadems are found no more, but the parallel lines between the hair and the back of the animal found, for instance, on figure 8.9 no. 6 may be considered as the last survivals of the diadem ties. Nor do the heads at the nape appear in Group H4. They may have been replaced by the *animal or bird heads at the forehead* (Fig. 8.9 no. 6) which, like the *breath-signatures at the nose* (Fig. 8.9 no. 6), belong primarily to Group H4.

Much more could be said about the occurrence and development of other seriated details, like eyes, ears, contours, infilling of hair-style, etc., but this must be

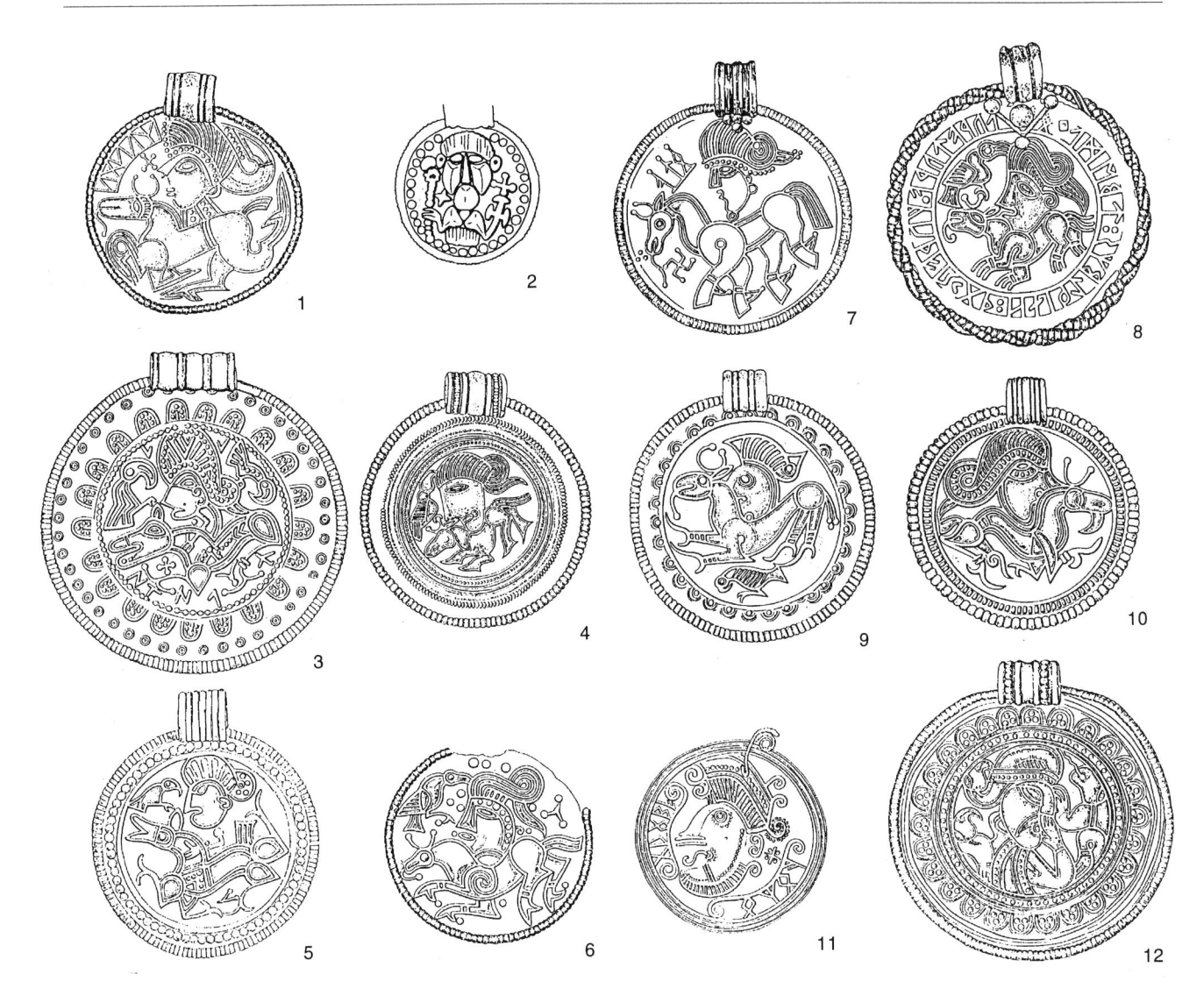

Fig. 8.8 Group H3: Selection of dies. 1: IK 244 Fredrikstad-C, Østfold (M 8:3). 2: IK 391 Gudme II-B, Fyn (Axboe 1987). 3: IK 25 Bjørnsholm-C, North Jutland (M 7:3). 4: IK 220 Near Böja-C, Västergötland (M 13:22). 5: IK 96,1 Kläggeröd-C, Skåne (M =7:2). 6: IK 19 Bakkegård-C, Bornholm (M 14:9). 7: IK 185 Tjurkö (II)-C, Blekinge (M 14:17). 8: IK 377,1 Near Vadstena-C, Östergötland (M 14:11). 9: IK 178,1 Sojs-C, Gotland (M 14:20). 10: IK 4 Åkarp-C, Skåne (A 238/2). 11: IK 76 Wurt Hitsum-A, Friesland (M 4:12). 12: IK 187 Tossene-A, Bohuslän (M 4:22); this die is placed in the transitional zone H3/H4.
Ausgewählte Brakteaten der Gruppe H3.

reserved for Hauck et al. (1985-) vol. 4. Generally speaking my investigation seems to conform to Mackeprang's intuitive chronology (1952): a development from Roman-inspired pieces with soft, low relief to a Germanic style with a frequent use of chip-carving. I have not compared these results with Malmer's chronology (1963) as his seriation uses other elements, though the basic principle of gradual change remains the same.

As shown in figures 8.6–8.9, A-, B- and C-bracteates occur in all four groups (*contra* Bakka 1973, 54f., who considered that 'not a single A-bracteate' should be referred to Mackeprang's Period 3; similarly Hines 1984, 201f) This is further demonstrated in figure 8.10. As definitely the most numerous of the seriated types (see IK 3,1 p. 76) the C-bracteates naturally dominate the picture, apart from Group H1 where the A-type is predominant. It should, however, be remembered that the A-type is the most likely to be affected by the 'renaissance'-phenomenon mentioned, where renewed inspiration from Roman prototypes may result in anachronistic 'typologically early' bracteates. Thus the number of A-bracteates in Group H1 may be too high, although a larger share here than later is to be expected. For the rest of the diagram the proportions between the three types seems surprisingly stable; it is not the case that the A- or B-bracteates disappear much earlier than the C-bracteates.

I shall not comment in detail on the local seriations for

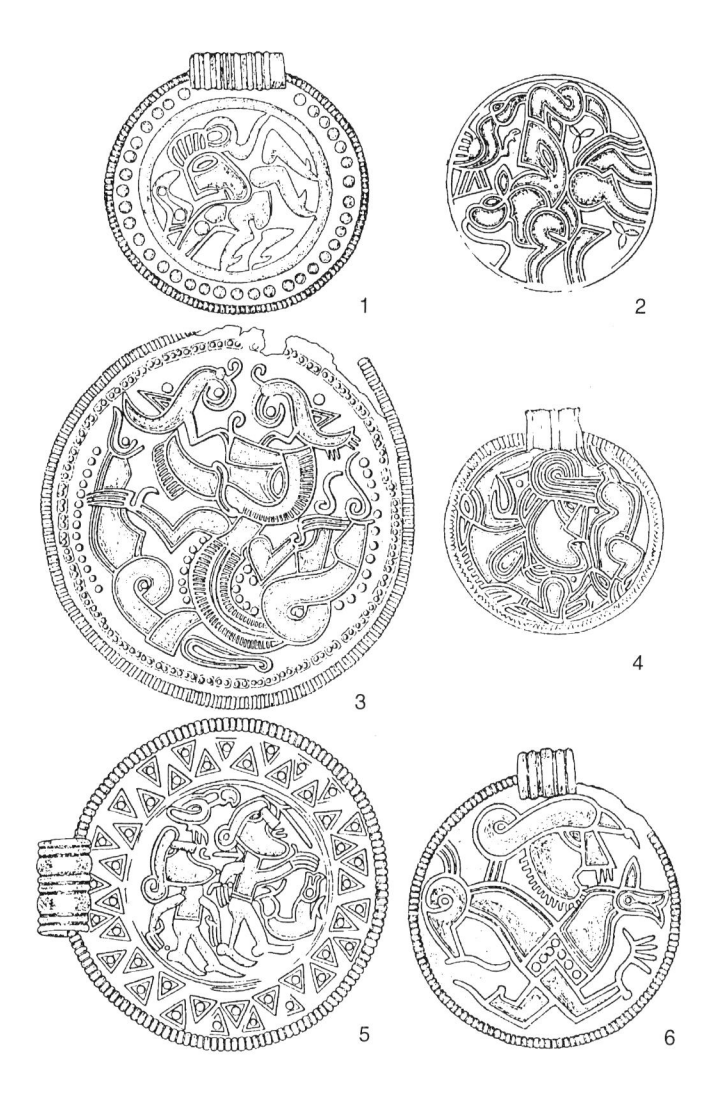

Fig. 8.9 Group H4: Selection of dies. 1: IK 280 Holmgårds Mose-C, North Jutland (M 13:34). 2: IK 334 Silleby Mellangården-C, Södermanland (M 14:10). 3: IK 176 Söderby-B, Uppland (M 5:3). 4: IK 120,2 Haugan-A, Vestfold (M 4:23). 5: IK 6 Års-B, North Jutland (M 6:6). 6: IK 46 Dokkum(?)-C, Friesland (M 7:16).
Ausgewählte Brakteaten der Gruppe H4.

Norway and southern Scandinavia (Fig. 8.3–8.4) but just draw attention to a few points. As already mentioned, the curves of the two correspondence analyses are slimmer than that of the general seriation, as was to be expected. But it is also evident that most of the southern Scandinavian dies lie in the left (i.e. early) half of the curve in figure 8.3, while the right (late) part is much more sparsely filled. The Norwegian dies in figure 8.4 show the opposite situation: few early pieces and more later. This would indicate that bracteate production started in southern Scandinavia, as already claimed by Mackeprang (1952, 29f.) and suggested by Malmer (1963, 195). It also fits well with the fact that 16 of the 22 dies in Group H1 come from southern Scandinavia. On the other hand southern Scandinavia is still represented by 9 of the latest 37 dies (Group H4 proper, excluding the transitional zone H3/H4), so that production here probably continued to the end of the bracteate period.

The material from southern Scandinavia, like the general seriation, permits a division in four groups (S1–S4). As almost half of the total material comes from southern Scandinavia the divisions in these two seriations can be coordinated quite well; however the groups appeared somewhat clearer in the smaller and more homogeneous body of material so that it appeared reasonable to let Groups S2 and S3 overlap slightly. In the Norwegian seriation only the transition between the two latest groups could be synchronized with the transition H3/H4, and the earlier Norwegian dies could only be divided into two groups. We thus get the groups N1–N3.

The relationship between the general and the local seriations is illustrated in figure 8.11. The basis for this comparison is the general seriation, with Group H1–H4 at the top of the diagram. Below is shown where the dies of Group S1–S4 and N1–N3 are placed in the general seriation. It can be seen that the divisions for southern Scandinavia correspond pretty well with the general seriation (the scattering at the group divisions is due to some extent to the seriation method), and that dies from southern Scandinavia are present from start to end of the general seriation. In respect of Norway, it is evident that productions starts late: there are no Norwegian dies in Group H1 and relatively few in H2. The bulk of the Norwegian bracteates are placed in the later half of the sequence.

One more thing should be said about the Norwegian seriation. To obtain the parabola in figure 8.4 it was necessary to remove four dies from the database: IK 24 Bjørnerud-A (M 4:17), IK 146 Røgenes-C (M 9:25), IK 244 Fredrikstad-C (M 8:3) and IK 331 Selvik-A (M 4:19). Before that the curve was much more blurred and the dies in question tended to form a third arm at the top of the parabola. This may indicate that the shape of the curve is not determined by chronological factors alone, or that the objects in question do not conform to the same line of development as the rest (pers. comm. Karen Høilund Nielsen). In this case the explanation could be that the four bracteates were imported, perhaps from western Sweden/ southern Scandinavia. None of them were, however, placed very early in the Norwegian development, neither in the general seriation nor in the Norwegian, so that it seems unlikely that they should be considered as early imports which might have given an impetus to indigenous Norwegian bracteate production.

In the general seriation the human dot-and-circle eyes (e.g. Fig. 8.6 nos. 5–6) were concentrated at the early and late ends of the sorted matrix and thus disturbed the correspondence analysis. They were therefore omitted as chronologically insignificant and are not included in the diagrams in figures 8.1–8.2. But they fit nicely into the local diagrams: for southern Scandinavia in the early part (with one late exception), for Norway in the late part. Here, then, we may have an example of local variation.

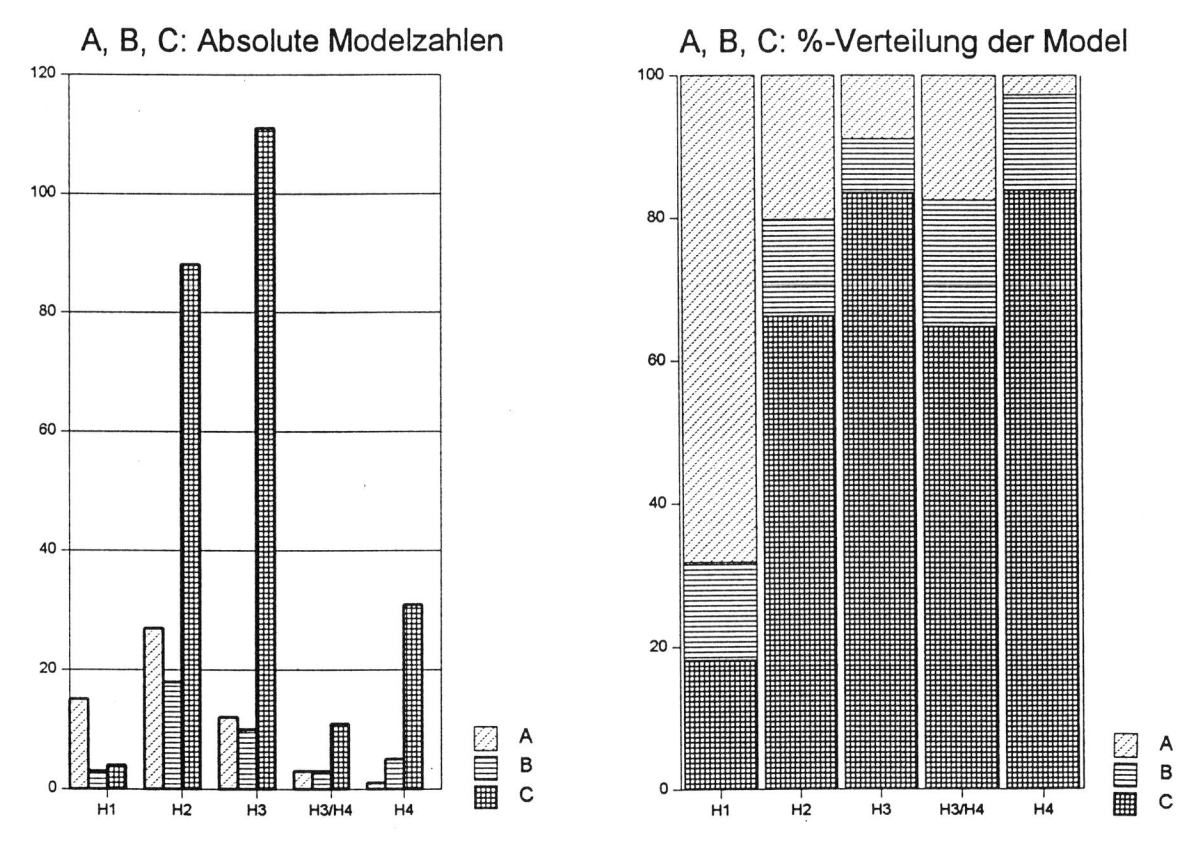

Fig. 8.10 The occurrence of A-, B- and C-dies in Group H1–H4. Left: absolute numbers, right: percentages.
Vorkommen von A-, B- und C-Modeln in den Gruppen H1 – H4. Absolute Häufigkeiten links, Prozente rechts.

The correlation of the D-bracteates with my seriation is an important matter, but unfortunately there is no straightforward solution. Though animals occur on all types of bracteates, the details of the animals are mostly so different from those of the A- to C-bracteates that it seems impossible to analyse them together, as D-bracteate animals are mostly executed in one of the developed variants of Style I (Haseloff's Stilphase D), in some cases even with tendencies to symmetry and ribbon-band style pointing forward to Style II (see Haseloff 1981, 216–30). No secure technical links, like tool-identities in punch-marks or loops have yet been established (on a few possible loop-identities, see Axboe 1982, 37f.). Find-combinations with A- to C-bracteates occur practically only in hoards, which are to be treated with some care for detailed chronological purposes. However, some points can be made.

An examination of the degree of wear in finds combining C- and D-bracteates seems to indicate that the D-bracteates begin later and also may continue longer than the C-bracteates. In Norwegian graves where A- to C-bracteates occur relatively frequently, only three single D-bracteates have been found. This may be another indication of a later dating for the D-type (Bakka 1973, 59), at least for the start of its production. In my seriation there is a concentration of A- to C-bracteates from graves in the first half of Group H3. Some bracteates from graves are placed later, but only one after the end of H3. This may indicate that the D-bracteates start when the custom of placing bracteates in graves was in decline, i.e. in the later part of H3. The combinations in Norwegian hoards may also, with some caution, be taken to indicate that the D-bracteates started somewhere in H3.

From a more general stylistic point of view can it be noted that the D-bracteates all show rather high relief, often even chip-carving. On the A- to C-bracteates high relief occurs from the beginning of Group H2 but becomes dominant only in H3, while chip-carving is diagnostic of Group H4. Square-headed brooches with band-shaped animals of Haseloff's Stilphase C–D, which must be considered both a parallel and a *sine qua non* for the style of many D-bracteate animals, start in Nissen-Meyer's Stadium 5 (Nissen Meyer 1934). In Norwegian graves, bracteates from 8 A- to C-dies are combined with such brooches: 2 from Group H2, the rest from around the middle of H3 to H4.

Taken together, these indications may point to the production of the D-bracteates beginning some time during Group H3, while to judge from their wear they may have continued later than the A- to C-bracteates.

The animals of some F-bracteates have their closest relatives on D-bracteates while others seem related to animals on C-bracteates. The occurrence of runes and swastikas on some F-bracteates also provides a link to the C-bracteates, and not to the very latest as runes become

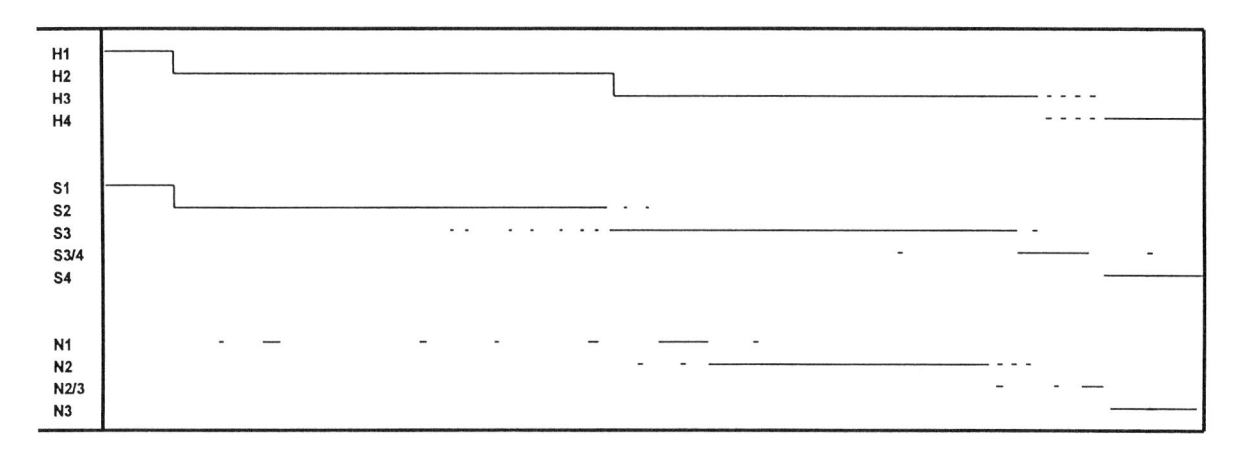

Fig. 8.11 The local seriations for southern Scandinavia (Groups S1–S4) and Norway (Groups N1–N3), compared with the general seriation of the large human heads (Groups H1–H4).
Die regionalen Seriationen für Südskandinavien (Gruppen S1 – S4) und Norwegen (Gruppen N1 – N3) im Vergleich zur Gesamtseriation der großen menschlichen Häupter (Gruppen H1 – H4).

rare in the later part of the development of the A- to C-bracteates and not are found in Group H4 (Axboe, forthcoming). Thus the F-bracteates can be supposed to run parallel with (at least part of) the C- and D-bracteates.

As for the external chronology it is evident that the gold bracteates in Scandinavia belong to the Migration Period (Danish: *ældre germansk jernalder*; Swedish and Norwegian: *folkevandringstid*). It is, however, a problem that in Scandinavia graves with gold bracteates are found almost exclusively in Norway (Bakka 1973) with only a few finds in central Sweden and on Gotland.

Thus most of the Scandinavian bracteates are single finds or come from hoards. Some of the hoards include other datable objects like brooches, scabbard mounts, or even solidi. The latter are evidently datable, but there is much debate over when and how they reached Scandinavia, how they were used here, and how long it may have taken before they were buried (e.g. Fagerlie 1967; Kyhlberg 1986; Fonnesbech-Sandberg 1989). Malmer has pointed out that the combinations of solidi and D-bracteates would allow for the latter to begin as early as c. A.D. 475 (Malmer 1977), the date otherwise proposed for the beginning of Style I (Haseloff 1974a, 14). I would not follow Malmer that far but instead point to the hoard from Bostorp on Öland, with 3 C-bracteates and a gold spiral ring, which were deposited in a pot together with a rather unusual collection of 6 solidi (IK 221–223; A 189a; Fagerlie 1967, find no. 90b): 1 of Libius Severus (461–65), 1 of Anthemius (467–72), 1 of Glycerius (473–74) and 3 of Leo I (457–74). Fagerlie considers this (and some other Öland finds) to have been buried soon after c. 475 (1967, 153f.), and although her theses of devastating attacks as the reason for the deposition of the Öland hoards may be challenged, the period after c. 475 may still be a reasonable date for the formation of the Bostorp hoard. A similar date of deposition is proposed by Herschend (1980, 252f.) and Kyhlberg (1986, 67f. and Tab. 44, Find No. 90b). Both the coins and the bracteates are somewhat worn, which may be an indication that they were of about the same age when deposited. All three bracteates are placed in the first part of my Group H2. Here we also find the C-bracteate from Rynkebygård on Fyn (IK 147, M 6:30, catalogue no. 64), which was found together with five solidi: 1 of Valentinian III (425–455), 1 of Marcian (450–457) and 3 of Leo I (457–474), a gold spiral ring and a gold ingot. The bracteate is rather worn. Kyhlberg considers this hoard to have been deposited at around the same time as Bostorp (1986, Tab. 44, Find No. 193).

Thus both the Bostorp and the Rynkebygård bracteates seem to have been deposited during the years after 475, after a period of use resulting in distinct wear.

Salin and Bakka have pointed out the close resemblance between the animals of the relief brooch from Gummersmark and the C-bracteate from Fyn (Bakka 1973, Fig. 1); both, for instance, have what Salin labelled 'Vimose-heads' with a marked nostril separated from the eye through two or three transverse lines. The transverse lines with an eventual nostril can be found on some other bracteates, which in my seriation occur from the beginning of Group H2 to the transition H3/H4 (e.g. Fig. 8.8 no. 3). On relief brooches this detail is common in Haseloff's Stilphase A and B and he considers it diagnostic of early Style I (1974a, 13), but it does also occur, though more rarely, in Stilphase C and D. The Gummersmark brooch is itself typical of his Stilphase A (1981, 176); the Fyn bracteate (IK 58; M 6:19) is placed among the first dies of my Group H2. Other features from early Style I which occur on bracteate animals are the use of contour lines not only as outer contour but also around the separate parts of the animals (thighs etc.), which is diagnostic of Style I as opposed to the Nydam Style (Haseloff 1984, 111f.); and the closely set parallel lines across the bodies of the animals, characteristic of Stilphase B (Haseloff 1974a, 9), although on the bracteates usually only part of the body can be filled this way (e.g. IK 33 British Museum-C M 6:21, IK 79 Hjørlunde-C M 8:21, IK 82 Højgård-C M 10:12, IK 122 Gummersmark-C M 8:4, IK 289–290 Kjellers Mose-C M 10:10–11). Most of these dies are placed in Group H2, a few in H3. The pointed-oval transverse lines in the body of the animal on IK 59 Fyn-C (M 8:18) may be compared with the similar lines occurring in Stilphase A (Haseloff 1981, 108ff.).

The use of rounded relief surrounded by contour lines is diagnostic of Haseloff's Stilphase A; in the Nydam Style contour lines (especially the internal ones) are not yet common, in Stilphase B the relief is flat (Haseloff 1984, 112f.). Both the human heads and the animals of the A- to C-bracteates have mostly more or less rounded relief, but in Group H4 many human heads show chip-carving, and this tendency is found on the animals too. This may be a parallel to the development in Stilphase C–D, albeit expressed in a slightly different way. The ribbon-like bodies of Stilphase D-animals are found on many D-bracteates, while the only attempt to adapt this style on a C-bracteate would seem to be the strange piece from Chippenham in Cambridgeshire (IK 228, A 307e), placed in Group H4.

When discussing the style of the gold bracteates one should, however, bear in mind that their pictures are not merely ornamentation. Both the A- to C-bracteates and the D-bracteates had an amuletic character beside their value as ornaments, as already acknowledged by Haseloff (1981, 217) and elaborated by Karl Hauck in numerous papers (most recently Hauck 1994a–b). They were executed in the style of the period, but it is conceivable that their stylistic development may have been impeded by their amuletic function. Nor could the horse-like animals of the C-bracteates simply develop into D-bracteate beasts, for although it may seem strange to discuss the 'zoology' of fabulous animals, they are quite different sorts of creatures and the D-bracteate monsters can be traced back to several different Roman prototypes (Hauck et al. 1985- Vol. 3,1, 15–36).

The 'ornamentation' of the relief brooches and other Nydam Style and Style I objects also includes Germanic adaptions of different Roman monsters (Haseloff 1986) and may have had symbolic or apotropaic functions, though this subject has been less thoroughly explored. I shall not venture further into this, but simply draw attention to the complicated interplay between the style and its spiritual contents, which may influence any examination of prehistoric 'ornamentation' and its development.

It must also be kept in mind that the stylistic development of the Migration Period is anything but straightforward. The Sösdala and Nydam Styles are considered to be at least partly contemporaneous, and so are Nydam Style and early Style I. Haseloff stresses that although he has labelled his stylistic phases A–D, they are not to be taken strictly as successive steps in a development (e.g. 1984, 112f.). Only Stilphase A is considered to be the substratum for the rest of the phases (1981, 174) while these may be more or less parallel developments. Stilphase C and D especially seem closely related, and as a matter of fact Haseloff, in a lecture of 1971, did not yet distinguish between these two (published as Haseloff 1974b, 379; cf. 1974a, 10, where C and D do appear as separate phases). We have quite a few examples of elements from more than one Stilphase (or even the Nydam Style) being combined in the same find or even on the same object: the Overhornbæk brooch, Snartemo V (sword, buckle, gold foil on glass), Högom mound 2 (Ramqvist 1992, 144ff.), to name but a few.

But let us return to the problems of the gold bracteates and their place in the chronology of the Migration Period. I have found no elements on the gold bracteates proving a connection with the Nydam Style as defined by Haseloff (1981, 8ff.; 1984, 112f.). As already stated, the animals of the C-bracteates are closely connected with early Style I (Stilphase A and to some extent B). Although many A-bracteates lack animals to be compared with the general development of the animal styles, and one might thus postulate that they could have been manufactured before the beginning of Style I, my seriations give me no reason to suppose that the A-bracteates should start significantly earlier than the C-bracteates. Thus I would presume that the start of the production of the gold bracteates was contemporary with the development from the Nydam Style to early Style I. It would fit nicely with the Bostorp and Rynkebygård hoards mentioned above if this transition should be dated somewhat earlier than c. 475; Näsman has argued for a start of Early Style I around the middle of the 5th century, running for a time parallel with the later part of the Nydam style (1984, 70). Moreover several weapon graves containing Style I have recently been dated surprisingly early, to around the middle of the 5th century (Bemmann and Hahne 1994).

But almost no matter how the early bracteates are dated there will be a problem with any hypothetical continuity from the medallion-imitations, several of which are securely datable to the Late Roman Iron Age. It is open to discussion whether the gap can be bridged by the few stray finds of medallion-imitations and the related pieces which were excluded from my seriation. A possible example could be the Gotlandic series IK 286,1–4, with three medallion-imitations, two of which were found in Late Roman Iron-age graves, and one 'bracteate' (A 212a) struck with the avers die. Unfortunately the latter has no secure context but may, according to Nerman (1935, 7 and find no. 126), have belonged to an early Migration-period cremation.

It is also worth noticing that, although solidi showing the emperor's head in profile were still being issued through the first half of the 5th century (Fagerlie 1967, Pl. I–IV, VI, XIII, XIV) and the first bracteate makers thus may have used relatively contemporary prototypes, it is evident that at least some bracteates were copying coins of the first half of the 4th century, like figure 8.6 no. 1 with its recognizable Constans-inscription. As it is impossible that the bracteates should have started so early, this illustrates the long survival of the Roman coins as possible sources of inspiration for the bracteate makers.

The date of the end of the bracteate production in Scandinavia must coincide with the transition to the Vendel Period (Danish: *yngre germansk jernalder*; Swedish: *vendeltid*; Norwegian: *merovingertid*), assuming that at least the D-bracteates with their harbingers of Style II (Haseloff 1981, 216–30) continued until the end of the Migration Period. This transition has been dated to c. 550/75 (Ørsnes 1966, 207) or 560/70 (Hines 1984, 31f., 1992, 83), but in the later years there has also been a tendency to an earlier dating c. 520/30 (Lund Hansen 1992, 185; Jørgensen 1990, 50f.). This earlier dating is based on a synchronization of the Scandinavian weapon graves with the Continental chronology, starting with Birgit Arrhenius' revision of the dates of the rich Vendel graves (1983). She parallelled the earliest Vendel graves X, XI, XII, and XIV with their ornamentation in Vendel Style A and B to Ament's AM III (560/70–600; Ament 1977; Arrhenius 1983, 64) and would place the transition phase to the Vendel Period around 550, corresponding to AM II (520/30–560/70; Arrhenius 1983, 68). The weapon sets of the Vendel-period graves on Gotland have subsequently been investigated by Anne Nørgård Jørgensen (1992), who aligned the beginning of her earliest phase with the transition AM I/AM II, i.e. c. 530 (Nørgård Jørgensen 1992, 18ff.). This absolute date has apparently been transferred to the general chronology for the transition from the Migration to Vendel Period, including the brooches and other jewellery (see for example Jørgensen 1994, Abb. 122–124), thus also affecting the gold bracteates and the general style history of the time.

A key problem in this discussion is obviously the equation 'start of Vendel Period = start of AM II' which must underlie the early date of the transition from Migration to Vendel Period. This equation is based on the weaponry, which in this period appears to have developed uniformly over large areas (Nørgård Jørgensen 1992, 20) in contrast to the more local female dress accessories which traditionally have formed the backbone of the chronology in Scandinavia (Ørsnes 1966; Høilund Nielsen 1987). The synchronization of the male and female chronologies has been attempted through horizontal stratigraphy on Bornholm, 'with all the necessary reservations for this uncertain method of dating' (Jørgensen 1990, 44) and Gotland (Nørgård Jørgensen 1992, 22ff.). But especially in respect of the Bornholm material I have not been able to find the basis for the repeated claims for a transition Migration/Vendel Period around 520 or in the following decade. On the contrary, Lars Jørgensen's diagram (Jørgensen 1990, Fig. 41), based upon the weaponry and the horizontal stratigraphy mentioned above, equates Høilund Nielsen's Phase 1B with AM III. Phase 1A he has (in my opinion correctly) criticized for being a mixture of types from both the Migration and the Vendel Periods, partly caused by the seriation of mixed finds (1990, 23ff.). Thus the start of Phase 1B, the first phase containing purely Vendel-period material, is dated to 560/70, in good agreement with Ørsnes (1966) and Arrhenius (1983, 68). Even if some of the Vendel-period material included in Phase 1A actually should predate Phase 1B would it correspond to Arrhenius' date of the 'transitional phase', c. 550. Meanwhile, Anne Nørgård Jørgensen's examination of the Kobbeå 1–Vendel XIV-horizon seems to correlate the early Vendel Period with Schretzheim Stufe 3 (565–590/600) and thus AM III (Nørgård Jørgensen 1991, 227).

The equipment of Phase 1 of the Vendel-period weapon graves on Gotland is compared with Schretzheim Stufe 1–2 (525–565/70), but for some reason the Gotlandic Phase 1 is dated only to 530/40–560/70 (Nørgård Jørgensen 1992, 20). There seems no reason to question the Vendel-period character of this material; some mounts even seem to display ornamentation in Style B (Nørgård Jørgensen 1992, Fig. 4). Here, too, a synchronization between male and female chronologies through horizontal stratigraphy is attempted and an acceptable agreement reached, but unfortunately only for finds later than the crucial early phases of the Vendel Period. And again, there appears no justification for letting the Vendel Period start already 520/30.

How can the gold bracteates contribute to this discussion? Within Scandinavia there are only limited possibilities, as already mentioned, but we should consider the more numerous bracteates found in England and on the Continent. Just like the relief brooches, which were also both imported and imitated, they represent Scandinavian influence which can be dated in English and Continental graves. Karen Høilund Nielsen has summarized this evidence in a diagram (1987, Fig. 16), but the bracteates can be examined in greater detail, as seen in figure 8.12. Some of the bracteates may be exports from Scandinavia while others are local products inspired from these, e.g. the light-weight bracteates found on the Continent (Axboe and Hauck 1985, 101f.; Axboe 1987, 80f.), the English silver bracteates, and the silver mounts of bracteate design from Bradstow School and Rhenen. I have not distinguished between imports from Scandinavia and local products as the latter must be dependent on the former, and as the Scandinavian pieces do not seem to have initiated local developments of any lasting importance, unlike, for instance, the acceptance of Style I among the Alamanni and Langobards. The two groups of bracteates will be broadly contemporaneous, and in many cases the origin of specific pieces will be debatable.

Figure 8.12 records the datable grave finds of bracteates of Scandinavian make or tradition in England and on the Continent. The diagram includes 56 bracteates and 3 examples of *Preßblech* (stamped foil) made from bracteate dies, as well as 2 loops possibly from bracteates. 42 different dies are represented. Generally I have used other authors' datings of the finds, for the Continental finds

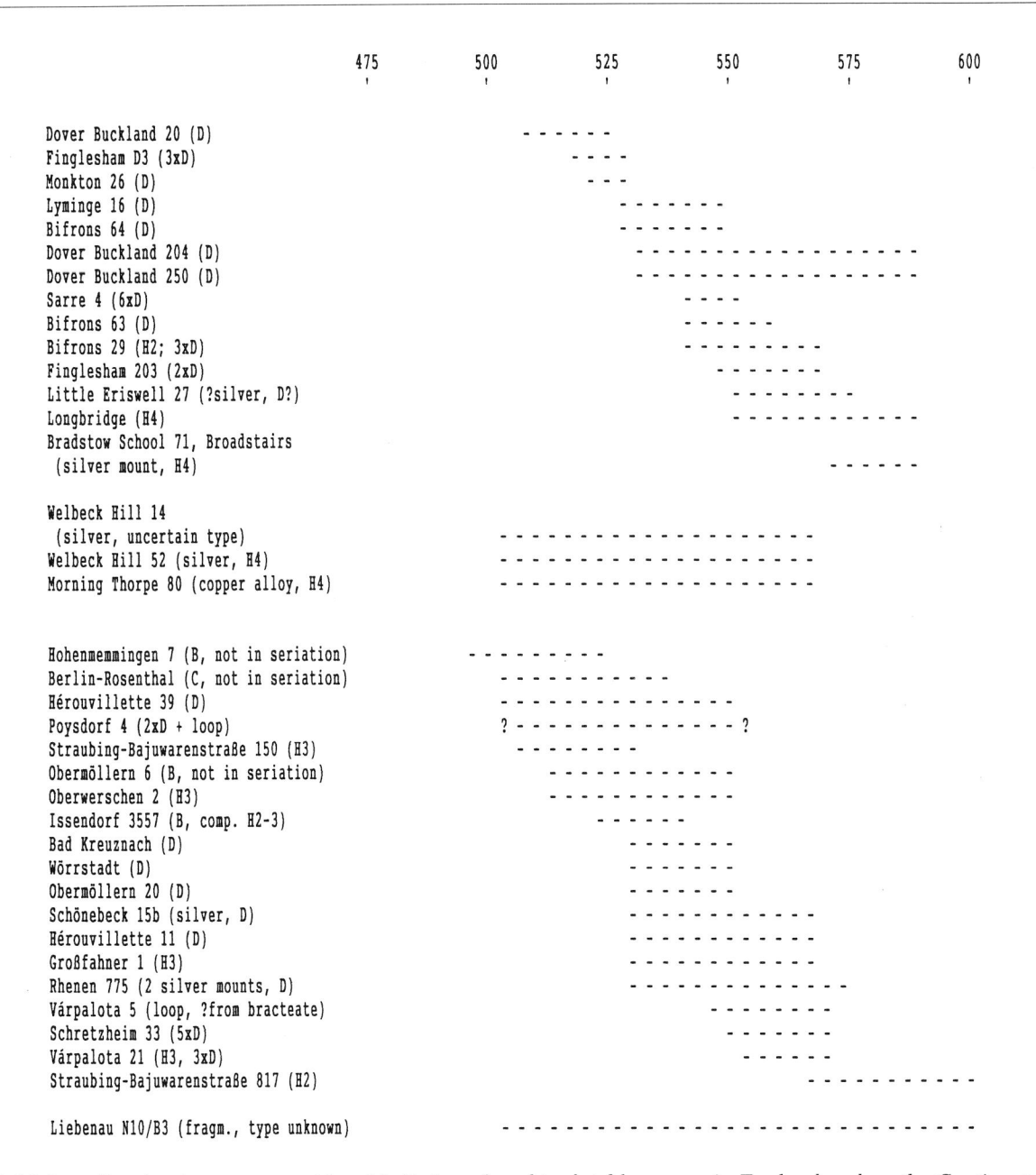

```
                           475      500      525      550      575      600
                            |        |        |        |        |        |
     Dover Buckland 20 (D)                    - - - - - -
     Finglesham D3 (3xD)                      - - - -
     Monkton 26 (D)                           - - -
     Lyminge 16 (D)                              - - - - - - -
     Bifrons 64 (D)                              - - - - - - -
     Dover Buckland 204 (D)                      - - - - - - - - - - - - - - - - - -
     Dover Buckland 250 (D)                      - - - - - - - - - - - - - - - - - -
     Sarre 4 (6xD)                                  - - - -
     Bifrons 63 (D)                                 - - - - - -
     Bifrons 29 (H2; 3xD)                           - - - - - - - - -
     Finglesham 203 (2xD)                           - - - - - - -
     Little Eriswell 27 (?silver, D?)               - - - - - - - - -
     Longbridge (H4)                                - - - - - - - - - -
     Bradstow School 71, Broadstairs
       (silver mount, H4)                              - - - - - -

     Welbeck Hill 14
       (silver, uncertain type)            - - - - - - - - - - - - - - - - - - - - -
     Welbeck Hill 52 (silver, H4)          - - - - - - - - - - - - - - - - - - - - -
     Morning Thorpe 80 (copper alloy, H4)  - - - - - - - - - - - - - - - - - - - - -


     Hohenmemmingen 7 (B, not in seriation)    - - - - - - - - - -
     Berlin-Rosenthal (C, not in seriation)    - - - - - - - - - -
     Hérouvillette 39 (D)                      - - - - - - - - - - - - - -
     Poysdorf 4 (2xD + loop)                 ? - - - - - - - - - - - - - - - ?
     Straubing-Bajuwarenstraße 150 (H3)        - - - - - - - -
     Obermöllern 6 (B, not in seriation)       - - - - - - - - - - -
     Oberwerschen 2 (H3)                       - - - - - - - - - - -
     Issendorf 3557 (B, comp. H2-3)                   - - - - - -
     Bad Kreuznach (D)                                - - - - - -
     Wörrstadt (D)                                    - - - - - -
     Obermöllern 20 (D)                               - - - - - -
     Schönebeck 15b (silver, D)                       - - - - - - - - - - -
     Hérouvillette 11 (D)                             - - - - - - - - - - -
     Großfahner 1 (H3)                                - - - - - - - - - - -
     Rhenen 775 (2 silver mounts, D)                     - - - - - - - - - - - - -
     Várpalota 5 (loop, ?from bracteate)                    - - - - - - - -
     Schretzheim 33 (5xD)                                   - - - - - - -
     Várpalota 21 (H3, 3xD)                                    - - - - -
     Straubing-Bajuwarenstraße 817 (H2)                           - - - - - - - - -

     Liebenau N10/B3 (fragm., type unknown)    - - - - - - - - - - - - - - - - - - - - - - - - -
```

Fig. 8.12 Scandinavian bracteates and local imitations found in datable graves in England and on the Continent. For details, see text and Appendix C.

Skandinavische Brakteaten und lokale Imitationen aus datierbaren Gräbern in England und auf dem Kontinent (vgl. Anhang C).

adjusted with information kindly provided by Claudia Theune; references and datings can be found in Appendix C. When indicated in the literature I have used dates of deposition, without considering the degree of wear etc. This must be kept in mind when comparing with my bracteate chronology, as the latter is fundamentally a typologically based chronology of production. For each find, it is indicated which group the bracteates are referred to in my seriation or if they are of D-type.

Although the datings of the finds are indicated in absolute terms, the general chronological framework is Ament's system (1977), which many finds are directly or indirectly referred to in the literature. Several English finds have only been dated '6th century' or the like (Welbeck Hill etc.). They have therefore been given proportionate room in figure 8.12. The same applies to the Continental finds not attributed to just one specific phase (Rhenen 775, Liebenau N 10/B 3, Hérouvillette 39).

It is evident from figure 8.12 that most of the datable finds outside Scandinavia belong to AM II. Only a few finds may be earlier (hardly by much) and very few later. The bulk of the finds are D-bracteates (38 bracteates and 2 *Preßblech*, from 27 different dies), but B- and C-types do occur with 10 bracteates from 8 dies, and 6 bracteates and 1 *Preßblech* from 6 dies, respectively.

It is no surprise that B- and C-bracteates from the small Group H1 are absent, but perhaps noteworthy that only 2 dies belong to Group H2, while H3 and even the rather small H4 each are represented by 4 dies – perhaps another indication that the latter two groups are more or less parallel with the D-bracteates.

Thus the bracteate influx is being recorded almost exclusively in AM II-graves, including finds dated rather late in the period. This makes it difficult for me to believe that their production ceased in Scandinavia before 520/30 or even 520, as is to be inferred from Lars Jørgensen's dating of the transition Migration/Vendel Period (520/30: Jørgensen 1988, 20 and 1990 Fig. 41 p. 51; 520: Jørgensen 1989, 181 and 1994, 533 with Abb. 122–124). And if Martin's revision of the transition AM I/II is accepted (Martin 1989), would that imply a date of c. 510 instead of 520/30 for the end of the Migration Period in Scandinavia?

I have not been able to see whether the datings for the transition in Scandinavia are to be regarded as dates of production or dates of deposition. I would expect the difference to be no more than at most 20 years or so on average and would find it difficult to see the majority of the bracteate finds outside Scandinavia as long-treasured heirlooms, exported only in the final phase of bracteate production, or as local derivatives.

On the other hand the English and Continental deposition of bracteates of Scandinavian make or tradition seems, by and large, to end in the period around 550/60. This would make an end of production at say 530/40 acceptable and thus accord with the datings of Arrhenius (1983) and Høilund Nielsen (1987) for the beginning of the Vendel Period proper.

Acknowledgements

Thanks to Deutsche Forschungsgemeinschaft, the Danish Research Council for the Humanities and the National Museum of Denmark for financing my work; to Claudia Theune and Birte Brugmann for information on their dating of Continental and English finds; and to Cathy Haith for information on the Dover-Buckland finds.

DEUTSCHE ZUSAMMENFASSUNG

Für den Auswertungsband von 'Die Goldbrakteaten der Völkerwanderungszeit' habe ich eine innere Chronologie der A-, B- und C-Brakteaten erstellt, und zwar mit Hilfe von Korrespondenzanalysen und Kombinationsdiagrammen der Details der großen menschlichen Häupter. Dadurch konnten vier typologische Gruppen H1–H4 umrissen werden (Abb. 8.6–8.9). Diese sind zwar ein typologisches Konstrukt, geben aber sicherlich die allgemeine stilistische Entwicklung zuverlässig wieder. Die Brakteatenproduktion begann in Südskandinavien, während die ältesten norwegischen Brakteaten erst der Gruppe H2 angehören (Abb. 8.11). Die späten A-, B- und C-Brakteaten sind dagegen in Norwegen relativ häufiger als in Südskandinavien. Die D-Brakteaten setzen nach Fundkombinationen und stilistischen Überlegungen während der Gruppe H3 ein.

Eine absolute Datierung des Beginns der Brakteatenproduktion bleibt wegen des Mangels an Grabfunden und der allgemeinen Probleme der skandinavischen Chronologie schwierig. Sie kann mittels stilistischer Vergleiche mit dem frühen Stil I sowie einiger Münzdatierungen auf die Zeit um oder nach 450 n.Chr. geschätzt werden. Späte Brakteaten kommen in englischen und kontinentalen Gräbern zahlreich vor, wo sie mittels der lokalen Chronologien in die erste Hälfte des 6. Jahrhunderts eingeordnet werden können, einige sogar noch später (Abb. 12 mit Anhang C). Dadurch werden nicht nur die Brakteaten datiert, sondern zugleich wird ein Beitrag zur absoluten Datierung der späten Völkerwanderungszeit Skandinaviens gewonnen. In den letzten Jahren wurde der Übergang zur Vendelzeit in die Zeit um 520/530 n.Chr. angesetzt. Dies erscheint nach der Brakteatenchronologie als zu früh; eher liegt dieser Übergang im zweiten Drittel des 6. Jahrhunderts.

APPENDICES

Appendix A. Descriptions of the types in figure 8.2, with the German equivalents of the type names

The types are listed in seriated order as in the diagram, figure 8.2. For each type the German name, an English translation, references to examples in figures 8.6–8.9 and in some cases also to type-specimens in IK are given, together with a short description in most cases. Definitions and discussions of problematic pieces must be found in Hauck et al. (1985-) vol. 4.

Type-name, with description if appropriate	German type-name
B-shaped ear (Fig. 8.6 nos. 1 left, 3).	B-förmiges Ohr
Central jewel in diadem (Fig. 8.6 nos. 1–2,4; IK 59, 240). The frontal part of the diadem is specially emphasized, e.g. with a rosette or extra beads, which must, however, be an integral part of the diadem.	Stirnjuwel im Diadem
Multiple-string diadem (Fig. Fig. 8.6 nos. 2,4,5; Fig. 8.7 no. 7). Two or more rows of beads; no contour lines.	Diadem: Mehrfache Perlreihen
Single-string diadem (Fig. 8.6 no. 3; IK 45, 299). Single row of beads; no contour lines.	Diadem: Perlreihe ohne Kontur

Type-name, with description if appropriate	German type-name
"De luxe" diadem (Fig. 8.7 nos. 1,4; IK 54). The diadem is decorated with circles, crosses, leaves, etc., and may have contour lines.	Prachtdiadem
Rounded hair-style (Fig. 8.6 nos. 1–6; Fig. 8.7 nos. 2,3,6 left). Hair at the back may be indicated, but not turned upwards.	Kalottenförmige Frisur
Comma-shaped ear (Fig. 8.6 nos. 2,4; Fig. 8.7 nos. 2, 4,7,12). C-shaped, but not symmetrical around the horizontal axis.	Kommaförmiges Ohr
Dotted hair contour alone (Fig. 8.6 no. 6; Fig. 8.7 nos. 1, 4,5). No contour line.	Haar: Perlsaum allein
Crescent-filled hair-style (IK 62,1).	Frisur mit Bogen-Füllung
Hair-style with (3 or more) alternately hatched sections (Fig. 8.7 nos. 1,4,8; IK 89, 186).	Strähnenwechsel
Linear diadem dividing hair (Fig. 8.6 no. 1; 7 no. 6; IK 33, 251). No indication of beads. Hair signature is found both over and under (at least) the front part of the diadem.	Strichdiadem im Haar
Eyebrow (Fig. 8.7 nos. 6,9,10; IK 50, 140, 376, 2). Line or accentuated relief area over the eye.	Augenbraue
Plait (Fig. 8.7 nos. 5,10; IK 49, 58).	Zopf
C-shaped ear (Fig. 8.6 no. 6; IK 49, 159). Symmetrical around the horizontal axis. The ends may curve slightly back.	C-förmiges Ohr
Flat relief as general impression of the head (IK 233).	Flaches Relief
Pretzel-shaped ear (IK 58, 349).	Brezelförmiges Ohr
Cock'd hat (IK 348). The shape of the hair-style is determined by the diadem with two loop-shaped ends.	Admiralshut
Decorated top of hair (IK 57,2).	Frisur mit Federrand
Extended eye contour line (Fig. 8.6 no. 2; Fig. 8.7 nos. 5,11; Fig. 8.8 no. 1). The rear corner of the eye is prolonged.	Lidstrich
Oval ear (Fig. 8.6 no. 1 right; Fig. 8.7 nos. 10,11; Fig. 8.8 no. 5). Oval-, crescent- or kidney-shaped ear, drawn with a closed contour line. Infills of different shapes may occur.	Ovales Ohr

Type-name, with description if appropriate	German type-name
Hatched hair-style (e.g. Fig. 8.6 nos. 1–6; IK 57,2, 59, 68). Hair shown with straight, curved or angled lines rising vertically or radially above the face. Excluded are dies with *Sweeping hair* or *Horizontal hair-style*.	Haarsträhnen
Oval eye with pupil (e.g. Fig. 8.6 nos. 1,2,4; Fig. 8.7 nos. 1,2,4–12). Eye with an oval or pointed oval contour line and marked pupil or eyeball.	Auge: Oval mit Pupille
Animal/bird head at nape (Fig. 8.7 nos. 6–8; Fig. 8.8 no. 12). The hair-style ends with an animal's, snake's or bird's head.	Tierkopf im Nacken
Dots outside the hair contour line (Fig. 8.7 nos. 11,12; Fig. 8.8 no. 1; IK 13,1, 93).	Haar: Äußerer Perlsaum
Contoured beaded diadem dividing hair (Fig. 8.7 no. 2; Fig. 8.8 no. 11; IK 48). The diadem is shown as a string of beads with a contour line above and below it. There must be a hair signature below the whole length of the diadem.	Diadem: Perlreihe mit Kontur im Haar
D-shaped ear (Fig. 8.8 nos. 1,7; IK 37, 330). The ear is shown as one or more concentric crescents resting at both ends on the contour line of the head.	D-förmiges Ohr
Spiral-shaped ear (Fig. 8.6 no. 5; Fig. 8.7 nos. 5,6; Fig. 8.8 nos. 6,8). Curved or spiral-shaped line, one end of which is based on the contour of the head while the other end is free.	Schnörkelförmiges Ohr
Upturned hair-style (Fig. 8.7 nos. 6–8,11; Fig. 8.8 nos. 1,12). The hair is bent up at the nape, but not so far that it touches the top of the head or its contour line.	Aufgebogenes Nackenhaar
Volute-shaped ear (Fig. 8.8 nos. 3,11; IK 45, 380). C-shaped, symmetrical around the horizontal axis, with ends strongly curved.	Volutenförmiges Ohr
Half-mask (Fig. 8.8 nos. 3,5; IK 84, 110). Both front and rear corner of the eye are prolonged, and each of the ensuing lines ends at a contour line.	Maskenbinde
Dots inside the hair contour line (Fig. 8.7 no. 10; Fig. 8.8 no. 3; IK 34, 75,3).	Haar: Innerer Perlsaum
Contoured beaded diadem below hair (Fig. 8.7 no. 8; Fig. 8.8 nos. 6,7,10,12). Placed between face and hair, at least in the front half of the head.	Diadem: Perlreihe mit Kontur unter dem Haar
Smooth hair-style (IK 88). The interior of the hair-style shows neither hatching nor relief. Beads etc. may occur.	Haar: Glatte Fläche

Type-name, with description if appropriate	German type-name
High relief as general impression of the head, but without the ridges of chip-carving (IK 264, 276).	Hohes Relief
Linear diadem below hair (Fig. 8.8 no. 9; IK 15, 158). Two or more parallel lines dividing the face from the hair, in some cases continuing under the back hair. No indication of beads.	Strichdiadem unter dem Haar
Hair-style with contour line (e.g. Fig. 8.6 no. 3; Fig. 8.7 nos. 3,6–12). Is also recorded where there is an additional beaded contour.	Haar: Konturlinie
Contour around face, as a contour line or a row of beads (e.g. Fig. 8.7 nos. 3–6,7–11).	Haupt: Kontur
Sweeping hair (Fig. 8.8 nos. 6,8; IK 37, 63, 313). The lines of the hair rise vertically from the brow and run parallel from there to the tip of the hair-style.	Geschwungene Haarsträhnen
Dotted lower eyelid (Fig. 8.8 nos. 10,11; IK 72,1, 296,1).	Punktiertes Unterlid
Triangular ear (IK 120,1).	Dreieckiges Ohr
Open oval-shaped eye (Fig. 8.8 nos. 5,10; Fig. 8.9 nos. 1,2). Oval or pointed oval contour line with no indication of eye-ball or pupil.	Offenes Oval-Auge
Framed nose/eyebrow-curve (Fig. 8.8 no. 12; IK 191, 383, 386). Nose and eyebrow are drawn as one continuous relief line, framed by a contour line.	Nasen/Augenbrauen-Bogen

Type-name, with description if appropriate	German type-name
Knotted hair-style (Fig. 8.7 nos. 1,4,9,12; Fig. 8.8 nos. 5–10; Fig. 8.9 nos. 1,2,4–6). The hair is bent so far up at the nape that it touches or crosses the hair-style or its contour line.	Eingerolltes Haar
Breath at mouth (Fig. 8.6 no. 4; Fig. 8.7 nos. 4,7,12; Fig. 8.8 nos. 6,11; Fig. 8.9 no. 4).	Mundatem
Lower eyelid (Fig. 8.8 nos. 7,12; Fig. 8.9 nos. 4,6).	Wangenbogen
Horizontal hair-style (Fig. 8.9 no. 4; IK 170). The lines of the hair start horizontally over the brow.	Waagerechte Haarsträhnen
Breath at nose (Fig. 8.8 no. 11; Fig. 8.9 nos. 3,6; IK 116, 327).	Nasenatem
Animal/bird head at forehead (Fig. 8.9 no. 6; IK 180, 327).	Protome
Triangular eye (Fig. 8.9 nos. 4–6; IK 119a, 201, 210). Two sides of the eye are an angle formed by the nose- and hair contours.	Dreiseitiges Auge
Chip-carving with sharp ridges as general impression of the head (IK 103, 151).	Kerbschnitt
Relief hair-style (Fig. 8.8 no. 12; Fig. 8.9 nos. 2,5–6). The interior of the hair-style is rounded or ridged relief.	Haar: Relieffüllung

Appendix B. Die-links etc. in figure 8.5.

a. IK 162,2 Darum-A (M 3:12; H1)
   IK 162,1 Skonager-A (M 3:13; early H2)
   After the production of the Darum bracteate the die was reworked and a bird's head added at the nape.

b. IK 162,2 Darum-A (M 3:12; H1)
   IK 43 Darum-C (IK 43; H2)
   Linked through the dot-and-circle punch.

c. IK 49 Espelund-C (M 15:12; H3)
   IK 60 Furulid-C (M 15:26; H2)
   The same volute-shaped punch has been used in the border zones. The loops may be also tool-linked (Axboe 1982, 37f.).

d. IK 279 Holmetorp-A (M 4:18; H2)
   IK 144,1 Ravlunda-C (same die as M 11:3; H2)
   Both the volute-shaped and the two hatched triangle punches occur on both bracteates.

f. IK 312,1 Overhornbæk-A (M 3:13; H1)
   IK 140 Overhornbæk-C (M 6:23; H2)
   IK 110 Lindkær-C (A 86a; H3)
   Linked through the unique runic borders with bird's head terminations. The two C-bracteates appear at first sight to be mirrored versions of the same motif, but the Lindkær piece has a diminutive Protome which puts it in a late place in the seriation. The A-bracteate is an obvious candidate for fresh inspiration from a Roman coin.

g. IK 119b Madla-C (loop, A 157/1; H3)
   IK 119a Madla-C (M 13:9; H4)
   The loop of IK 119a was made from the picture field of another C-bracteate (IK 199b).

Appendix C. References to the datings used in figure 8.12.
The finds are listed in alphabetical order for England and the Continent respectively. Bracteates are referred to with their IK-number, type and, if possible, their group in my seriation.

Theune: Datings according to the chronology for south-west Germany kindly supplied by Claudia Theune (pers. comm. 16.4.1996 + 28.11.1996; cf. Roth and Theune 1988). However, for Straubing-Bajuwarenstraße 817 and Várpalota 21 I have not used the datings offered by Theune as they were based mainly on the beads found in the graves. The chronology of Merovingian beads seems still to be under construction, with both the contents and the datings of the groups currently subject to revision: see Theune-Vogt 1990; 1991; Sasse and Theune 1996).

A revised version of this survey of Continental finds will appear in Hauck et al. (1985–) vol. 4.

England
Bifrons 29 (M 311)
IK 23-B (H2), extremely worn. IK 410-D, 2 ex., both worn. IK 412,2-D, very worn; die-linked to Bifrons 64.
Deposited around the middle of the 6th century or possibly later (Hawkes and Pollard 1981, 346).
Brugmann, this vol.: Kentish Phase III.

Bifrons 63 (M 312)
IK 411-D, hardly worn.
Probably buried around the middle of the 6th century (Hawkes and Pollard 1981, 347).

Bifrons 64 (M 313)
IK 412,1-D, slightly worn; die-linked to Bifrons 29.
Deposited nearer c. 530 than 550 (Hawkes and Pollard 1981, 347).

Bradstow School, Broadstairs, grave 71 (A 314b)
IK 224: Silver mount with C-bracteate design (H4); hardly worn.
Weapon grave with a Merovingian *tremissis* (A.D. c. 575) as Charon's obol.

Dover Buckland 20 (A 314d)
IK 421-D, very worn.
Bakka dates the grave to late Böhner II (1981, 24, 27). Evison 1987, 137: Late Phase 1 (475–525).

Dover Buckland 204
IK 580-D, worn (according to photo).
Kentish Phase IV or perhaps III (Preliminary dating, based on information from Cathy Haith, British Museum).

Dover Buckland 250
IK 582-D, very worn (according to photo).
Kentish Phase III or IV (Preliminary dating, based on information from Cathy Haith, British Museum).

Finglesham D3 (M 314)
IK 425-D, only slightly worn. IK 426,1-D, 2 ex., according to S. Hawkes both considerably worn; die-linked to Finglesham 203.
The deposition of the grave is dated c. 520–30 (Hawkes and Pollard 1981, 331f.; Bakka 1981, 27: Late Böhner II; Hines 1984, 21).

Finglesham 203 (A 314a)
IK 426,2-D, 2 ex., according to Hawkes and Pollard 1981, 356f. moderately worn; die-linked to Finglesham D3.
Possibly deposited not long after c. 550 (Hawkes and Pollard 1981, 338f.
Brugmann, this vol.: Kentish Phase IV/V.

Little Eriswell 27 (A 307a)
IK 293-D?. Vierck recorded fragments of a silver D-bracteate in the grave.
Probably buried after 550 (Hines 1984, 217).

Longbridge (M 306)
IK 114-C (H4), worn. Punch-linked with the associated disc pendant and silver armlet.
Hardly buried before c. 550, but on the other hand earlier than the end of the 6th century (Hines 1984, 214).

Lyminge 16 (A 314e)
IK 462-D, moderately worn.
Possibly second quarter of the 6th century (Hawkes and Pollard 1981, 347).

Monkton 26
IK 467-D, very worn.
Weapon grave, dated by Sonia Hawkes to c. 520/30 (Perkins and Hawkes 1984, 102ff.).

Morningthorpe 80 (A 307f)
IK 306-C: Two C-bracteates of copper alloy. Both damaged, but hardly markedly worn.
6th century (Hines 1984, 214).

Sarre 4 (M 309)
IK 493-D, 3 ex., all slightly worn. IK 494-D, hardly worn. IK 495-D, hardly worn. IK 496-D, distinctly worn.
Possibly deposited c. 540–550 (Hawkes and Pollard 1981, 350).
Brugmann, this vol.: Kentish Phase III

Welbeck Hill 14 (A 305c)
IK 388, of uncertain type, silver; damaged, but also worn.

Welbeck Hill 52 (A 305d)
IK 387-C (H4), silver; very worn.
Both graves probably 6th century (Hines 1984, 218).

Continent
Bad Kreuznach (A 334c)
IK 408-D, worn.
Böhner IIIa (AM II; Clauß 1978, 136; Bakka 1981, 23).
Theune: Radiate brooches SW II D (530–550).

Berlin-Rosenthal (A 329a)
IK 322-C, not in seriation; only slightly worn.
Schmidt: Gruppe IIIa (AM II; 1961, 130).
Theune: The Thuringian *Zangenfibel* is typologically earlier than those in Schretzheim 197 and 219 (Schretzheim Stufe 1, 525/30–545/50; SW I C2 (510–530). Alignment with Werner's *norddanubische Phase* (489–526/57) seems reasonable.

Großfahner 1
IK 259-B: 3 bracteates, showing a woman en face (H3).
Schmidt: Late group IIIa or early IIIb (1961, 127).
Theune: Square-headed brooch SW II D (E) (530–550 (-570)).
Cfr. Nieveler and Siegmund, this vol., Rheinland Phase 4 (520/30–550/60).

Hérouvillette 11 (A 315b)
IK 440-D, worn.
Middle third of the 6th century (Hawkes and Pollard 1981, 347).

Hérouvillette 39 (A 315c)
IK 492,2-D; extremely worn; die-linked to IK 492,1 Sarre 90 and IK 492,3 Kent.
First half of the 6th century (Hawkes and Pollard 1981, 343).

Hohenmemmingen 7 (A 335a)
IK 278-B (not seriated), only moderately worn.
The grave contained two S-shaped brooches of Werner's Poysdorf-type, which is considered diagnostic for his *norddanubische* phase (489–526/27; Werner 1962, 77f.). According to information from the excavator W. Kettner the brooches were only very slightly worn.
Theune: Possibly c. 500/early 6th-century. The S-shaped brooches have parallels in Weingarten 509 which, unfortunately, has too few types to be included in the Weingarten seriation. Both the brooches and the beads in the Weingarten grave would fit well into SW I C1 (490–510), and so would the string of beads including some large specimens in Hohenhemmingen 7.

Issendorf 3557
IK 574-B, showing a man en face; worn. Not in seriation; the human face can however be compared to the en face-bracteates of the Oberwerschen type, placed in Group H2 and H3.
According to Häßler the grave was constructed c. 520–40 (Häßler et al. 1997, 116).

Liebenau N10/B3 (A 324a)
IK 292: Bracteate fragment with rim and punched border; uncertain type, worn.
6th century (Häßler et al. 1997, 115)

Obermöllern 6 (M 332)
IK 132-B, not in seriation; very worn.
Schmidt: Gruppe IIIa (525–560), possibly early (1961, 14, 122, 130).
Theune: Kidney-shaped buckle, brooches, beads SW I C2 / II D (510–530/550).

Obermöllern 20 (M 333)
IK 477-D, moderately worn.
Schmidt: Gruppe IIIa (525–560; 1961, 139).
An identical Thuringian brooch was found in Schretzheim and a pair of similar small garnet disc brooches in Schretzheim 197 combined with two *Zangenfibeln*; both graves are dated to Stufe 1 (525/35–545/50; Koch 1977, 16).
Theune: Thuringian brooches, small garnet disc brooches SW II D (530–550).

Oberwerschen 2
IK 311-B, showing a woman en face (H3), worn.
Schmidt: First half to middle of 6th century (1961, 122, 139. The type of bow brooches found in the grave is placed in Gruppe IIIa Abb. 49 p. 90).

Häßler dates the grave to Schmidt Gruppe IIIa (Häßler et al. 1997, 116).
The two animal brooches (Type Herpes) have a parallel in Schretzheim 472, dated to Stufe 1 (525/35–545/50; Koch 1977, 16).
Theune: Animal brooches, silver pin, bow brooches SW I C2 / SW II D (510–530/550).

Poysdorf 4 (A 335c)
IK 484-D: Two unusually naturalistic D-bracteates + a loop for bracteate or pendant.
Early 6th century (Werner 1962, 58f., 75).
Theune: According to brooch and pot possibly SW II (530–570).
An earlier date might, however, be indicated by the combination in Alternerding 319 (Sage 1984) of a similar brooch with two *Fünfknopffibeln*, which Roth and Theune date to SW I C1–2 (490–530; 1988, 30 Type 18).

Rhenen 775 (A 319a)
IK 486-D: Two silver mounts with D-bracteate design. Both damaged, but hardly worn.
The grave is only rather broadly dated to Böhner III; the shield boss is said to belong „probably to the middle to third quarter of the 6th century" (Ypey 1983, 471).

Schönebeck 15b (A 331a)
IK 497-D silver; damaged but apparently only slightly worn.
Schmidt refers the grave to his Gruppe IIIa (1961, 139); it is not evident how far this dating is based on the bracteate.
Theune: Buckle with thickened tongue base, beads of Combination Group C: c. SW II D (E) (530–550 (-570)).

Schretzheim 33 (M 335)
IK 500-D; 5 D-bracteates, all very worn.
Schretzheim Stufe 2 (545/50–565/70; Koch 1977, 18–21).
Theune: SW II E (550–570).

Straubing-Bajuwarenstraße 150
IK 347-B (H3), rather worn.
Dated to the Irlmauth phase, possibly around the beginning of the 6th century (Geisler and Hauck 1987, 132).
Theune: 2 *Vierpaßfibeln*, 2 *Fünfknopffibeln*, D-shaped buckle SW I C2 (510–30).

Straubing-Bajuwarenstraße 817
IK 348-C (H2), extremely worn.
Second half 6th century, cf. Schretzheim Stufe 3 (565–590/600; Geisler and Hauck 1987, 138f.).

Várpalota 5 (A 336b)
Loop from gold bracteate or pendant.
c. 530? (Werner 1962, 45).
Deposited 546–68 (Martin 1989, 134).
Theune: 2 bow brooches, 2 garnet disc brooches, shield-on-tonge buckle SW II E (550–570).

Várpalota 21 (A 336c)
IK 206-B (H3). IK 559-D: 3 D-bracteates.
Possibly deposited before 568 (Werner 1962, 36f., 45f.).

Wörrstadt (A 334b)
IK 566-D, very worn.
Brooches similar to Schretzheim Stufe 1 (525/35–545/50; Clauß 1978, 135f.).
Theune: 2 small garnet disc brooches, 2 bow brooches, shield-on-tongue buckle SW II D (530–550).

ABBREVIATIONS

A: Axboe 1982 (Catalogue numbers)
IK: Hauck et al. (Catalogue numbers)
M: Mackeprang 1952 (Plate numbers: M 8:13, etc.; catalogue numbers: M 314, etc.)

BIBLIOGRAPHY

Ament, H. 1977. 'Zur archäologischen Periodisierung der Merowingerzeit'. *Germania* 55, 133–40.

Arrhenius, B. 1983. 'The chronology of the Vendel graves'. In J. P. Lamm and H.-Å. Nordström (eds), *Vendel Period Studies*, 39–70. Stockholm.

Axboe, M. 1982. 'The Scandinavian Gold Bracteates. Studies on their manufacture and regional variations. With a supplement to the catalogue of Mogens B. Mackeprang'. *Acta Archaeol* 52, 1–100.

Axboe, M. 1987. 'Die Brakteaten von Gudme II'. *Frühmittelalterliche Stud* 21, 76–81.

Axboe, M. 1992. 'Skizze der Forschungsgeschichte zur Chronologie der Goldbrakteaten'. In K. Hauck (ed.), *Der historische Horizont der Götterbild-Amulette aus der Übergangsepoche von der Spätantike zum Frühmittelalter*, 103–10. Göttingen.

Axboe, M. 1993a. 'Gold bracteates and correspondence analysis'. In J. Andresen, T. Madsen and I. Scollar (eds), *Computing the Past. Computer Applications and Quantitative Methods in Archaeology*, 333–42. Århus.

Axboe, M. 1993b. 'Korrespondensanalyse og kombinationsdiagrammer – en nyttig kombination'. *KARK Nyhedsbrev* 1993 Nr. 2, 7–12.

Axboe, M. forthcoming. 'Die innere Chronologie der A-C-Brakteaten und ihrer Inschriften'. In K. Düwel (ed), *Runeninschriften als Quellen interdisziplinärer Forschung*, 231–52. Berlin (1998).

Axboe, M. and Hauck, K. 1985. 'Hohenmemmingen-B, ein Schlüsselstück der Brakteatenikonographie' (Zur Ikonologie der Goldbrakteaten XXXI). *Frühmittelalterliche Stud* 19, 98–130.

Axboe, M. and v. Padberg, L. 1977. 'Ein C-Brakteat als Ösenröhre'. *Acta Archaeol* 48, 239–42.

Bakka, E. 1973. 'Goldbrakteaten in norwegischen Grabfunden: Datierungsfragen'. *Frühmittelalterliche Stud* 7, 53–87.

Bakka, E. 1981. 'Scandinavian-type gold bracteates in Kentish and continental grave finds'. In V. I. Evison (ed), *Angles, Saxons, and Jutes. Essays presented to J. N. L. Myres*, 11–35. Oxford.

Bemmann, J. and Hahne, G. 1994. 'Waffenführende Grabinventare der jüngeren römischen Kaiserzeit und Völkerwanderungszeit in Skandinavien'. *Berichte Römisch-Germanische Komm* 75, 283–640.

Benner Larsen, E. 1982–83. 'Værktøjsspor/På sporet af værktøj. Identifikation og dokumentation af værktøjsspor, belyst ved punselornamenterede genstande fra Sejlflod'. *Kuml* 1982–83, 169–80.

Benner Larsen, E. 1985 'The Gundestrup Cauldron. Identification of Tool Traces'. *ISKOS* 5, 561–74.

Benner Larsen, E. 1987. 'SEM-Identification of Tool Marks and Surface Textures on the Gundestrup Cauldron'. In J. Black (ed.), *Recent Advances in the Conservation and Analysis of Artifacts*, 393–408. London.

Clauß, G. 1978. 'Ein neuer Grabfund mit nordischem Goldbrakteaten aus Wörrstadt, Kr. Alzey-Worms'. *Archäol Korrespondenzblatt* 8, 133–40.

Evison, V. I. 1987. *Dover: The Buckland Anglo-Saxon Cemetery*. London.

Fagerlie, J. M. 1967. *Late Roman and Byzantine Solidi Found in Sweden and Denmark*. Numismatic Notes and Monographs 157. New York.

Fonnesbech-Sandberg, E. 1989. 'Münzfunktionen in der Kaiserzeit und Völkerwanderungszeit Dänemarks'. *Frühmittelalterliche Stud* 23, 420–52.

Geisler, H. and Hauck, K. 1987. 'Zwei Frauengräber von Straubing-Bajuwarenstraße mit Goldbrakteaten aus dem Norden' (Zur Ikonologie der Goldbrakteaten XXXIV). *Frühmittelalterliche Stud* 21, 124–46.

Haseloff, G. 1974a. 'Salin's Style I'. *Medieval Archaeol* XVIII, 1–15.

Haseloff, G. 1974b. 'Der germanische Tierstil. Scine Anfänge und der Beitrag der Langobarden'. In *Atti del Convegno Internazionale sul tema: „La civiltà dei Longobardi in Europa" 1971*, 361–86. Accademia Nazionale dei Lincei, Quaderno N. 189. Roma.

Haseloff, G. 1981. *Die germanische Tierornamentik der Völkerwanderungszeit. Studien zu Salin's Style I*. 3 vols. Berlin.

Haseloff, G. 1984. 'Stand der Forschung: Stilgeschichte Völkerwanderungs- und Merowingerzeit'. In *Festskrift til Thorleif Sjøvold på 70-årsdagen*, 109–124. Universitetets Oldsaksamlings Skrifter, Ny rekke 5. Oslo.

Haseloff, G. 1986. 'Bild und Motiv im Nydam-Stil und Stil I'. In H. Roth (ed.), *Zum Problem der Deutung frühmittelalterlicher Bildinhalte*, 67–110. Sigmaringen.

Häßler, H.-J., Axboe, M., Hauck, K. and Heizmann, W. 1997. 'Ein neues Problemstück der Brakteatenikonographie: Issendorf-B, Landkreis Stade, Niedersachsen' (Zur Ikonologie der Goldbrakteaten LIV). *Stud zur Sachsenforschung* 10, 101–75.

Hauck, K. 1994a. 'Gudme als Kultort und seine Rolle beim Austausch von Bildformularen der Goldbrakteaten' (Zur Ikonologie der Goldbrakteaten L). In P. O. Nielsen, K. Randsborg and H. Thrane (eds), *The Archaeology of Gudme and Lundeborg*, 78–88. Copenhagen.

Hauck, K. 1994b. 'Altuppsalas Polytheismus exemplarisch erhellt mit Bildzeugnissen des 5.-7. Jahrhunderts' (Zur Ikonologie der Goldbrakteaten LIII). In H. Uecker (ed.), *Studien zum Altgermanischen. Festschrift für Heinrick Beck*, 197–302. Hoops Reallexikon Ergänzungsbände 11. Berlin.

Hauck et al. 1985-. M. Axboe, K. Düwel, K. Hauck, L. v. Padberg et al, *Die Goldbrakteaten der Völkerwanderungszeit. Ikonographischer Katalog*. Münstersche Mittelalter-Schriften 24/1-3. Vol. 1,1–3 1985, Vol. 2,1–2 1986, Vol. 3,1–2 1989, Vol. 4, in prep. Munich.

Hawkes, S. C. and Pollard, M. 1981. 'The gold Bracteates from sixth-century Anglo-Saxon graves in Kent, in the light of a new find from Finglesham'. *Frühmittelalterliche Stud* 15, 316–70.

Herzog, I. and Scollar, I. 1987. 'Ein 'Werkzeugkasten' für Seriation und Clusteranalyse'. *Archäol Korrespondenzblatt* 17, 273–9.

Hines, J. 1984. *The Scandinavian Character of Anglian England in the pre-Viking Period*. BAR British Series 124. Oxford.

Hines, J. 1992. 'The Seriation and Chronology of Anglian English Women's Graves: a critical assessment'. In L. Jørgensen (ed.), *Chronological Studies of Anglo-Saxon England, Lombard Italy and Vendel Period Sweden*, 81–93. Copenhagen.

Høilund Nielsen, K. 1987. 'Zur Chronologie der jüngeren germanischen Eisenzeit auf Bornholm. Untersuchungen zu Schmuckgarnituren'. *Acta Archaeol* 57, 47–86.

Jørgensen, L. 1988. 'Family Burial Practices and Inheritance Systems. The Development of an Iron Age Society from 500 BC to AD 1000 on Bornholm, Denmark'. *Acta Archaeol* 58, 17–53.

Jørgensen, L. 1989. 'En kronologi for yngre romersk og ældre germansk jernalder på Bornholm'. In L. Jørgensen (ed.) *Simblegård-Trelleborg. Danske gravfund fra førromersk jernalder til vikingetid*, 168–87. Copenhagen.

Jørgensen, L. 1990. *Bækkegård and Glasergård. Two Cemeteries*

from the Late Iron Age on Bornholm. Copenhagen.
Jørgensen, L. 1994. 'Fibel und Fibeltracht: Römische Kaiserzeit und Völkerwanderungszeit in Skandinavien'. In *Reallexikon der Germanischen Altertumskunde* 8, 523–36. Berlin.
Koch, U. 1977. *Das Reihengräberfeld bei Schretzheim.* Germanische Denkmäler der Völkerwanderungszeit A 13. Berlin.
Kyhlberg, O. 1986. 'Late Roman and Byzantine Solidi. An archaeological analysis of coins and hoards'. In A. Lundström and H. Clarke (eds), *Excavations at Helgö X*, 13–126. Stockholm.
Lund Hansen, U. 1992. 'Die Hortproblematik im Licht der neuen Diskussion zur Chronologie und zur Deutung der Goldschätze in der Völkerwanderungszeit'. In K. Hauck (ed.), *Der historische Horizont der Götterbild-Amulette aus der Übergangsepoche von der Spätantike zum Frühmittelalter*, 183–94. Göttingen.
Mackeprang, M. B. 1952. *De nordiske Guldbrakteater.* Jysk arkæologisk Selskabs Skrifter 2. Århus.
Malmer, M. P. 1963. *Metodproblem inom järnålderns konsthistoria.* Lund.
Malmer, M. P. 1977. 'Chronologie der Solidi und Goldbrakteaten'. In G. Kossack and J. Reichstein (eds), *Archäologische Beiträge zur Chronologie der Völkerwanderungszeit*, 107–11. Bonn.
Martin, M. 1989. 'Bemerkungen zur chronologischen Gliederung der frühen Merowingerzeit'. *Germania* 67, 121–41.
Näsman, U. 1984. 'Zwei Relieffibeln von der Insel Öland'. *Prähistorische Zeitschrift* 59, 48–80.
Nerman, B. 1935. *Die Völkerwanderungszeit Gotlands.* Stockholm.
Nissen Meyer, E. 1934. *Relieffspenner i Norden. Bergens Mus Årb* 1934 Hist-ant rekke 4.
Nørgård Jørgensen, A. 1991. 'Kobbeå Grab 1 – ein reich ausgestattetes Grab der jüngeren germanischen Eisenzeit von Bornholm'. *Stud zur Sachsenforschung* 7, 203–39.
Nørgård Jørgensen, A. 1992. 'Weapon sets in Gotlandic grave finds from 530–800 A.D.: a chronological analysis'. In L. Jørgensen (ed.), *Chronological Studies of Anglo-Saxon England, Lombard Italy and Vendel Period Sweden*, 5–34. Copenhagen.
Ørsnes, M. 1966. *Form og Stil i Sydskandinaviens yngre germanske jernalder.* Copenhagen.
Perkins, D. R. J. and Hawkes, S. C. 1984. 'The Thanet Gas Pipeline Phases I and II (Monkton Parish), 1982'. *Archaeol Cantiana* 101, 83–114.
Ramqvist, P. H. 1992. *Högom. The excavations 1949–1984.* Neumünster.
Roth, H. and Theune, C. 1988. *SW ♀ I–V: Zur Chronologie merowingerzeitlicher Frauengräber in Südwestdeutschland.* Archäologische Informationen aus Baden-Württemberg 6. Stuttgart.
Schmidt, B. 1961. *Die späte Völkerwanderungszeit in Mitteldeutschland.* Halle.
Steuer, H., forthcoming. 'Datierungsprobleme in der Archäologie'. In K. Düwel (ed), *Runeninschriften als Quellen interdisziplinärer Forschung*, 129–49. Berlin (1998).
Theune-Vogt, C. 1990. *Chronologische Ergebnisse zu den Perlen aus dem alamannischen Gräberfeld von Weingarten, Kr. Ravensburg.* Kleine Schriften aus dem Vorgeschichtlichen Seminar 33. Marburg.
Werner, J. 1962. *Die Langobarden in Pannonien. Beiträge zur Kenntnis der langobardischen Bodenfunde vor 568.* Munich.
Wright, R. 1985. 'Detecting pattern in tabled archaeological data by principal components and correspondence analysis: Programs in BASIC for portable computers'. *Science and Archaeol* 27, 35–8.
Ypey, J. 1983. 'Ein Männergrab mit D-Brakteatenbeschlägen des fränkischen Gräberfeldes bei Rhenen, Provinz Utrecht, Niederlande'. *Frühmittelalterliche Stud* 17, 460–78.

9. A peaceful discussion of a martial topic: the chronology of Scandinavian weapon graves

Anne Nørgård Jørgensen

INTRODUCTION

Weapons are well suited to broad-based studies across time and space because, as a natural consequence of their function, they developed under 'international' or – to use a more appropriate term for this period – cross-regional influence. Weapons constitute one of the largest groups of finds from the Iron Age. For more than a century, archaeologists have worked on improving the weapon chronology from shortly before the birth of Christ down to the Viking Period. The results make it possible for archaeology to trace the frequent changes in weapon-types, which in some periods evidently took place at intervals of only a generation or two.

The number and distribution of weapon graves varies across the Iron Age. The men's graves of the Late Germanic Iron Age on Bornholm constitute 95% of the Danish weapon graves from this period but have not hitherto been subjected to relative or absolute chronological analysis. The reason for this can be found in the poor state of preservation of the iron objects that were excavated in the late 19th century and are now kept in the National Museum in Copenhagen (Jørgensen 1990, 45). Only with the excavation of the Nørre Sandegård Vest cemetery was there a new opportunity to typologize the weapons as the basis for an independent weapon chronology. During the publication of the large quantity of finds from the 166 grave groups at Bækkegård, 34 of which were weapon graves, a preliminary sketch of the chronological development of weaponry on Bornholm was made. This was done by using the excavator, Emil Vedel's, drawings of the 'old' weapon finds alongside the newly excavated Nørre Sandegård Vest material (Jørgensen 1990, 44ff.).

The chronology of the Germanic Iron Age has up to now been based exclusively on the richly varied and well-preserved women's articles – first with Ørsnes's system from 1966, followed by Høilund Nielsen's system from 1987, and most recently with the Nørre Sandegård Vest women's graves and the improved fine chronology presented in Nørre Sandegård Vest publication (Ørsnes 1966/1970; Høilund Nielsen 1987; Jørgensen and Nørgård Jørgensen 1997). A prerequisite for further studies of the men's graves from Nørre Sandegård Vest is that the material should be placed in a chronological context that includes the other weapon finds from Denmark of the Late Germanic Iron Age.

The total number of weapon graves from Denmark now comprises about 105 graves, of which 100 are from Bornholm. The weapon grave finds overwhelmingly comprise ordinary, undecorated iron objects that can be regarded as functional weapons before they were used in the funerary ritual and not, as is the case with a large proportion of the weapon finds from Gotland and central Sweden, as both parade and practical weaponry. The weapons of the Germanic Period vary greatly in form, and this must reflect the frequent replacement of weapon-types. Detailed studies have shown that ordinary weapons obtained their form under the indirect influence of weapon-manufacture in other areas outside of Denmark. Saxes, shields and spears, for instance, have super-regional features typical of northern and central Europe in this period. This gives us an excellent opportunity to establish a detailed absolute chronology.

In 1993 a weapon chronology for the Scandinavian equipment in the period A.D. 520/30–800 was put forward by the author (Nørgård Jørgensen 1993). The chronology concerning the Bornholm weaponry which will be presented here is based upon a Ph.D. thesis entitled *Skandinaviske Våbengrave, kronologiske, regionale og kulturhistoriske studier* (Scandinavian weapon graves, chronological, regional and culture-historical studies) (Nørgård Jørgensen 1989; 1991a; 1991b; 1992a; 1992b; 1993; 1996a; 1996b; 1997a; 1997b; in print).

For the chronological seriation of the Bornholm weapon finds and the extremely sparse material from western Denmark, computerized seriation and correspondence analysis have been used, as with the women's grave goods (Scollar 1995). The typological classification has been based on the form of the weapons, something which is crucial in determining their function. The goal of this seriation analysis has been to find the inter-relationships of the various weapon-grave combinations, shown on a two-dimensional matrix in which the central range of the occurrence of the types is arranged along the diagonal of the graph. It is thus possible to trace the introduction of new weapon-types as well as to find clusters in the material.

PHASING

Through seriation it transpired that 46 graves and 32 types were linked. The low number of linked graves was due to the severe degradation of the finds from Bornholm, rendering several objects unclassifiable. This causes problems in phasing, especially with the late finds when there are often only two objects in the grave: a large one-edged sword (SAX4) and a small knife (MD). In many cases with the older excavated material it is no longer possible to identify or classify this small knife, so that many of these grave finds are eliminated from the seriation. Emil Vedel reported the knives from these grave groups as being small, but did not give their measurements. Furthermore, only knives which it is still possible to analyse have been included, which means overwhelmingly the Nørre Sandegård Vest knives together with some of the better preserved knives from the other grave finds. All of the uncertain knives have been excluded from the seriation.

Upon the first seriation, the sequence revealed a steady development in the grave furnishings. Types that were obviously of no chronological significance or were simply misleading were removed from these series: combs, wooden vessels, beads and nails. These objects could have been overlooked in excavations in the 19th century. After the second seriation whetstones were removed, as these appear in the majority of graves and are of a single type for most of the Late Germanic Iron Age. In the final seriation, 46 graves and 32 types were put in order. It must be noted, however, that all of the 'lost' graves, up to a total of 100 grave groups, include one main type that does occur in the seriation, and that in most cases this is a one-edged sword (SAX1–5).[5–9] The term *main type* means a distinct and prominent artefact-type which is specifically linked to a particular phase but which has successors in the following phase(s). This is in contrast to a *leading type*, which is exclusively linked to a single phase. The seriation in figures 9.1 and 9.2 shows a clear development in the grave goods, dividing into five phases, with Phase I as the earliest.

Phase I (Figs. 9.3–9.4) has as its main types the two-edged sword with a small copper-alloy pommel (SP2), a shield boss with a spike on the tip (SBA) and a long, slender spearhead with copper-alloy rivets on the socket (L3d), both of which latter types continue in Phase II.[1] Other types that occur in Phase I are the two-edged sword with no preserved metal pommel or hilt (SP), a belt buckle with a large oval backplate (GU3), square iron or copper-alloy belt mounts with domed rivets (KR) and finally wool shears (SC).[2] A distinctive feature of Phase I is the absence of one-edged swords. What can be described as a very large knife may occur (MF), but this type is not included in the seriation.

Phase II has as its main types square strap-distributors (RF1A), tongue-shaped strap ends (ZR), rectangular belt mounts (RR1), small ring bits (R1) and tweezers (P1). The main types of weapon in Phase II are the short one-edged sword (SAX1) and the shield boss with a spike on the tip (SBA). There are also two-edged swords with no preserved pommel or hilt (SP). New types appearing for the first time in Phase II are the set of two small knives (MC) and the short, wide-bladed spearhead (L2), bridle harness of simpler forms (GU1) and (R2). Old types from Phase I which cease in Phase II are the long lance (L3d), the shield boss (SBA) and the shears (SC). Those graves which include the rarely deposited pottery of the Late Germanic Iron Age (TO) appear in Phase II and the following phase.

Phase III has as its characteristic main types the medium-sized one-edged sword (SAX2) and the short, wide-bladed spearhead (L2), the dominant weapon-types of this phase. New types of Phase III beside SAX2 are the shield boss with a concave neck and no top spike (SBB), the two-edged sword with a large pommel and possibly some of the metal hilt preserved (SP3a). Artefact-types that apparently disappear in Phase II are the two-edged sword with no metal hilt or grip (SP) and the short one-edged sword (SAX1). Bridle harness and mounts are of simpler forms (RR2, R2, GU1), and are associated with large ring bits (R3), and square strap-distributors (RF1A), tongue-shaped strap ends (ZR), wool shears (SC), the set of two small knives (MC) and pottery (TO).

Phase IV has as its leading types the large one-edged sword (SAX3), the short and narrow spearhead (L3c) and the sugar loaf-shaped 'Galgenberg' shield boss (SBD). New types are the single small knife (MD) and the short and very slender one-edged sword (SAX5). Types that continue from Phase III are the two-edged sword with pommel and hilt (SP3a), the shield boss with concave neck and no top spike (SBB) and the small, wide spearhead (L2), together with the set of a large and a small knife (MC) and an offshoot of the one-edged sword (SAX2). Horse gear now comprises only the large ring bit (R3) and bridle harness mounts of simpler forms (RR2), while an iron ring buckle (GU6) also occurs in Phase IV. The spearhead (L5) occurs in just one example and is therefore not represented in the seriation.

Phase V has as its characteristic main type the long and heavy one-edged sword (SAX4). There are also boxes (K), copper-alloy ring pins (GU5) and iron ring buckles (GU6). There is also one example of a hemispherical shield boss (SBC) which, however, is not represented in the seriation itself. The same is the case with the set of two small knives (MA) and the large knife (ME) which occur in Phase-IV and Phase-V contexts but which are not represented in the seriation. The small knife (MD) continues into Phase V. Phase V, like Phase I, cannot have any leading types as its types can appear in both the preceding and the following periods.

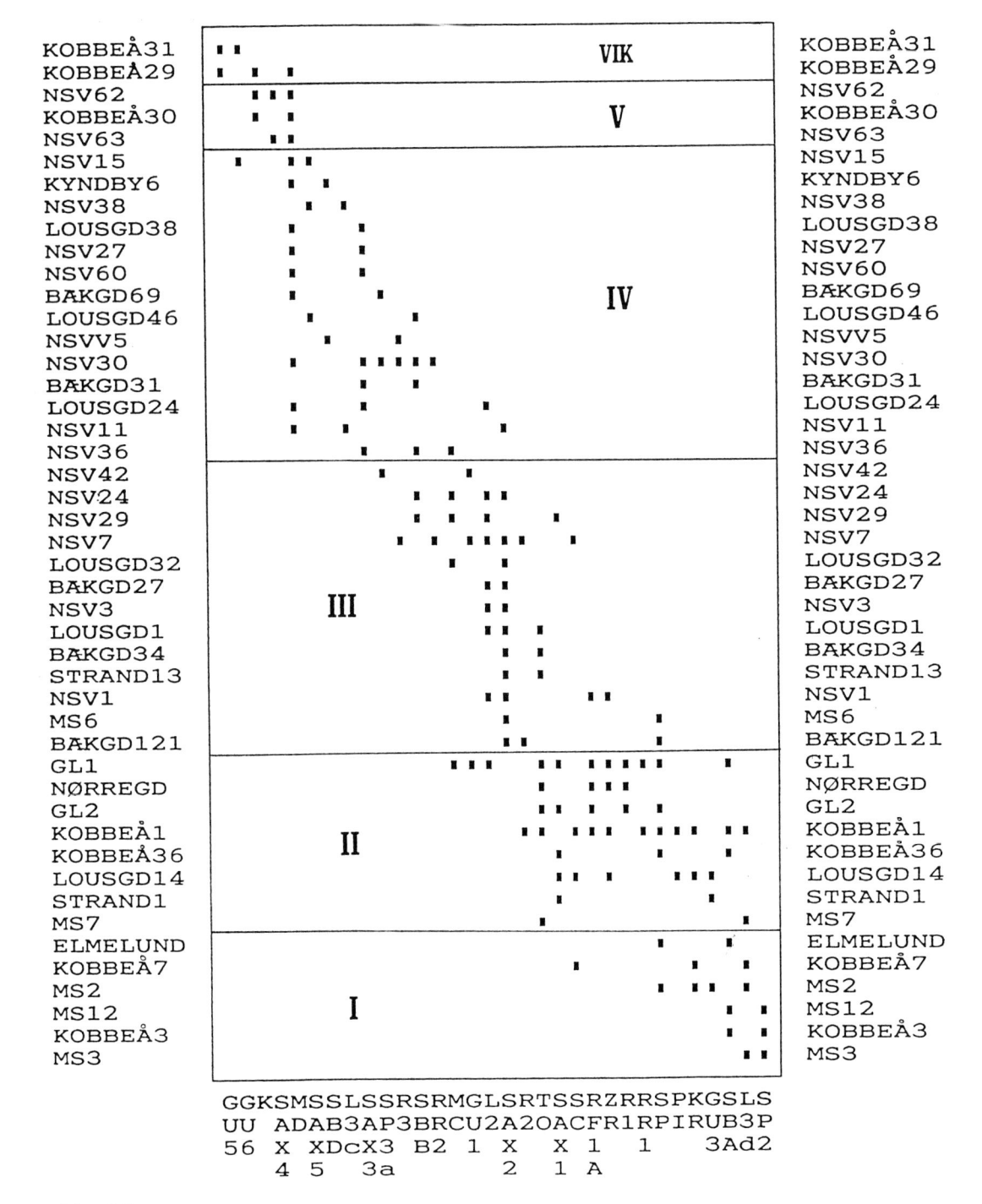

Fig. 9.1 *The seriated matrix of the Bornholm weapon graves with 46 graves and 32 types.*
Geordnete Matrix der Bornholmer Waffengräber.

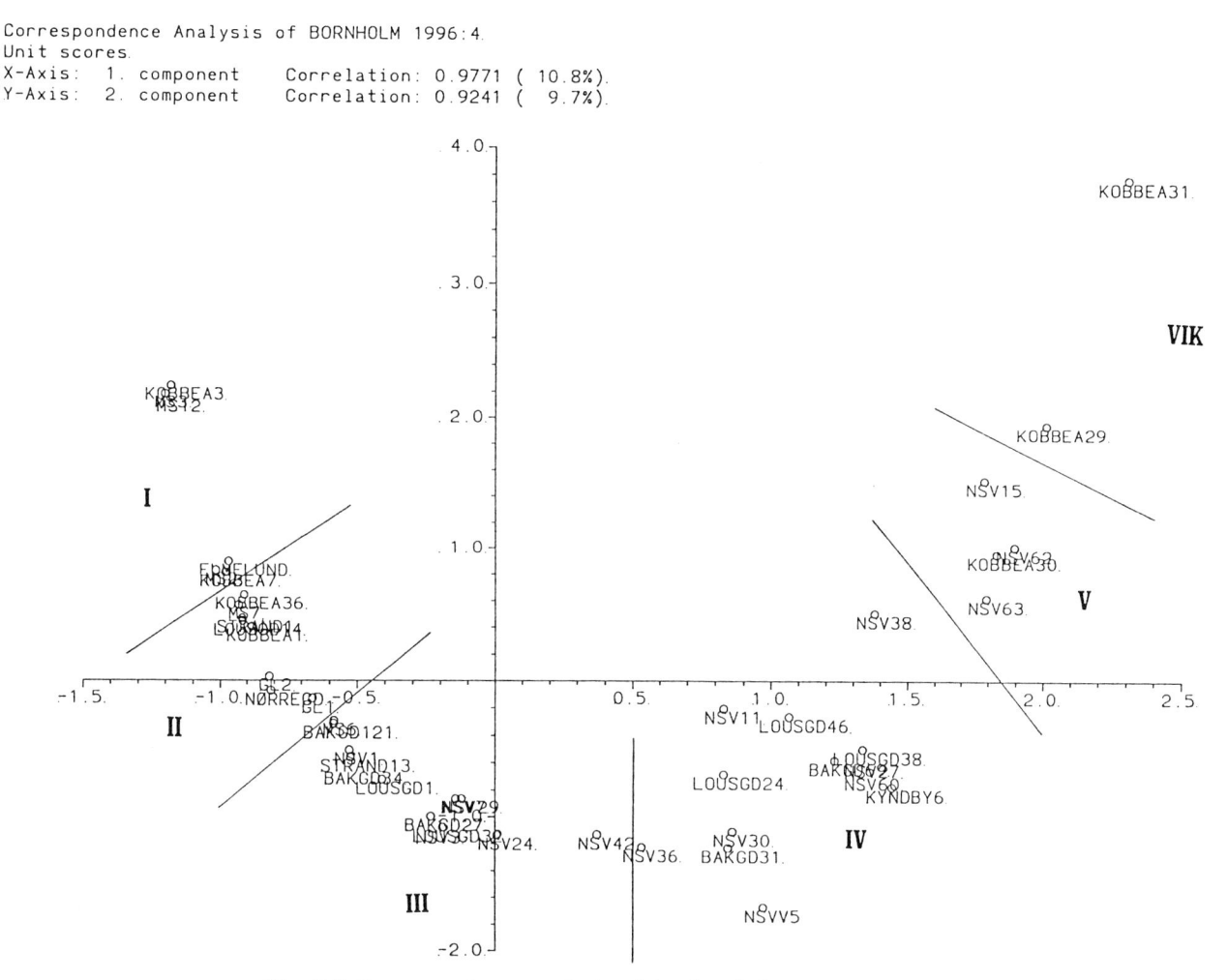

Fig. 9.2 Correspondence analysis of the Bornholm weapon graves.
Korrespondenzanalyse der Bornholmer Waffengräber.

ABSOLUTE CHRONOLOGY

Phase I is equivalent to the trasition from the Early Germanic Iron Age to the Late Germanic Iron Age in Danish terminology. The dating of the beginning of Late Germanic Iron Age has been discussed by Arrhenius, who showed that some of the earliest Vendel-period graves in Sweden could be aligned with Ament's phase *ältere Merowingerzeit* II (AMII) in the Continental chronology, 520/30–560/70 A.D. (Ament 1976; 1977; Arrhenius 1983, 39ff.; see also Nørgård Jørgensen 1991a, 227; 1991b, 157ff.).[3] Some artefact-types of Phase I belong to the Early Germanic Iron Age/Migration Period, for instance the two-edged sword (SP2) (Nerman 1935, Taf. 55:586), but new types appear in the form, for instance, of belt fittings and spearheads.[4] The Continental *Schmalsax*, however, which appears on Bornholm as the short one-edged sword (SAX1), is completely absent from Phase I.[5] This indicates that this period on Bornholm lies *before* the introduction of the Schmalsax, which belongs to Ament's phase AMIII of the late sixth century (Böhner 1958, 135; Ament 1976, 285ff.; 1977, 133ff.). The inventory of Phase I corresponds to a period that is dated 525–565/70 A.D. at the Alamannic cemetery of Schretzheim (Koch 1977, Abb.8b:37). Phase I is therefore dated to 520/30–560/70 A.D. The combination of the short sax and the two-edged sword is characteristic of the earlier Merovingian-period grave finds, and this holds also in parts of Scandinavia – for instance in the chronology of the Gotlandic Vendel Period (Nørgård Jørgensen 1992a, 5ff.). In Bemmann and Hahne's chronological work on the latest Norwegian material of the Migration Period, a Nerhus Group also comprises weapon combinations with short saxes (Bemmann and Hahne 1994, 329ff.; see Fig. 9.5). The dating of the Nerhus Group is suggested to cover the period from 510/25 to the beginning of the Vendel Period at 565/70 A.D. This means that we have to accept Phase I as being a transitional period from the Early Germanic Iron Age to the Late Germanic Iron Age.

The inventory of Phase II can be paralleled with other Scandinavian chieftainly graves which contain horse harness, a two-edged sword, a narrow sax (SAX1) and a shield boss with a top spike. This horizon corresponds to Ament's *ältere Merowingerzeit* III graves on the Continent (= 560/70–600 A.D.) (Nørgård Jørgensen 1991a, 219ff.).

152 *Anne Nørgård Jørgensen*

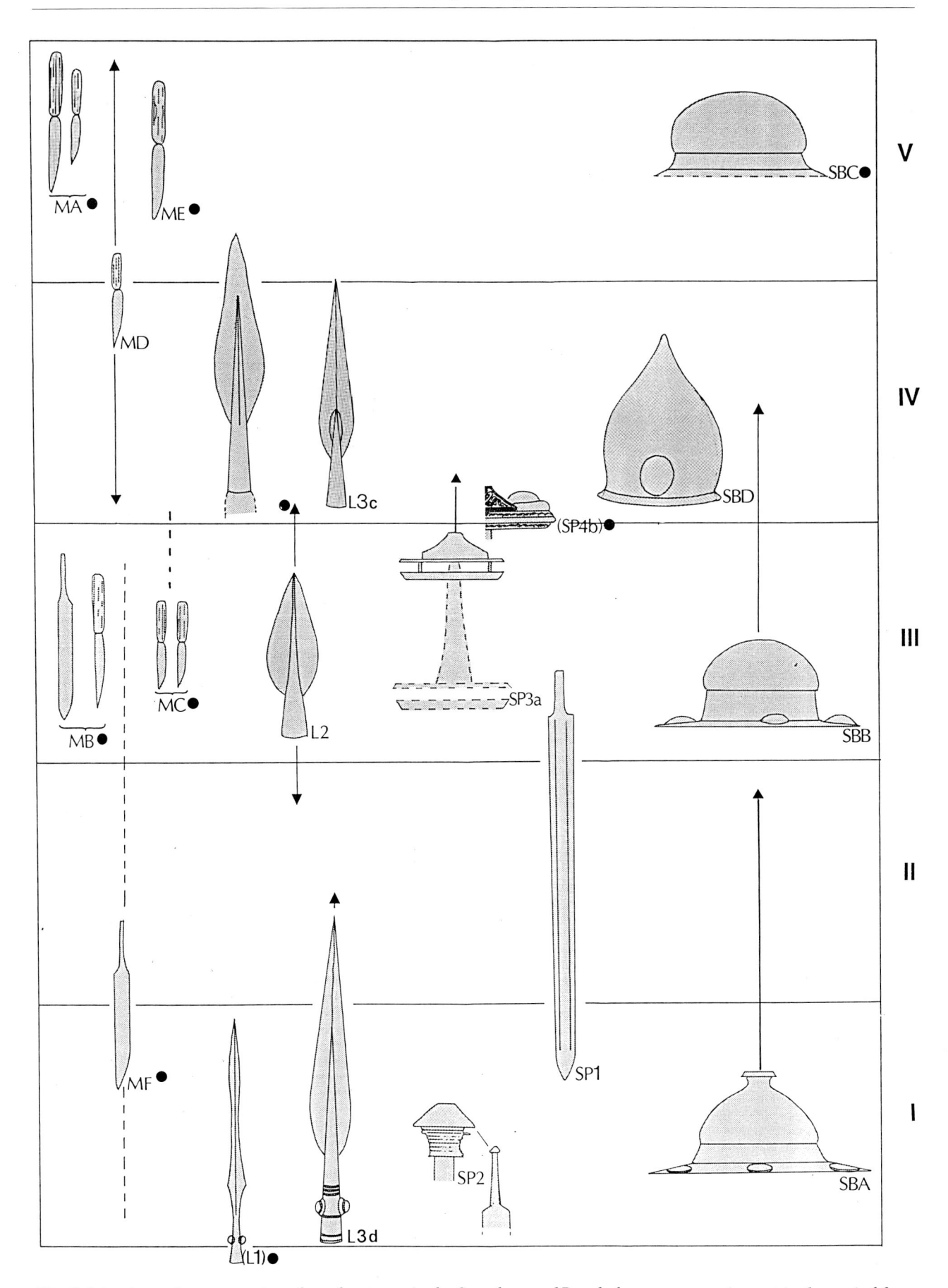

Fig. 9.3 A schematic presentation of artefact-types in the five phases of Bornholm weapon equipment in the period from 520/30 to ca. 800 A.D. Phase I is dated to 520/30–560/70 A.D., Phase II: 560/70–610/20 A.D., Phase III: 610/20–ca.680 A.D., Phase IV: ca. 680–740/50 A.D., Phase V: 740/50 to shortly after 800 A.D. * Not in the seriation.
Schematisierte Darstellung der Typen in den fünf Phasen der Bornholmer Waffenausstattungen zwischen ca. 520/530 und 800 n.Chr.

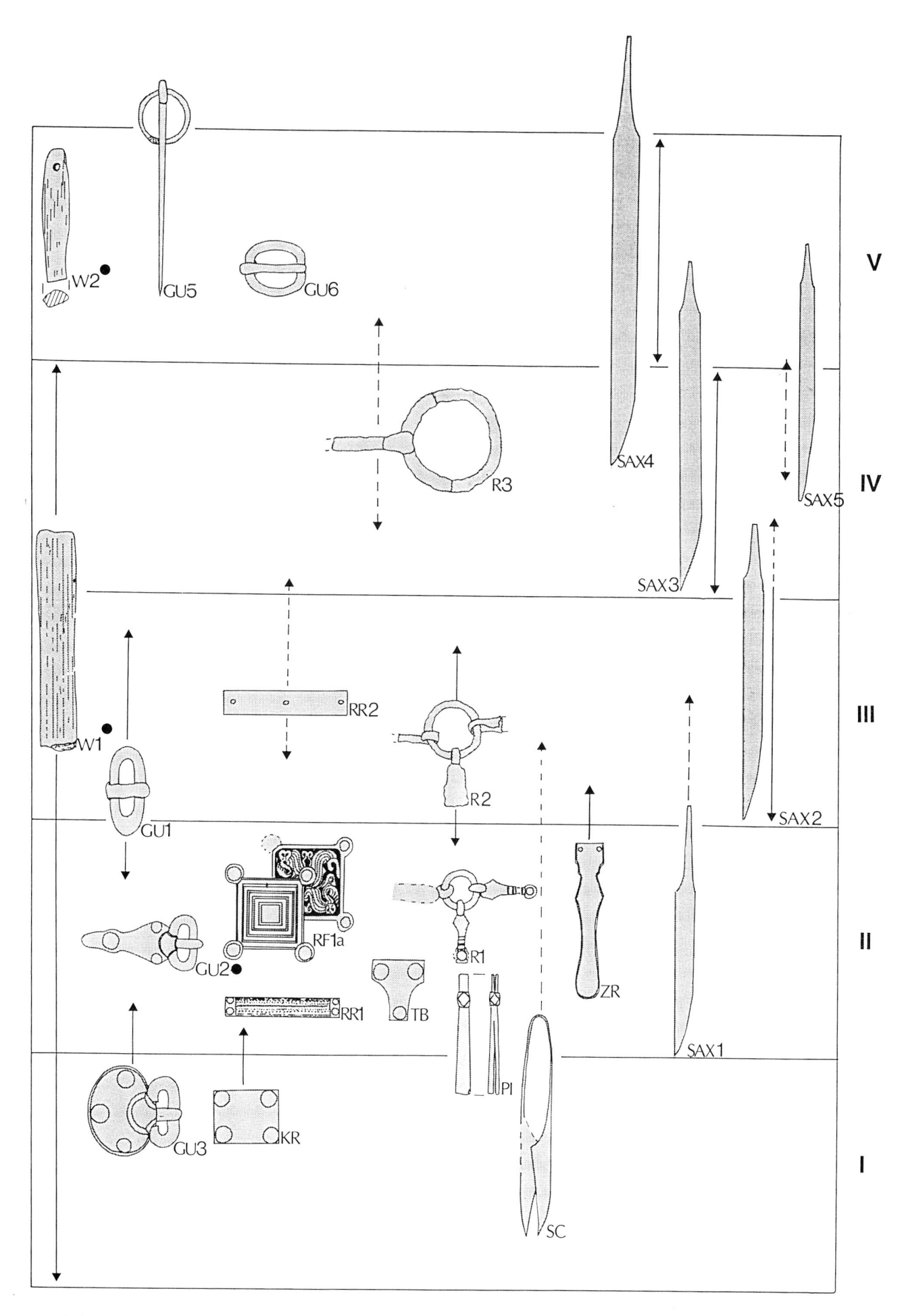

*Fig. 9.4 A schematic presentation of artefact-types in the five phases of Bornholm weapon equipment in the period from 520/30 to ca. 800 A.D. Phase I is dated to 520/30–560/70 A.D., Phase II: 560/70–610/20 A.D., Phase III: 610/20–ca.680 A.D., Phase IV: ca. 680–740/50 A.D., Phase V: 740/50 to shortly after 800 A.D. * Not in the seriation.*
Schematisierte Darstellung der Typen in den fünf Phasen der Bornholmer Waffenausstattungen zwischen ca. 520/530 und 800 n.Chr.

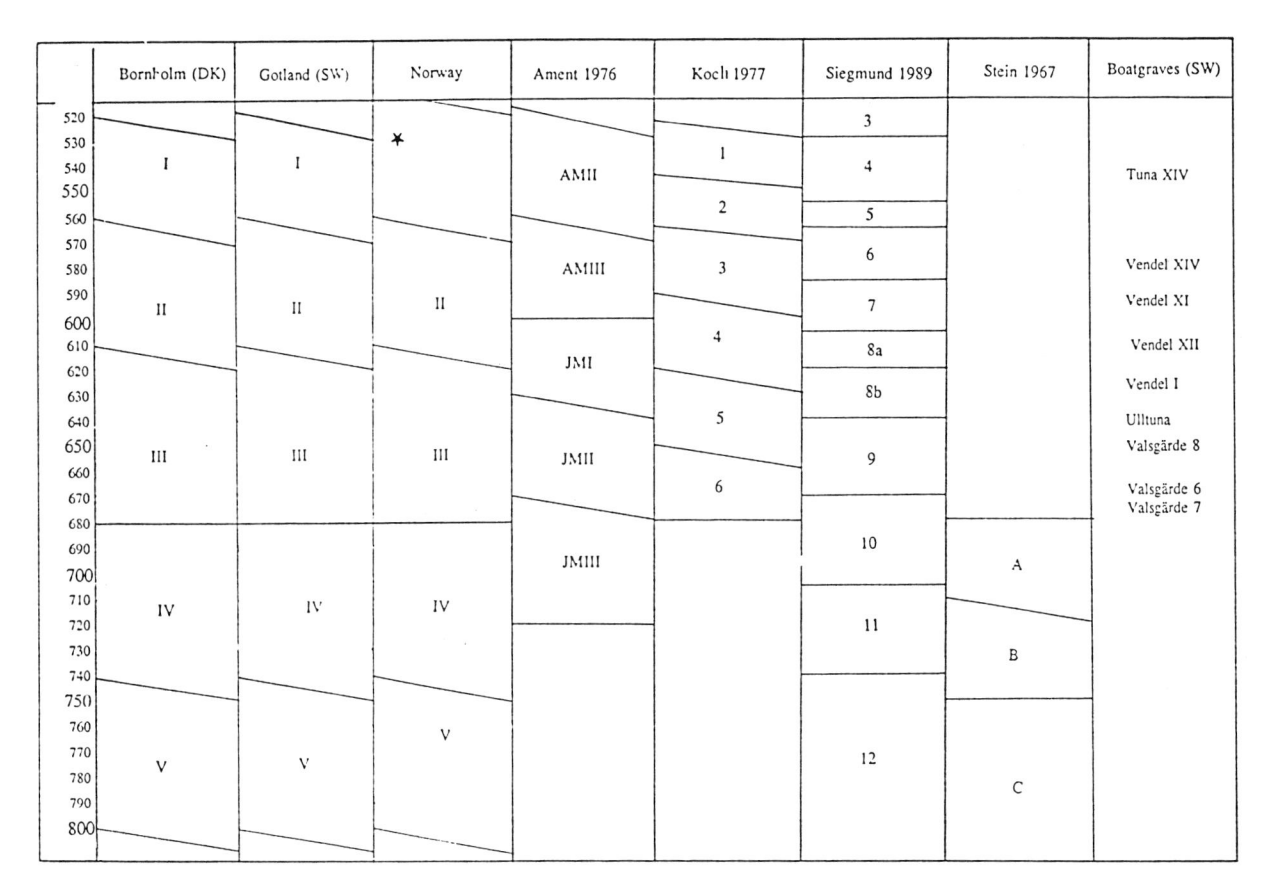

	Bornholm (DK)	Gotland (SW)	Norway	Ament 1976	Koch 1977	Siegmund 1989	Stein 1967	Boatgraves (SW)
520						3		
530	I	I	*	AMII	1	4		Tuna XIV
540								
550					2	5		
560								
570				AMIII	3	6		Vendel XIV
580								
590	II	II	II			7		Vendel XI
600					4	8a		Vendel XII
610				JMI				Vendel I
620						8b		
630					5			Ulltuna
640	III	III	III	JMII		9		Valsgärde 8
650					6			
660								Valsgärde 6
670								Valsgärde 7
680				JMIII		10	A	
690								
700	IV	IV	IV			11		
710							B	
720								
730								
740			V					
750								
760	V	V				12		
770							C	
780								
790								
800								

Fig. 9.5 The results of the chronological studies on Bornholm, Gotland and in southern and central Norway linked to Continental chronological systems. *The Nerhus Group is identified and defined in Bemmann and Hahne's chronological work on Norway (1994, 329ff.).
Chronologien für Bornholm, Gotland, Süd- und Zentralnorwegen in Korrelation zu kontinentalen Systemen.

Several circumstances, however, render an end date of around 600 for Phase II impossible. First and foremost, the horizontal stratigraphic datings of the Bornholm finds indicate that the inventory of this phase must continue for more than one generation and thus into the 7th century, with men's graves with Phase-II combinations being most closely associated with women's graves of both Phases 1B2 and 1C (Fig. 9.6). Phase II must therefore continue until after 600, and is consequently dated to the period 560/70–610/20 A.D. In respect of the synchronization of Scandinavian weaponry with the Continental, Phase II is certainly the most important phase, as there is no doubt as to the close contact between the Continental weapon graves of AMIII in the late 6th century and the series of well-furnished graves from central Sweden (Vendel XIV etc.), central Norway (Torgård), Bornholm (Kobbeå 1 etc.) and the rich Phase-II graves of Gotland (Nørgård Jørgensen 1991a, 218ff.; 1991b, 157f.). The weapon set comprising a narrow sax and a two-edged sword and a shield boss with a top spike which recurs in the more regularly furnished graves is thus an international one.

The Phase-III inventory cannot be paralleled directly on the Continent as we do not have in Scandinavia the Continental sword-types *leichter Breitsax* or *schwerer Breitsax* (Böhner 1958, 138ff.; Koch 1977, 107; Hübener 1989a, 81, Abb. 4). Instead the Scandinavian one-edged sword – Type SAX2 – becomes heavier than the Schmalsax although at the same time slenderer and longer than the Continental Breitsax.[6] Neither does the long, two-handed hilt that is sometimes found on the Continental Breitsax appear in Scandinavian finds. According to the horizontal stratigraphy, the men's graves with SAX2 are predominantly linked to women's graves of Phase 1D1–1D2 (Fig. 9.6). SAX2 was probably superseded by SAX3 before 700 according to the date of the beginning of Phase IV.[7] Phase III is therefore dated to 610/20–ca. 680 A.D. It is not possible to confirm the absolute dates in this case but it can be assumed that the replacement of the Scandinavian type of narrow sax, SAX1, by SAX2 took place at the beginning of the 7th century, approximately at the same time or shortly after the shift from the Schmalsax to the Breitsax on the Continent, a feature of the transition between Ament's phases AMIII and JMI (Ament 1976, 334). The two-edged sword becomes rarer in Phase III, although a few specimens are found of a type that remained widespread on Gotland (SP3).

In Phase IV, a weapon set including the heavier and

A peaceful discussion of a martial topic: the chronology of Scandinavian weapon graves

Jørgensen 1996 ♀   Nørgård Jørgensen 1996 ♂

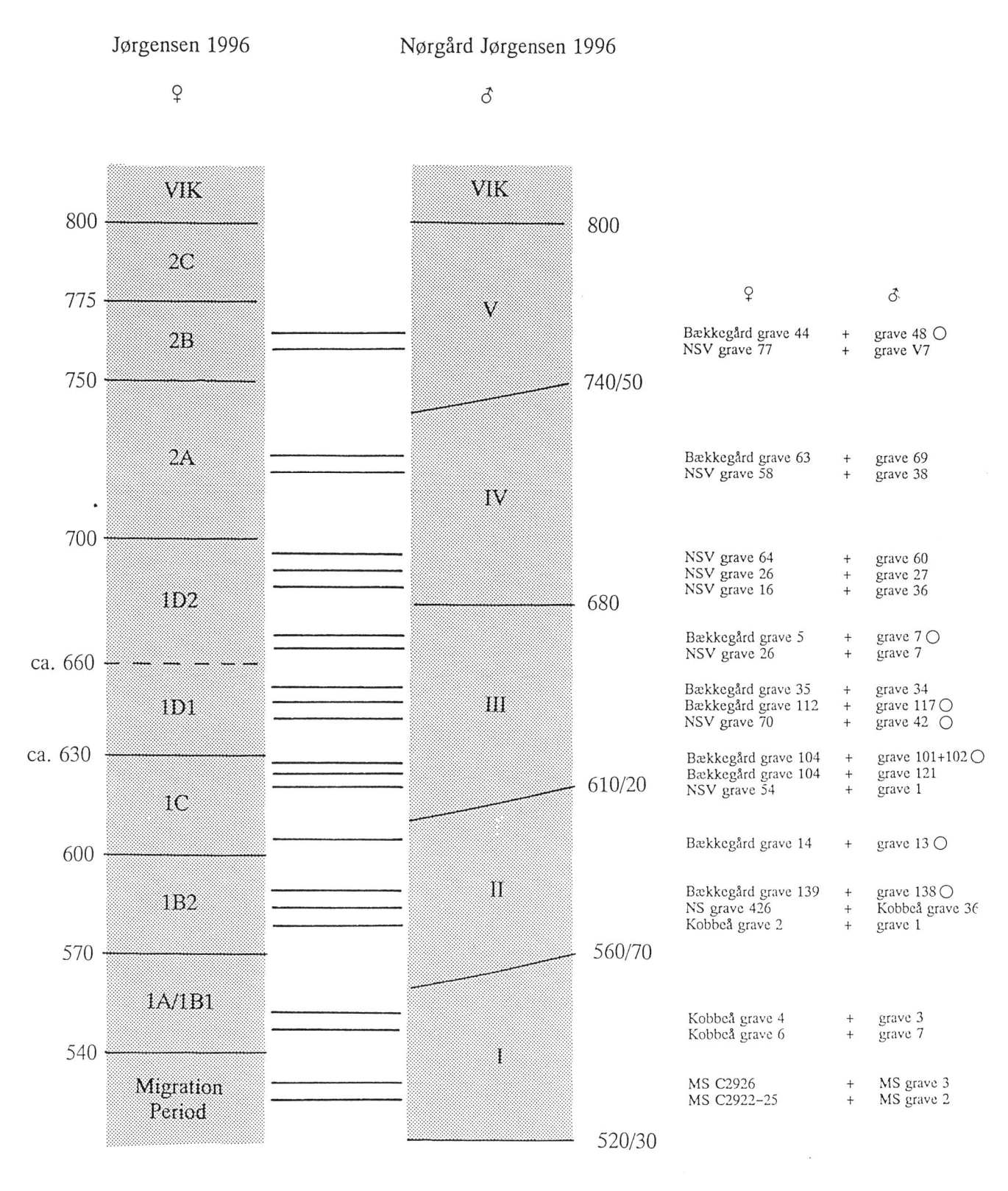

♀	♂
Bækkegård grave 44 +	grave 48 ○
NSV grave 77 +	grave V7
Bækkegård grave 63 +	grave 69
NSV grave 58 +	grave 38
NSV grave 64 +	grave 60
NSV grave 26 +	grave 27
NSV grave 16 +	grave 36
Bækkegård grave 5 +	grave 7 ○
NSV grave 26 +	grave 7
Bækkegård grave 35 +	grave 34
Bækkegård grave 112 +	grave 117 ○
NSV grave 70 +	grave 42 ○
Bækkegård grave 104 +	grave 101+102 ○
Bækkegård grave 104 +	grave 121
NSV grave 54 +	grave 1
Bækkegård grave 14 +	grave 13 ○
Bækkegård grave 139 +	grave 138 ○
NS grave 426 +	Kobbeå grave 36
Kobbeå grave 2 +	grave 1
Kobbeå grave 4 +	grave 3
Kobbeå grave 6 +	grave 7
MS C2926 +	MS grave 3
MS C2922-25 +	MS grave 2

Fig. 9.6 Horizontal-stratigraphical synchronization of the female and male chronology on the basis of the closest graves. The phase-datings of the pairs of graves used are linked by a line. It is clear that the relative chronological sequence amongst the women's graves is matched by a sequence amongst the men's graves but that the phases are not completely synchronous. A circle indicates that the man's grave in question is not included in the seriation and that its assignation to a phase is based solely upon the type of sax. NSV = Nørre Sandegård Vest; NS = Nørre Sandegård.
Chorologische Synchronisierung der Frauen- und Männerchronologie auf Grundlage der nächstgelegenen Bestattungen.

more powerful type of one-edged sword that was practically a slashing or chopping weapon appears (SAX3). The dating of the beginning of Phase IV falls, through comparison with the Continent, around 680 A.D. The basis for this is that the one-edged sword-type SAX3 can be regarded as a parallel to the Continental *Langsax* and the classic dating of the introduction of that in Continental grave finds is 680 A.D. (Böhner 1958, 144; Ament 1976, 334: the transition from JMII to JMIII). Men's graves with SAX3 are linked to a series of women's graves, in this case one which includes Phases 1D2 and 2A and which may represent more than one generation (Fig. 9.6). The lower dating limit is very difficult to fix, but the replacement of SAX3 by SAX4 may have taken place around 740/750 A.D.[8] The proposed dating of Phase IV is therefore to the period ca. 680–740/50 A.D.

The inventory of Phase V is difficult to align with other finds. Several of the types are also known from Viking-period finds, such as the shield boss (SBC), and not least the large and heavy one-edged sword (SAX4) which continued in the late Merovingian Period and on to the Viking Period in Norwegian grave finds in slightly different forms classified as SAX5, SAX6, SAX7 and SAX8 (Nørgård Jørgensen 1993).[9] Associated with the large sword (SAX4) we often find only a single slender knife, more rarely a set of them, and occasionally a shield boss of the hemispherical type with a narrow collar (SBC). These features are common to Norway, Gotland and Bornholm. There are no two-edged swords in Bornholm's finds from this phase. The lower date limit of Phase V cannot immediately be determined. The grave goods of Phase V are poor because of the partial abandonment of the practice of depositing weapons in graves. Nørre Sandegård Vest is represented in this phase, and everything indicates that these graves are linked to women's graves of Phase 2B–2C (Fig. 9.6). Kobbeå graves 29 and 31, by contrast, belong to the latest part of 8th century and beginning of the 9th century, and these graves are linked to the women's graves of beginning of the Viking Period. The transition from weapon graves to graves with only horse harness and possibly a shield boss thus appears most clearly in the Kobbeå cemetery. There we find the latest type of weapon grave with sword in Kobbeå 9, situated in association with women's graves dated to the late 8th and early 9th centuries. Phase V thus comprises artefact-types from both groups of men's graves in this long period, which probably runs from 740/50 down to around 800 A.D. It is therefore proposed that Phase V covers the period from 740/50 to shortly after 800 A.D., but there is no evidence on which to base a closing date.

In summary, we have five phases dated as follows:

Phase I: 520/30–560/70 A.D.
Phase II: 560/70–610/20 A.D.
Phase III: 610/20–ca. 680 A.D.
Phase IV: ca. 680–740/50 A.D.
Phase V: 740/50 to shortly after 800 A.D.

CONCLUSION

It was a real bonus to encounter, at the symposium in Ålborg, a forum of specialists which could serve as a critical audience for this chronological study a short time before it was to be published as part of the Nørre Sandegård Vest report in 1997. The chronology for Bornholm in the Germanic Iron Age forms part of a general Scandinavian chronology which is at present being prepared for publication under the title of *Waffen und Gräber: Skandinavische Waffengräber 520/30–900 n.Chr. Typologische und Chronologische Studien*. It was thus of particular importance to have the less certain points discussed at just that time.

There appears to be broad agreement that seriation using, for instance, the Bonn programme is of great value in archaeological chronology. Only a few years ago the view was relatively sceptical, but gradually, as most scholars have come to perceive that combination analyses using seriation are simply a modern form of the most traditional archaeological methods, such attitudes have softened. There was, in consequence, relatively little discussion of the seriation within the present chronology itself.

A second point over which there appears to be general agreement is the overriding need for the synchronization of chronological systems. Various contributions to this symposium (see especially Høilund Nielsen, this volume) touched upon this theme, and most opted to seek synchronization with the Continental chronological systems, especially that of Ament (1976; 1977). This provoked an immediate debate, with the German participants pointing out the weaknesses of this chronological scheme, which comprises the phases AM I–III and JM I–III but is not based upon substantially documented archaeological material. The terms *ältere Merowingerzeit* and *jüngere Merowingerzeit* are, however, familiar to most scholars, and easy both to remember and to use. It is also sensible to have a common international terminology for the various subdivisions of every distinct culture-historical age. Nevertheless, if a chronological scheme is synchronized with Ament's phases, it is essential that at the same time attention is drawn to the relationship with other Continental chronologies, such as those in the synchronized chronology diagram in figure 9.5. In this way one can get past the problem while we are still waiting for the further work of Continental scholars on a comprehensive chronology of the Merovingian Period, such as the great project on the publication of the Weingarten cemetery and the chronology of south-western Germany (Roth and Theune 1988; forthcoming).

The questions which were taken up as a direct result of this chronological study naturally included the date of the transition between the Early Germanic Iron Age and the Late Germanic Iron Age (as they are called in Denmark; *alias* the Migration Period and the Vendel Period), which is here equated with Phase I. The start date of Phase I is 520/530 A.D. The beginning of the Late Germanic Iron Age has traditionally been placed around 565/570 A.D. Looking,

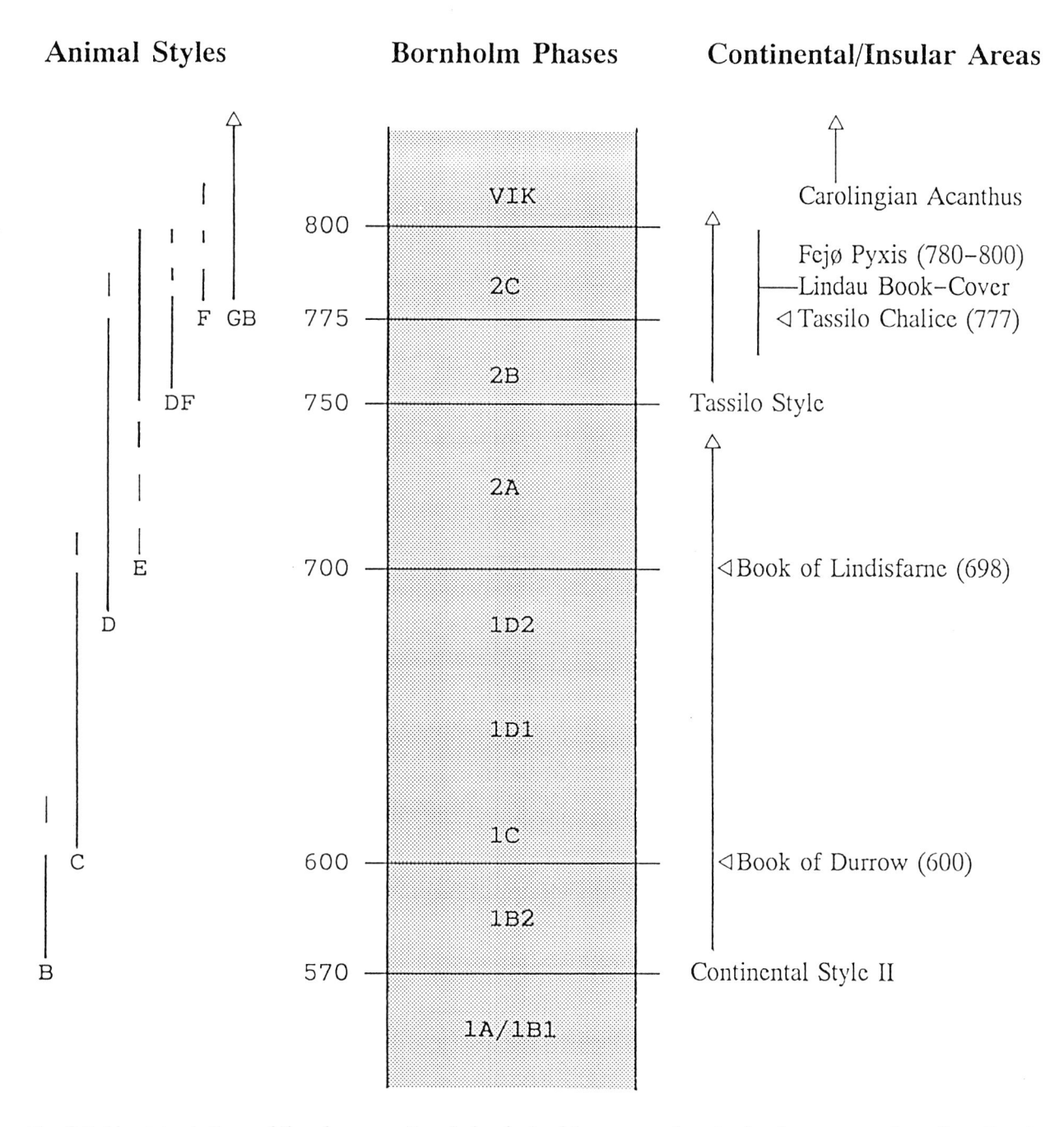

Fig. 9.7 Absolute datings of the phases on Bornholm derived from a synchronization between southern Scandinavian Styles B–F and GB (the gripping beast style) and Insular and Continental material. The date of ca. 600 A.D. for the Book of Durrow is adopted from Roth 1987.
Absolute Datierung der Bornholm-Phasen. Grundlage ist eine Synchronisierung mit den südskandinavischen Stilen B–F und dem Greiftierstil sowie insularem und kontinentalem Material.

for comparison, at Bemmann and Hahne's chronological study of the Norwegian material from the Migration Period, one finds their latest phase, the Nerhus Group, covering the period from 510/525 to 565/570 A.D. (Bemmann and Hahne 1994, 329ff.). The weapon combinations in this phase include two-edged swords and short saxes, both in Phase I of the present chronology and in the Nerhus Group. There is no doubt that the phases are to some degree contemporary.

Moreover the later combination of two-edged sword and narrow sax is quite absent from this period.

In order to explain a change in the period 520/530–560/570 in the weapon chronology we have to turn to cultural history. What we find reflected in this phase are burials of those persons who participated in what, from the perspective of military history, was a comprehensive reform during and immediately after the conquests of

Chlodevic. The military equipment changes in AM II, *alias* Phase I, for instance with the disappearance of the two-spear, lance plus javelin, combination; the symbols of the military elite change as the earliest of the massive gold ring knobs appear in the first half of the 6th century, both in Scandinavia and on the Continent; and there is much to indicate that the methods and order of battle may have undergone certain changes: as implied, for instance, by the increase in riding gear and the use of the lance, together with the obligatory use of the sword in most weapon sets beginning with the narrow sax which appears both in Scandinavia and on the Continent in Phase II, around 560/570 A.D. This appears to have happened with some degree of synchronization between parts of southern Scandinavia and western Europe. It is thus reasonable to characterize Phase I as a transitional phase between the Early Germanic Iron Age and the Late. There is no point in placing a chronological dividing line in the middle of a phase of steady new developments. Furthermore, I can see no purely Early Germanic Iron-age features in Phase I such as Style-I elements in male equipment which would justify us in describing the phase as belonging purely to the Early Germanic Iron Age or Migration Period. This does not deny that Style I may still appear on female accoutrements in the middle of the 6th century.

The chronological watershed in the female range apparently occurred a little later (cf. Fig. 9.6), as we discuss in the Nørre Sandegård Vest report (Jørgensen and Nørgård Jørgensen 1997, 24ff.). Here the male and female chronologies were worked out independently. They were each separately synchronized and discussed in relation to other comparable chronological schemes. The female chronology was also discussed on the basis of the most recent stylistic datings from Europe. Finally a horizontal-stratigraphical analysis to coordinate male and female graves was undertaken (Fig. 9.6). The result of these analyses is that female material can scarcely be described as Late Germanic Iron-age before ca. 540 A.D, Phase 1A/1B1 (Fig. 9.6).

Finally, the Ålborg meeting also found room for a detailed discussion of individual artefact-types such as belt buckles with oval backplates and shield-tongues (type GU3). These very large buckles are known from both Gotlandic and Danish finds, and are dated to Phase I. This is a point we need to take proper note of, and to revise our perceptions of long-distance contacts to include mutual influence.

There is a general tendency for there to be a time-lag between the western European and Scandinavian material, on the one side, and that from England on the other. The question is whether this dislocation is real or if there is a general chronological misalignment. On the basis of both the regional and the trans-regional analyses of combinations in weapon graves between these areas, it is concluded that in certain respects England does stand somewhat apart, while there is a high similarity between Scandinavia and south-western Germany (Nørgård Jørgensen 1997a, 159). This debate will, it is hoped, be carried forward in future symposia. I should personally like to take this opportunity to express my thanks for a good and profitable discussion, and for the material sent to me by colleagues after the Ålborg meeting.

DEUTSCHE ZUSAMMENFASSUNG

Die hier vorgestellte Chronologie für die Waffengräber Bornholms beruht auf meiner noch unpublizierten Kopenhagener Dissertation: Anne Nørgård Jørgensen, 1993, *Skandinaviske Våbengrave: Kronologiske, regionale og kulturhistoriske studier* (Waffen und Gräber. Skandinavische Waffengräber 530–900 n.Chr. Typologische und Chronologische Studien). Grundlage sind die Waffengräber Bornholms und das sehr spärliche Material aus dem westlichen Dänemark, die einer Korrespondenzanalyse unterzogen wurden. Als Ergebnis kann das Material in fünf Phasen gegliedert werden: Phase I ca. 520/530 – 560/570 n.Chr., Phase II ca. 560/570 – 610/620 n.Chr., Phase III ca. 610/620 – 680 n.Chr., Phase IV ca. 680 – 740/750 n.Chr., Phase V ca. 740/750 bis kurz nach 800 n.Chr. Auf dem Symposium in Ålborg ergab sich eine sehr intensive Diskussion insbesondere über die Phase 1 und den Übergang von der Älteren zur Jüngeren Germanischen Eisenzeit.

NOTES

1 Type SP2 is only found in Phase I of the Late Germanic Iron Age. However it also occurs in the Early Germanic Iron Age on Gotland, and therefore cannot be a leading type.

2 'SP' is not a distinct type; it covers all types of two-edged sword with organic grips or hilts.

3 Max Martin (1989) has revised the dating of the transition from AMI to AMII to 510 A.D., but this proposition is not generally accepted. The transition from the Early Germanic Iron Age to Late Germanic Iron Age was discussed at the working seminar in Ålborg in March 1996. On that occasion Phase I of the weapon chronology was generally accepted as a transitional phase. The discussion concentrated on special types such as the belt buckle with a large oval backplate (GU3), which appears in Scandinavia, on Gotland and Bornholm, in Phases I and II. This dating is earlier than the Continental datings of a smaller belt buckle very similar to the type GU3. The equipment in this period, however, must be regarded as subject to mutual influence and not just 'one-way' contacts.

4 The Early Germanic Iron-age finds in the male graves Melsted-Sandhuset graves 2 and 3 are found in burial mounds together with female equipment of pure Early Germanic Iron-age date.

5 The saxes are defined on a supra-regional level in Scandinavia as 'Nordic'-types. The regional differences derives from the local production of saxes in each area (Bornholm, Gotland and southern and central Norway). The Nordic SAX1 is defined as follows: total length 27 – 52 cm, length of the blade 22 – 38 cm and width of the blade 2.2 – 3.6 cm.

6 The Nordic SAX2 is defined as follows: total length 40.5 – 69 cm, length of the blade 36 – 53 cm and width of the blade 3.5 – 4.3 cm.

7   The Nordic SAX3 is defined as follows: total length 55 – 81 cm, length of the blade 43.8 – 67 cm and width of the blade 4.3 – 4.9 cm.
8   The Nordic SAX4 is defined as follows: total length 68 – 99 cm, length of the blade 51 – 85 cm and width of the blade 5 -5.9 cm.
9   The Nordic SAX5 is defined as follows: total length 32 – 55 cm, length of the blade 22.5 – 45 cm and width of the blade 2.2 – 3.0 cm.

This chronology has now been revised and is about to be published in: *Waffen und Gräber. Skandinavische Waffengräber 530–900 n. Chr. Typologische und Chronologische Studien*. Nordiske Fortidsminder.

BIBLIOGRAPHY

Ament, H. 1976. 'Chronologische Untersuchungen an fränkischen Gräberfeldern der jüngeren Merowingerzeit im Rheinland'. *Berichte Römisch-Germanische Komm* 57, 285–336.

Ament, H. 1977. 'Zur archäologischen Periodisierung der Merowingerzeit'. *Germania* 55, 133–40.

Arrhenius, B. 1983. 'The Chronology of the Vendel graves'. In J. P. Lamm and H.-Å. Nordström (eds), *Vendel Period Studies*, 39–70. Stockholm.

Bemmann, J. & Hahne G. 1994. 'Waffenführende Grabinventare der jüngeren römischen Kaiserzeit und Völkerwanderungszeit in Skandinavien. Studie zur zeitlichen Ordnung anhand der norwegischen Funde'. *Berichte Römisch-Germanische Komm* 75, 283–640.

Böhner, K. 1958. *Die fränkischen Altertumer des Trierer Landes*. Germanische Denkmäler der Völkerwanderungszeit B 1. Berlin.

Høilund Nielsen, K. 1987. 'Zur Chronologie der jüngeren germanischen Eisenzeit auf Bornholm. Untersuchungen zu Schmuckgarnituren'. *Acta Archaeol* 57, 47–86.

Hübener, W. 1989. 'Die Langsaxe der späten Merowingerzeit'. *Acta Praehistorica et Archaeol* 21, 75–84.

Jørgensen, L. 1990. *Bækkegård and Glasergård. Two Cemeteries from the Late Iron Age on Bornholm*. Arkæologiske Studier VIII. Copenhagen.

Jørgensen, L. & Nørgård Jørgensen, A. 1997. *Nørre Sandegård Vest. A Cemetery from the 6th–8th Centuries on Bornholm*. Nordiske Fortidsminder 14. Copenhagen.

Koch, U. 1977. *Das Reihengräberfeld bei Schretzheim*. Germanische Denkmäler der Völkerwanderungszeit A 13. Berlin.

Martin, M. 1989. 'Bemerkungen zur chronologischen Gliederung der frühen Merowingerzeit'. *Germania* 67, 121–41.

Nerman, B. 1935. *Die Völkerwanderungzeit Gotlands*. Stockholm.

Nørgård Jørgensen, A. 1989. 'Elmelunde – en våbengrav fra yngre jernalder på Møn'. In L. Jørgensen (ed.), *Simblegård – Trelleborg. Danske gravfund fra førromersk jernalder til vikingetid*, 208–17. Arkæologiske Skrifter 3. Copenhagen.

Nørgård Jørgensen, A. 1991a. 'Kobbeå 1 – ein reich ausgestattetes Grab der jüngeren germanischen Eisenzeit von Bornholm'. *Studien zur Sachsenforschung* 7, 203–39. Hannover.

Nørgård Jørgensen, A. 1991b. 'Kobbeågravpladsen – en yngre jernaldergravplads fra det nordbornholmske kystområde'. *Aarb Nordisk Oldkyndighed og Historie* 1991, 123–83.

Nørgård Jørgensen, A. 1992a. 'Weapon sets in Gotlandic grave finds from 530–800 A.D. A Chronological Analysis'. In L. Jørgensen (ed.), *Chronological Studies of Anglo-Saxon England, Lombard Italy and Vendel Period Sweden*, 5–34. Arkæologiske Skrifter 5. Copenhagen.

Nørgård Jørgensen, A. 1992b. 'Regional Studies of the Weapon Burial Practice in Scandinavia, 530–800 AD'. In *Medieval Europe, Pre-printed Papers vol. 4*. York.

Nørgård Jørgensen, A. 1993. *Skandinaviske Våbengrave, kronologiske, regionale og kulturhistoriske studier*. University of Copenhagen Ph.D. thesis. Unpublished.

Nørgård Jørgensen, A., with Crumlin Petersen, O. and Holmberg, B., 1996a. 'Kystforsvaret'. In O. Crumlin-Pedersen et al. (eds), *Atlas over Fyns Kyst i Jernalder, Vikingetid og Middelalder*, 182–93. Odense.

Nørgård Jørgensen, A. 1996b. 'Kriger og hird i germansk jernalder'. *Nationalmus Arbejdsmark* 1996, 84–98.

Nørgård Jørgensen, A. 1997a. 'Scandinavian Military Equipment and the Weapon-burial Rite, A.D. 530–800. Foreign Influence and Regional Variation'. In C. Jensen and K. Høilund Nielsen (eds), *From Burial to Society*, 200–9. Højbjerg.

Nørgård Jørgensen, A. 1997b. 'Introduction'. In A. Nørgård Jørgensen and B. Clausen (eds), *Military Aspects of Scandinavian Society AD 1–1300*, 7–9. PNM Stud Archaeol Hist 2. Copenhagen.

Nørgård Jørgensen, A. (in print): *Waffen und Gräber. Skandinavische Waffengräber 530–900 n. Chr. Typologische und Chronologische Studien*. Nordiske Fortidsminder.

Ørsnes, M. 1966. *Form og Stil i Sydskandinaviens yngre germanske jernalder*. Nationalmus Skrifter, Arkæol-hist række XI. Copenhagen.

Ørsnes, M. 1970. 'Südskandinavische Ornamentik in der jüngeren germanischen Eisenzeit'. *Acta Archaeol* 40, 1–121.

Roth, H. und Theune, C. 1988. *SW ♀ I–V. Zur Chronologie Merowingerzeitlicher Frauengräber in Südwestdeutschland*. Archäologische Informationen aus Baden-Württemberg 6. Stuttgart.

Roth, U. 1987. 'Early Insular manuscripts: ornament and archaeology, with special reference to the dating of the Book of Durrow'. In M. Ryan (ed.), *Ireland and Insular Art A.D. 500–1200*, 23–9. Dublin.

Scollar, I. 1991/1995. *The Bonn Seriation and Archaeological Statistics Package ver. 4.1*. Bonn.

Siegmund, F. 1989. *Fränkische Funde vom deutschen Niederrhein und der nördlichen Kölner Bucht*. Inaugural-Dissertation, Universität Köln.

Stein, F. 1967. *Adelsgräber des achten Jahrhunderts in Deutschland*. Berlin.

10. Female grave goods of southern and eastern Scandinavia from the Late Germanic Iron Age or Vendel Period

Karen Høilund Nielsen

The Germanic Iron Age in Scandinavia, comprising the Migration and Vendel Periods, has been studied for more than 150 years. Most chronological studies have focussed on the Vendel Period, from which most of the grave finds derive. In the Migration Period finds from Denmark are dominated by hoards and stray finds. In Sweden, and especially in Norway, more graves are known from this period, but they are often rather sparsely furnished.

An increasing interest in these periods, partially due to many recent metal-detector finds, has revitalized research and led to the excavation of large cemeteries such as Sejlflod (Nielsen 1980; Nielsen *et al* 1985), Hjemsted (Ethelberg 1986) and Nørre Sandegård Vest (Jørgensen and Nørgård Jørgensen 1997). To be able to include both new and old finds in modern studies of social and historical developments in the various areas the dating of finds has also to be confirmed. To some extent it has been possible to date very rich male graves in Scandinavia on the basis of comparative analyses both with other areas and with some of the female graves within their own local region. This, however, is not enough for comparative Scandinavian studies. Therefore chronology has been reintroduced, as it were, though in Denmark at least chronological analysis never really ceased but its methods changed. Nowadays computerized correspondence analysis is used mostly, to produce the seriations on which the grave chronologies are based. New analyses are necessary to meet the demands for more detailed chronological schemes and to establish supra-regional chronologies.

A lot of evidence is available from Gotland and Bornholm, whereas Denmark (excluding Bornholm) and mainland Sweden offer comprehensible problems. Mainland Sweden has a large amount of cremation cemeteries, but very few of the graves contain more than one identifiable object that can be dated; often not even that. This makes the amount of graves suitable for analysis of type-combinations very small. Furthermore there is no real female counterpart to the rich Uppland inhumation boat graves. The problem with the Danish material is that no firm Danish chronology based on material from the central part of the country can be established at all. For the Late Germanic Iron Age the only large cemetery is Lindholm Høje, which is a cremation cemetery (Høilund Nielsen 1996; Ramskou 1976) that suffers from the same lack of artefacts as the Anglian cremation cemeteries. At Lindholm Høje not even the pottery is of an artistic quality that might have provided a basis for a chronological analysis. The chronological scheme for Denmark therefore relies heavily on the chronological scheme achieved from Bornholm, which has a large amount of inhumation graves, and which has the same female artefact-types as are known in central Denmark. This leaves us, however, with a serious problem: how we assess the datings in Denmark will then depend on our view of the historical and political relationship between Denmark and Bornholm. If Bornholm is seen as politically close to the centre, the chronology in the one area should not differ from that in the other. But what if Bornholm is a border area far from the centre? Should its dating value still be equal to that in the centre, or does the distance from the centre imply a time-delay? Actually there is clear evidence that the difference in material culture between Denmark and Bornholm increased during the Late Germanic Iron Age (Høilund Nielsen 1992a). The same points must be considered when, for instance, we date the Swedish boat graves which contain equipment of a foreign character. This is not exactly like the old discussion of long or short chronologies – I think in the most cases the chronology is short – but there are clear indications that items were not simply scattered around and end up in the graves immediately afterwards. Sutton Hoo mound 1, for instance, contains artefacts more than a century old. We thus cannot rely on such rich graves when establishing chronologies. The risk that the grave goods cover a large time span is too great. Because of the large amount of well-furnished and well-preserved cremation and inhumation graves from Gotland, the results from this area have served as the chronological basis for mainland Sweden, although the Gotlandic types often differ considerably from those of the Swedish mainland.

The aim of this paper is to present some typological

analyses of dress-accessories from the Vendel Period, and analyses of grave groups from three regions of southern and eastern Scandinavia and their interrelationships, including discussions of the problems involved in these analyses.

RESEARCH HISTORY

Typological and chronological analyses have constituted a conspicuous area of research within Scandinavian archaeology for more then two centuries. H. J. Thomsen and later Sophus Müller and Oscar Montelius did their pioneer work within this field in the 19th century and thus, especially Montelius, established the main chronological division still in use. The many contributions to this field since then have to a high degree been refinements and adjustments to the Montelian system. In Sweden a properly defined 'Vendel Culture' was discussed at an early stage, whereas in Denmark the Middle Iron Age (*Mellemjernalderen*), an early name for the Late Germanic Iron Age, was seen as a stage in the continuously changing material culture. For this reason the period was only clearly defined at a rather late stage.

In 1865 J. J. A. Worsaae divided the Late Iron Age into 'the first time span of the Late Iron Age (450–700)' (*den yngre jernalders første tidsrum (450–700)*) and its 'second time span (700–1000)' (*andet tidsrum (700–1000)*). The objects placed in 'the first time span' all belongs to the Early Germanic Iron Age, whereas some objects, including the early tortoise brooches, from the Late Germanic Iron Age belong to his 'second time span'.

To avoid working with a bipartition of the Late Iron Age, Conrad Engelhardt in 1867 introduced the 'Middle Iron Age' (*Mellemjernalderen*). To this phase was assigned material from both the Early Germanic Iron Age and the early part of the Late Germanic Iron Age. This supposed contemporaneity between material from these two periods was to characterize views for many years.

The man behind the study of the Late Germanic Iron Age on Bornholm, Emil Vedel, presented his first contribution in 1878. In this he isolated most of the material belonging to the Germanic Iron Age. He divided the Middle Iron Age into an early and a late phase, with the early phase containing the later part of the Early Germanic Iron Age and phases 1–2 of Ørsnes (see below). The late phase covers the later part of Ørsnes's phases 2 and 3. In 1886 Vedel's principal work, *Bornholms Oldtidsminder og Oldsager* (= 'Monuments and Antiquities from Bornholm'), was published. This dealt with the whole prehistory of Bornholm, thus including the Middle Iron Age which he divided into a series of successive stages each characterized by one or more brooch-types:

1) Narrow bow-brooches and equal-armed brooches
2) Relief- and disc-on-bow brooches, beak brooches, bird-shaped brooches, oval and rectangular disc brooches
3) Circular disc brooches
4) Small tortoise brooches
5) Large tortoise brooches

The Early Germanic Iron Age, however, was still not isolated as an independent period, but partly included in the Late Germanic Iron Age. Like Worsaae and Montelius, he dated the entire period to 450–700. Vedel also studied the male graves, but not with the same refined results as in the case of the female graves.

In 1897 Sophus Müller published *Vor Oldtid* (= 'Our Antiquity'), in which he also discussed the questions of periodization. He used the term 'the post-Roman Time' (*Den efterromerske Tid*) for the period previously called the Middle Iron Age. He reported that these centuries were not distinctly different from the Viking Age, so that the Middle Iron Age was not to be seen as a main phase; rather that one had agreed on a division of the entire Iron Age into early and late sections. Thus, the Late Iron Age was subdivided in to 'the post-Roman Time and the Viking Age. To the post-Roman time belongs everything that lies between the rich grave and bog finds of the Migration Period and the large group of finds definitely belonging to the Viking Age' (Müller 1897, 594).

In 1905, Knut Stjerna published his large article *Bidrag til Bornholms Befolkningshistoria under Järnåldern* (= 'Contribution to the history of the population of Bornholm in the Iron Age'). This also included a discussion of typological and chronological problems. Stjerna typologized the material and studied typological development. Subsequently, on this basis, he studied the chronological development in order to find the order of the burials. In Stjerna's work for the first time there was a reasonable division between the material belonging to the Early and to the Late Germanic Iron Age from Bornholm in line with the chronology presented by Montelius in 1892 (see below).

Although ahead of his time in many ways, Stjerna was not without his critics. When Lindqvist, in 1926, in *Vendelkulturens ålder och ursprung* (= 'The Age and Origin of the Vendel Culture') went through the entire Continental comparative material from the late Migration Period and the Vendel Period, including absolute chronologies both for the Continent and for Scandinavia, he was very critical of previous chronological analyses. Stjerna (1905) and Salin (1904) were criticized especially for their typological dating method, which was inspired by Darwinism: 'The main emphasis is on a detailed typological analysis of the shapes and decorations of the objects occurring, a study which ... might actually lead to decisive conclusions at least for relative chronology, as, like in Darwin's biological system, many types may be arranged in sequence like natural links in a coherent, continuous successive series of development and – in contrast to the animal species – be assumed to die out by turns as they give life to new forms. Experience shows that really good typological series, fulfilling the specified criteria, can most easily be established for periods characterized by local isolation and stuntedness, for which reason these series rather ought to be named series of degeneration than series of development' (Lindqvist 1926, 116).

Lindqvist also criticized Stjerna for building on Salin's chronology, which was based on the typological part of Montelius's works. In contrast to most others, Lindqvist dated the transition between the periods VI and VII to circa 500, whereas they are usually dated to 550 or 600.

The term 'Germanic Iron Age' was first introduced for the period 400–800 by Johannes Brøndsted in 1934. He chose this name for the reason that it would be consistent with the Celtic and Roman Iron Age, names chosen to emphasize the main cultural groups influencing the Danish material at the given time (Brøndsted 1934, 175f.).

In the 1950's and 60's, the chronology of the Late Germanic Iron Age of Denmark and Skåne was thoroughly studied by Mogens Ørsnes (Ørsnes 1966; Ørsnes-Christensen 1956). He developed his typological analysis much further than Stjerna did, in accordance with current principles, with concrete definitions. He also based his chronological analysis on a sort of seriation of artefact combinations. Like Vedel, his emphasis lay on the female dress-accessories and much less on male equipment and weaponry. The result was a division of the Late Germanic Iron Age into three main phases with two or three sub-phases each. Not until the 1980's were further analyses done, and then, by the application of correspondence analysis, his results were tested and re-interpreted in some cases, with the basic development remaining the same although the phase-division is slightly changed as also the typological system (Høilund Nielsen 1987; 1988). A further refinement has been produced by Lars Jørgensen, by adding the new finds from Nørre Sandegård Vest and a re-division of some of the types and some of the phases (Jørgensen and Nørgård Jørgensen 1997). The male graves from Bornholm have also recently been studied and a more detailed chronology is now available (Nørgård Jørgensen 1992a). However it is still not possible to establish a weapon chronology from material from the other parts of Denmark.

In Sweden too, the Iron Age had been divided into three main periods, as in Oscar Montelius's *Svenska fornsaker* (= 'Swedish Antiquities') of 1872. The Middle Iron Age covered the period A.D. 450–700 and the Late Iron Age the period A.D. 700–1000/1100. When the first graves from Vendel were excavated in the 1880's and published (Stolpe 1884), they were assigned to the Middle and Late Iron Age.

The introduction of a new chronological scheme for the Nordic material by Oscar Montelius in 1892, which still holds, led to a division of the Iron Age into eight periods (I–VIII). The period 400–800 was called the Migration Period, but was divided into period VI, Early Migration Period (400–600), and period VII, Late Migration Period (600–800) (Montelius 1892; 1893). As a result, most of the finds from and related to the boat graves from Vendel in Uppland were placed in Period VII. The chronological scheme constructed for the Swedish material rested to a large extent on its *de luxe* elements. No attempt to use the more ordinary finds was seen until Åberg's publication of 1953 (see below).

The concept 'Vendel Period' first really became generally accepted in the 1930's. The previous name of the period was the Late Migration Period (cf. Olsén 1981), and the Montelian name for the period was Period VII. T. J. Arne introduced the concept 'Vendel Time' (*Vendeltid*: Arne 1919), whereas Nils Åberg was the one who made it into a name of a period (Åberg 1922). He also used the concept of 'Vendel Culture' – actually the only 'culture' referred to in the Scandinavian Late Iron Age.

For Gotland a fine chronology was produced at an early date. In 1878, the Gotlandic Iron Age was divided into six groups by Hans Hildebrand. The second group covers the Migration Period, whereas the third and the fourth groups cover the Vendel Period. Through Birger Nerman's work from 1919 onwards, the Montelian system was also introduced to the material from Gotland. Nerman divided Period VII into four phases. The absolute dates he gave for period VII, however, were changed to 550–800 (Nerman 1919). This dating of Period VII has survived to the present day. In 1969 (and 1975) Nerman presented a revised version of the chronology of the material from Gotland. This divided Period VII into five phases and is still in use.

Recently, a new analyses of the Gotlandic male chronology has been presented by Anne Nørgård Jørgensen (1992b). She divides the Vendel Period into five phases, and four of them into into further sub-phases each. Her absolute dates are slightly earlier than Nerman's. This chronology has also been related to the male chronologies of Bornholm and Norway (Jørgensen and Nørgård Jørgensen 1997; Nørgård Jørgensen, this volume).

ABOUT THE METHOD

The method of producing chronologies used here is based on the seriation principle, where the material is subject to a thorough typological analysis, and on basis of the combinations of types in the graves a serial structure is searched for. For the establishment of the series multivariate correspondence analysis has been used.

The seriation principle, in relation to chronology, builds upon the idea that types are successively replaced, and that a type first occurs sporadically, then increases in frequency to a maximum, and then declines, and that the contextual unit/grave is dated on basis of the average date of the objects within it (Gräslund 1974, 69ff.; Jensen and Høilund Nielsen 1997, 37f.; Ørsnes 1970, 11). It has been argued that this model is wrong, and that the 'seriation' ought to be sorted according to the latest object in the unit (Andresen and Ilkjær 1989; see also Høilund Nielsen 1992b). This difference in chronological model ought, however, to be judged in terms of the archaeological material which was the basis for the models. The former model was constructed into relation to female accessories found in graves, the latter to military hoards and a Europe-wide corpus of weapon burials. In the case of the latter model, how do we determine which type is latest before we make the chronological analysis? There is a real difference between weaponry from a battle and female dress accessories as grave goods, as the usefulness of the weapon may change radically when types change whereas a brooch still works as a dress-fastener even in such circumstances. The weapon hoards are also a sort of frozen picture, whereas female burials together represent a long and often slow development, and the single grave may in itself express the entire life of a human being unless it can be proven that the ornaments are all of the most recent fashion, which the often worn jewellery shows not to be the case. The conclusion is that the seriation model ought to work in the case of female ornaments, and that in this case there is hardly any possibility of stating beforehand which objects are the latest.

It must nevertheless be noted that a seriation chronology is a chronology based on an average date for the analysed object-types from the particular graves. Thus, two individuals whose grave goods have the same seriation-dating may have died at very different times, as two contemporaries – one who died as a young person and the other as a very old person – carrying the same ornaments given to them in their youth, on basis of their ornaments would be placed together in the seriation. These anomalies can only be dealt with if other data such as skeletal data or topo-chronology indicate this time-difference. If it were possible to argue that the lay-out of a cemetery is determined solely by the time of burial, it would be possible to some degree to re-establish the 'real' order in which people were buried, but this can rarely be done in detail and unambiguously. Actually this is an entirely different question, as a topo-chronology would give the order of the burials, which is of most interest at the local level, whereas the seriation shows the trends over a broader range of society, at the macro-level, where the isolated occurrence of an old person with out-dated ornaments is of minor importance.

The inclusion of larger geographical areas causes the risk of inconsistency to increase, as equipment, style, fashion and taste may vary both localy and regionally. In supra-regional chronologies it is therefore necessary to accept a coarser picture in order to eliminate regional variation.

Seriation usually ignores that the ornaments are not necessarily given to the deceased according to certain rules; that the order is a statistical approximation; and that a regional variation may blur the picture. To meet these problems minor irregularities are accepted. Furthermore, the result ought not to be seen as a rigid and frozen system, which can be used for ever, but one which can be used as a tool in relation to the body of material concerned. New finds may always change the chronological details — no result represents the entire truth as long our associated body of finds make up such a small amount of the pre-historic material culture: probably between 0.1 and 1 per cent of the material originally deposited (Steuer 1982, 68). In mainland Sweden especially, we must be prepared for changes caused by new finds, while Bornholm and Gotland are already much better covered, albeit less well at the period boundaries. Jørgensen's analysis has also shown that even here minor changes appear when a new cemetery is included (in Jørgensen and Nørgård Jørgensen 1997).

To produce the seriations, correspondence analysis was applied (Baxter 1994; Jensen and Høilund Nielsen 1997). The immediate result of a correspondence analysis is a plot, and – if it is a good seriation – the units and variables together form an approximate parabola shape. If this shape does not appear, it is not possible to make a good seriation of he material: in other words the criterion of continuity is not fulfilled (cf. Malmer 1963). On basis of the plot it is possible to sort the matrix that underlies the correspondence analysis, as the units (graves) and variables (object-types) are sorted according to their first coordinate in the plot. This produces a matrix with the graves and the object-types organized in such a way that the occurrences form a diagonal from one corner of the matrix to the other: the so-called main-diagonal. There is a difference between the plot and the sorted matrix. On basis of the plot of the types, it is possible only to see which types are most characteristic for which graves, but not the entire range of graves in which the types occur. By comparing the plot and the matrix it is possible both to find out when types are especially character-istic and their overall range. Both are important when suggesting a phasing of the chronological series. The chosen model also means that the concept of diagnostic leading types ('Leittypen') has no meaning at all as it is almost impossible to make exclusive phases; the more even the development, the more difficult it is to establish phases, to make a chronological scheme.

Let us turn to a more practical example of how a supra-regional or master chronology for an area can be established. Three main areas could be established on basis of the distribution of brooch-types. Material culture varies considerably between the three analysed areas (Høilund Nielsen 1991; 1992a).

With regard to the distribution of burials of the Late Germanic Iron Age in southern and eastern Scandinavia, the areas chosen for chronological analysis are Bornholm, Gotland and mainland Sweden. The other areas simply do not provide sufficient graves for any analysis. Even mainland Sweden is questionable, but has been analysed despite that. To avoid regional variance having too much influence on the chronology, and because the sets of brooches in the three areas are very different, detailed chronologies were made within each region, at which point mainland Sweden actually provides some problems partly because the area is too large, and partly because there are too few graves with sufficient recordable material for combinational analyses. A further basis for choosing these areas is an analysis of the regional variation within the area, which clearly separates the material from southern

Scandinavia, Gotland, and central Sweden respectively (Høilund Nielsen 1991). Following the regional chronological analyses an attempt is made to combine the results in a sort of supra-regional chronology for the entire area. This is based on the ornament types that after all were the same in the three regions, and on occurrences on 'contamination' of objects from the neighbouring areas.

TYPOLOGICAL ANALYSIS OF DRESS-FASTENERS AND ORNAMENTS

The area analysed includes Denmark and Sweden from Uppland southwards including Gotland.[1] The typology is based on the typology of Ørsnes (1966) with additions from my later revisions (Høilund Nielsen 1987). The inclusion of a much larger body of material resulted in an expansion of several parts of the typology and a different typology for certain types. It is not my intention here to go deeply into the typological framework; rather to refer to the appended typological lists. However, the beads, the disc brooches and the tortoise brooches are analysed here. These two brooch-types were subject to a time-governed change of size and decoration which has never been studied in detail before now.

Tortoise brooches

From the area analysed a total of 174 tortoise brooches, including also the animal-shaped examples, have been analysed. Ørsnes (1966) was of the opinion that the growing length of the tortoise brooches was chronologically significant. He divided the Danish material into two size-groups: 3.6–7.5 cm and 8.0–11.2 cm. With the Swedish specimens included this division was, however, insufficient, as many of the Swedish brooches are considerably smaller than the Danish ones. Hence a new measurement-based analysis is applied, based on all measurable brooches of non-Viking-age character. In this analysis no distinction has been made between the ordinary tortoise brooches and the animal-shaped tortoise brooches, as they are very closely related.

The recorded brooches are listed in figure 10.1, where also lengths and decoration are recorded. As a result of decomposition and wear the measurements are accurate to within a couple of millimetres. The distribution of measurements of length is illustrated in figure 10.2, and in figures 10.3–10.6 the length measurements are differentiated by the various types of decoration (plain; geometrical decoration; animal seen from above; intertwined animals). A comparison of the diagrams reveals a series of groups. The division between N1 and N2, however, is mainly based on decoration. The measurement groups are thus:

N1: 2.5 – 4.8 cm
N2: 4.9 – 5.6 cm
N3: 5.7 – 7.5 cm
N4: 7.8 – 9.9 cm
N5: 10.3 – 11.2 cm

The decoration is divided into four groups:

a) plain
b) geometrical decoration
c) animal seen from above
d) intertwined animals

Figures 10.3–10.7 show that the measurement groups clearly correlate with changes in decoration. *Plain* decoration is found exclusively on the smallest brooches (N1), and *geometrical decoration* is also most frequent here. The *animal seen from above* is predominantly found in N1–3 and is dominant in N2. *Intertwined animals* are nearly only seen in N3–5, where they are dominant in all three groups. A typological description is thus as follows:

N1: the length ranges from 2.5–4.8 cm. Dominant are brooches without decoration, although some with geometrical decoration and with an animal seen from above occur.
N2: the length ranges from 4.9–5.6 cm. Dominant are brooches with the animal seen from above, although geometrically decorated brooches also occur.
N3: the length ranges from 5.7–7.5 cm. Intertwined animals are dominant, although brooches with an animal seen from above are relatively frequent.
N4: the length ranges from 7.8–9.9 cm. Intertwined animals are nearly totally dominant.
N5: the length ranges between 10.3–11.2 cm. Intertwined animals are totally dominant. There are, however, only a few brooches in this group.

To give a detailed description of the brooches the type-names will be N1a–N5d; N means it is an ordinary tortoise brooch (O that it is an animal-shaped tortoise brooch); 1–5 decribes the length; a–d the decoration. Not all combinations of length and decoration exist.

As noted above, I have so far not distinguished between ordinary tortoise brooches and animal-shaped tortoise brooches. Ørsnes's division of the animal-shaped tortoise brooches is closely connected to the style of the animal, which has not been considered here. It is striking that although the type is presumed to derive from virtual animals seen from above, most of the brooches have far more complicated decoration. The brooch-type with only one animal is only found very rarely within types N1–4, most frequently in N4 (Figs. 10.8–10.9). It is also clear that the ordinary tortoise brooches with an animal seen from above are smaller than the animal-shaped ones with such an animal (Fig. 10.9). Furthermore intertwined animals are seen on ordinary tortoise brooches of all lengths, but most frequently in N3 (Fig. 10.10). Those of animal shape are only found in N3 (=O3). Animal-shaped tortoise brooches are thus found with an animal seen from above predominantly in N/O3 and with intertwined animals in N/O4 (with few exceptions, only O-types O1c, O2c, O3d, O4c occur in the data set).

	Length, cm	Plain	Plain geom. dec.	Geom.dec.	Animal	Intertwined animals	Type	Combinations
asrd2592	5,9					x	N3d	
åhm129x1230	5,4		x				N2c	
åhm129x1520	3		x				N1c	
åhm129x1725	3,5	x					N1a	R3b?
bmr1729	6,5				x		N3d	M3
bmr1730	6,5				x		N3d	M3
bmrjnr1002A	6,4			x	x		O3d	
bmrjnr1002B	6,4			x	x		O3d	
fsm7390	6,7				x		N3d	
Helgö150:54	4,2	x					N1b	
hm1319	4,1		?	?			N1d	
klm05853	7		x				N3c	
klm07565	4,5		x				N1c	
luhm10623	9				x		N4d	
luhm29055	>7				x		N4-5d	
luhm29226	4		x				N1c	
nm10774a	8,8		?	?			N4d	
nm10774b	9,6				x		N4d	
nm3413	8,5				x		N4d	
nm4970/83x24	4,2		x				N1c	
nm5267/83	8				x		N4d	
nm570/72	4,4				x		N1d	
nm5725/84d	4,2		x		(x)		N1b	
nm999x875	2,5		?				N1c	
nmc10269	5,5				x		N2c	I3B,P2C,Q2C,S1,R3B
nmc10270???	5,5				x		N2c	I3B,P2C,Q2C,S1,R3B
nmc10285	6,3		?	?			O?d	R3B
nmc10289	6,2		?	?	x		O3d	R3B
nmc12263	9,5				x		N4d	
nmc12264	8,5				x		N4d	
nmc12265	9,3				x		N4d	
nmc13494	6				x		N3d	
nmc1752	11,2				x		N5d	
nmc23186	5		x				N2c	
nmc2385	6,5		?		x		O3d	I3,R3B,Q3C
nmc2389	5,7		x				N3c	
nmc2394	7,5		?		x		O3d	R3B
nmc2395	6,5		?		x		O3d	R3B
nmc2396	8,8		?				O4c	R3B
nmc2408	7				x		O3d	R3B
nmc2409	9		?				O4c/d	R3B
nmc2454	9,4		?		x		N4d	R3B,Q3C
nmc2455	6,1				x		N3c	R3B,Q3C
nmc2456			x		?			R3B,Q3C
nmc25000	4,6				x		N1d	
nmc2504	8		?		x		N4d	I3,R3B
nmc2511	8,4				x		N4d	I3,Q3B,Q3D,R3B
nmc2512	8,4				x		N4d	I3,Q3B,Q3D,R3B
nmc2530	8,4		x				N4c	
nmc2534	6,5		?		x		O3d	I3,Q3C,R3B
nmc25519	10,6				x		N5c	Q3B,Q3C,R3D
nmc25520	10,6				x		N5c	Q3B,Q3C,R3D
nmc26581	6,1				x		O3d	R3C
nmc26582	6,1				x		O3d	R3C
nmc26583	4,8				x		N1c*	R3C
nmc27959	9,1				x		N4d	
nmc27995					?	?		
nmc2898	9,5		x		?		N4d	Q3C,R3D
nmc2903	11		x		x		N5d	
nmc2904	11		x		x		N5d	
nmc30664	3,6		x				N1c	
nmc30665	8,4		x		x		N4d	
nmc30730	5,4		x				N2c	
nmc30750	6,8		x				N3c	
nmc30830	6,5				x		N3d	
nmc30983	4,4				x		N1d	
nmc30998	4,9		x				N2c	
nmc3102	8,2				x		N4d	Q3B/C,R3B
nmc3103	8,5				x		N4d	Q3B/C,R3B
nmc4802	4,6	x					N1b	
nmc5339	4,2		x				N1c	Beads
nmc5362	8,5				x		O4c/d	R3C
nmc5405	6,5				x		O3d	
nmc5587a	7				x		O3d	R3B
nmc5587b	7				x		O3d	R3B
nmc5592	8,2				x		O4c	R3B
nmc5593	6,2		?		x		O3d	R3B
nmc5603	6,8		?		x		O3d	R3B
nmc5604	6,8		?		x		O3d	R3B
nmc5607	7,3				x		O3c	Q3B,Q3C,R3B
nmc5608a	4,7		x				N1c	Q3B,Q3C,R3B
nmc5608b	4,7		x				N1c	Q3B,Q3C,R3B
nmc5706?	10,3				x		N5d	R3D
nmc5707?	10,4				x		N5d	R3D
nmc6593a	9,7				x		N4d	Q3B,R3D
nmc6593b	9,7				x		N4d	Q3B,R3D
nmc6872	7,2				?	x	O3d	

	length, cm	plain	plain geom. dec.	geom.dec.	animal	intertwined animals	type	combinations
nmc7222	9,5				x		O4c	R3C
nmc7223	6,5				x		N3c	R3C
nmc7741	9,1					x	N4d	
nmc7756	5,2				x		N2c	
nmc965:b	5				x		N2c	R3B
nmc965a	5,2				x		O2c	R3B
privateje(Ohlsson)	5				x		N2c	
privateje(Paulsson)	3,8					x	N1d	
Raä141:51:a	5,9	x					N3a	
Raä141:51:b	6	x					N3a	
Raä141:51:c	3,8		x				N1b	
raä:10:10								R3B
raä:10:13	5,3				x		N2d	
raä:10:36	3,5	x					N1a	
raä:277:19	4,3		?	?			N1b	R3B,Q2B
raä:27:91:a	4,6	x					N1a	
raä:27:91:b	4,2	x					N1a	
raä:27:91:c	3,9		x				N1b	
raä:6693/68:a	6				x		N3d	R3C?
raä:6693/68:b	7,2				x	x	N3d	R3C?
shm 5208:33	6,1				x		N3d	
shm01304:1834:97	5,3		x				N2c	
shm01304:1835:98	5,6		x				N2b	
shm01304:1836:33	4	(x)					N1a	
shm01985:1848:11	7,2		x				N3c	
shm03414:c:3	8,8				x		N4d	
shm05042	5,1		?	?			N2d	
shm06745:1	6,4		x				N3c	E2C2
shm06745:2	6,4		x				N3c	E2C2
shm07584	6,7				x		N3d	
shm08800:28	3,7	x					N1a	A2E,R3B
shm08923:7:a1	4,8			x			O1c*	
shm08923:7:a2	6,8		x				N3c	
shm09521:b:5c	3,5	?					N1a	K1A
shm10033:14		x						
shm11229	9,9				x		N4d	three-lobed
shm11494	3,8		x				N1c	
shm12176	6,1		?	(?)			N3d	
shm12314:5	4,8		x				N1c*	
shm14723:a	3,8	(x)					N1a	
shm15174:11:a	3,8	x					N1a	
shm15174:11:b	3,8	x					N1a	
shm15174:18	4	x					N1a	
shm15618:12	3,1	x					N1a	
shm17906:10	4,7	(x)					N1a	
shm17906:11	5,2		?	?			N2d	
shm18355:36	3,8	x					N1a	
shm18876:6	4,5	x					N1b	R3E
shm20319	6,5				x		N3d	
shm20523:53:a	8-10cm				x		N4d	S2F
shm2076:15	4,9		x				N2d	
shm2076:16	4,1	x					N1b	
shm23304:17	4,4	x					N1b	
shm23648	5		x				N2c	
shm24198:2	7,8			x			N4b	I3B9,R3B,S2F
shm24362:a	7			x	?		N3d	
shm24362:b	7			x	?		N3d	
shm25848:86	6,3				x		N3d	R3D
shm25849:10	4,8	x					N1a*	R3C,S2F,P1B2
shm26042:172/24	4	x					N1a	R3B
shm26042:173/125	4,6	x					N1a	R3B
shm26042:180/121	4,3	x					N1a	R3B
shm26042:182/132	3,4	x					N1a	R3B
shm26481:60	3,5			x			O1c	R3B
shm26696	8						N4	
shm27124:stl						x		C/K,E2
shm27702:13								
shm27877								R3E?
shm28913	5,9				x		O3c	
shm30621:11	4,4	x					N1a	R3B
shm30621:36	3,5	x					N1a	R3B
shm31039:19	3,8	x					N1a	R3B
shm31039:6				x?				R3D
shm31277:10	4				x		N1c	R3B
shm31439:22:a	6,8					?	N3d	R3D
shm31439:22:b	7,2	x					N3b	R3D
shm31461:34	3,4	(x)					N1a	R3B
shm31461:6	4,4	x					N1a	R3B??
shmbj349	5,8				x	?	N3d	
shmbj485:a	8				x		N4d	M3
shmbj485:b	7,9				x		N4d	M3
shmbj602	3,6				x		N1d	M3
shmbj655:a	9,6				x		N4d	R3C
shmbj655:b	9,6				x		N4d	R3C
ssm:e3	4		x				N1c	
umf3294	9				x		N4d	
ym145	4,9				x	(x)	N2c	

Fig. 10.1 Catalogue of tortoise brooches. (Based on material from Nationalmuseet, Copenhagen; Bornholms Museum, Rønne; Fyns Stiftsmuseum, Odense; Den Antikvariske Samling, Ribe; Aalborg Historisk Museum; Horsens Museum; Statens Historiska Museum, Stockholm; Stockholms Stadsmuseum; Upplands Museet för Nordiska Fornsaker, Uppsala; Kalmar Läns Museum; Lunds Universitets Historiska Museum; Ystad Museum; and Riksantikvarieämbetets rapporter).
Katalog der Schildkrötenfibeln.

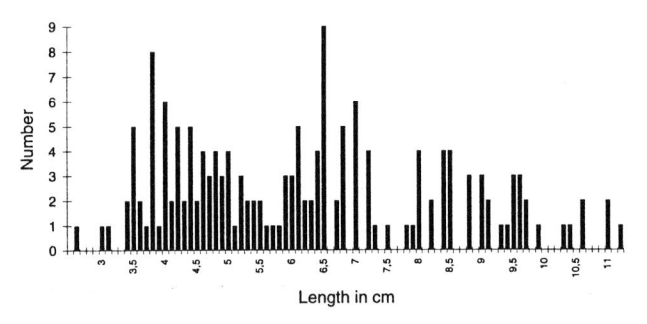

Fig. 10.2 Lengths of all measured tortoise brooches.
Längen aller vermessenen Schildkrötenfibeln.

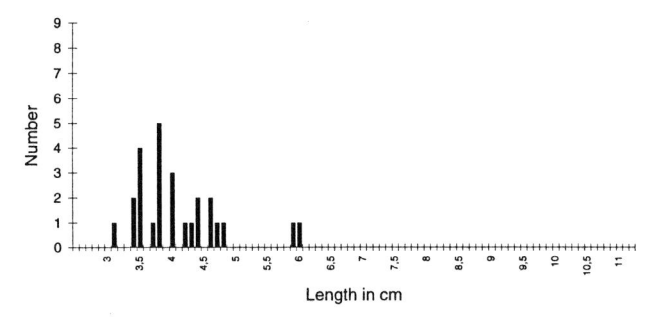

Fig. 10.3 Lengths of all plain tortoise brooches.
Längen der einfachen Schildkrötenfibeln.

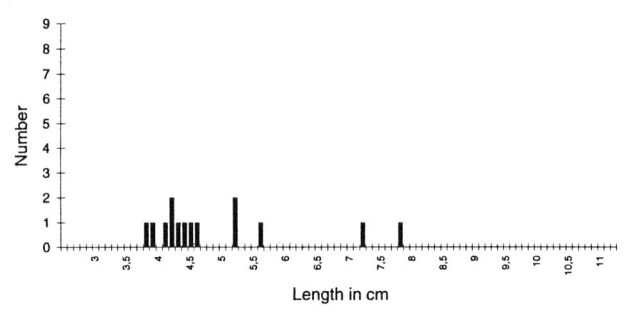

Fig. 10.4 Lengths of all geometrically decorated tortoise brooches.
Längen der geometrisch verzierten Schildkrötenfibeln.

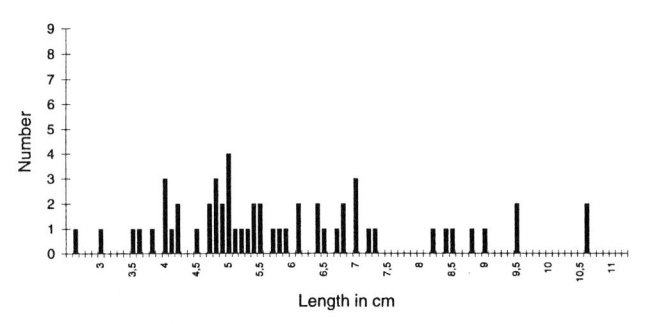

Fig. 10.5 Lengths of all tortoise brooches with animal seen from above.
Längen der Schildkrötenfibeln mit von oben gesehenem Tier.

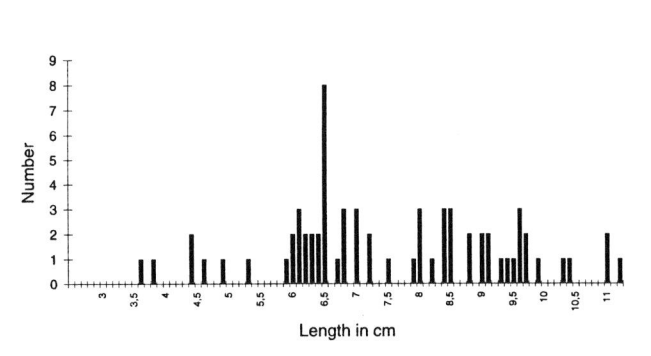

Fig. 10.6 Lengths of all tortoise brooches with intertwined animals.
Längen der Schildkrötenfibeln mit verflochtenen Tieren.

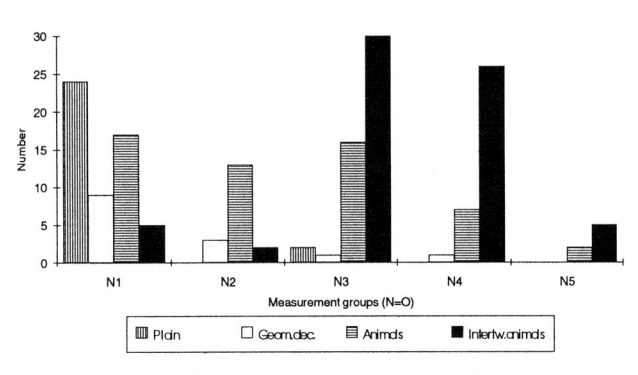

Fig. 10.7 Distribution of decoration in all measurement groups.
Verteilung der Verzierungen auf die Größenklassen.

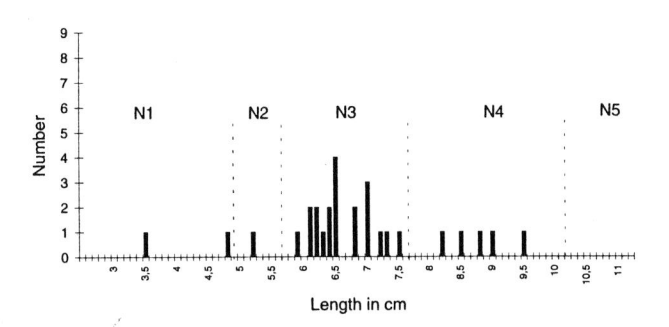

Fig. 10.8 Lengths of all animal-shaped tortoise brooches.
Längen der tierförmigen Schildkrötenfibeln.

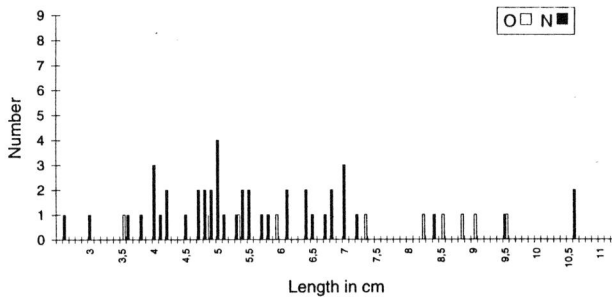

Fig. 10.9 Lengths of all tortoise brooches with animal seen from above. The animal-shaped tortoise brooches are marked.
Längen der Schildkrötenfibeln mit von oben gesehenem Tier. Die tierförmigen Fibeln sind besonders hervorgehoben.

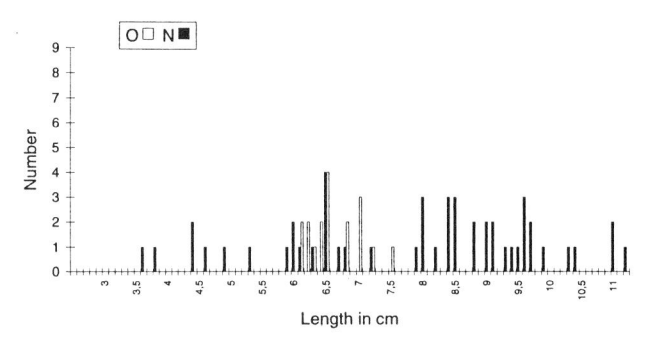

Fig. 10.10 Lengths of all tortoise brooches with intertwined animals. The animal-shaped tortoise brooches are marked.
Längen der Schildkrötenfibeln mit verflochtenen Tieren. Die tierförmigen Fibeln sind besonders hervorgehoben.

Circular brooches with a rim

An examination of the Gotlandic disc and early box brooches (type I) made it clear that the brooches are highly stereotyped, and it is possible to use the face decoration to divide the brooches into a series of clearly distinct groups (Fig. 10.11). Some are unique, however. Consultation of Nerman (1969) reveals a clear tendency to increasing rim sizes and a parallel development in the decoration of the rim from plain to zoomorphic. The following typology is therefore based on outer rim size and the type of decoration on the rim. The published Gotlandic brooches constitute the data set. A corpus of 121 brooches is included, and measurements, rim decoration and face-type are listed in figure 10.12. The measurements are accurate to *circa* 1 millimetre.

A clear tendency to four size groups with their own type of rim and face decoration is seen (Fig. 10.13):

1: no rim
2: 2–7 mm
3: 8–10 mm
4: 11–15 mm
5: 16–17 mm

Group 1 is not included in this analysis as it has no rim. On the basis of all rim sizes without respect to the decoration it is not possible to reveal any clustering: only that there are more brooches with narrow rims than with broad rims in the set analysed (Fig. 10.14).

It is possible to divide the rim decoration into six groups:

a: plain
b: horizontal lines
c: two twisted ribbons
d: interlace
e: chequer pattern
f: animals

From figures 10.15–10.17 it appears that the plain rim, horizontal lines and two twisted ribbons are found on brooches with a relatively narrow rim, whereas figure 10.18 shows that interlace is found on both narrow and broad rims and figures 10.19–10.20 show that the chequer pattern and animals are tied to the broad rims. The brooches have decoration chosen from a limited set of patterns (Fig. 10.11). Those are, as also mentioned above, highly dependent on rim size and rim decoration. A typological description can now be established:

I1: Not treated here.
I2: Rim sizes from 3–7 mm. The rim is plain or decorated with horizontal lines, two twisted ribbons or interlace. On the face types 1–2, 5–6 and 8–11 occur.
I3: Rim sizes from 8–10 mm. The rim is decorated with a chequer pattern or is plain. Very few are decorated with animals. On the face types 3, 7 and 12–13 occur.
I4: Rim sizes from 11–15 mm. The rim is decorated with interlace or the chequer pattern and animals in particular. On the face types 4 and 13–16 occur.
I5: Rim sizes from 16–17 mm. The rim is decorated only with animals. On the face types 14 and 17–18 occur.

It is thus possible to prove a close correlation between face-type, rim size and rim decoration. This relationship now means that more brooches are datable. The type names will be I2a1–I5f18, where 'I' indicates the main brooch-type, 1–5 the rim size, a–f the rim decoration and the final 1–18 the face-types. As with the tortoise brooches, not all possible types have been found.

Necklaces

The necklaces have already been subjected to thorough analysis for their typological and chronological significance on the basis of necklaces from Bornholm (Høilund Nielsen 1987; 1988; 1997). With the analysis of the Swedish female graves it proved to be necessary to include the necklaces. Chronological analysis would otherwise be impossible. On first impression, the Swedish necklaces appear to match contemporary necklaces from Bornholm. Some of the Swedish necklaces were in such bad condition, coming from cremations, that in these cases the typology was based on intuition conditioned by analysis of the material from Bornholm.

The evidence that could be recorded was registered at a level like the coarse one of those used for Bornholm (Høilund Nielsen 1987, 51ff.; Vedel 1890, 68ff.) and necklaces from both Bornholm and mainland Sweden were included. Correspondence analysis was applied to the Bornholm necklaces on their own and to all the necklaces from both Bornholm and mainland Sweden. The result of the analysis of the Bornholm necklaces, including about 3,900 beads, appears in figure 10.21. The result did not appear as the usual parabola and is thus not a traditional seriation. The distribution is not random, however, but a series of clusters with tendency towards seriation. A division into two groups is clearly visible along the first axis, while two further groups may be distinguished from the second axis. A sorting of the matrix according to the first axis is

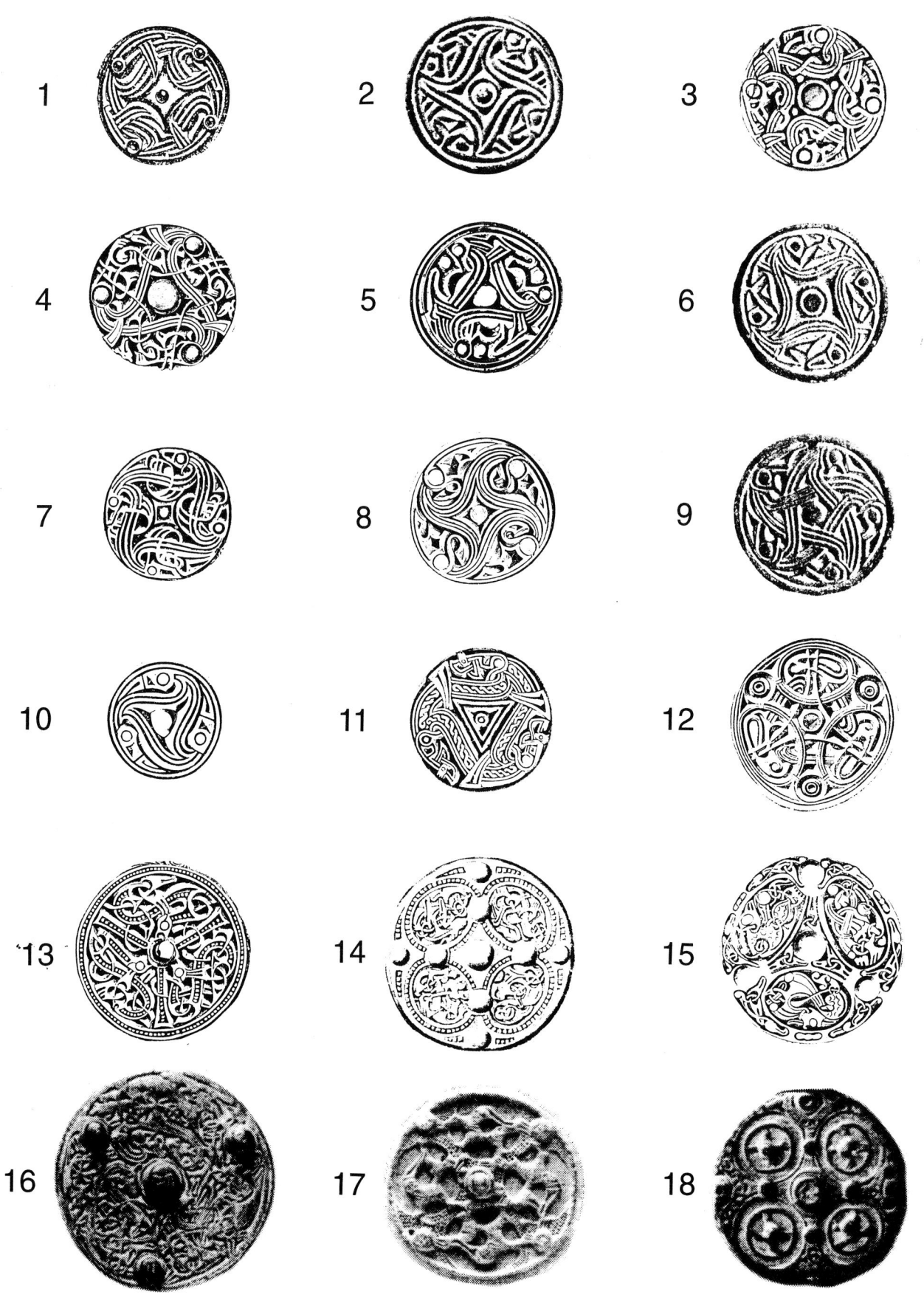

Fig. 10.11 Face-types of disc brooches. (After Nerman 1969).
Bildseitentypen der Scheibenfibeln.

Female grave goods of southern and eastern Scandinavia 169

Nerman fig.	Rim size, mm	Plain	Horizontal lines	Two twisted ribbons	Interlace	Chequer pattern	Animals	Face type
80	3	x						-
81	3	x						5
82	5	x						-
83	4	x						-
84	3	x						-
85	5		x					2
86	4	x						-
87	4	x						-
88	2	x						-
89	5	x						-
90	4	x						-
91	2	x						-
872	3	x						10
873	5		x					2
874	5	x						2
875	4		x					2
876	4	x						3
877	5		x					2
878	5			x				2
879	5			x				2
880	4	x						4
881	5	x						-
882	3	x						1
883	5	x						1
884	5	x						1
885	4		x					1
886	6		x					1
887	5		x					1
888	4			x				1
889	3	x						1
890	4	x						1
1380	5	x						5
1381	5	x						5
1382	5				x			5
1383	6				x			5
1384	6				x			5
1385	7					x		5
1386	7			(x)	x			5
1387	7				x			5
1388	6	x						5
1389	6	x						10
1390	7	x						5
1391	7		x					11
1392	7					x		11
1393	7				x			11
1395	5		x					1
1396	5		x					1
1397	4		x					1
1398	5				x			1
1399	6	x						1
1400	3	x						-
1401	5				x			6
1402	7					x		8
1403	5		x					8
1404	6				x			8
1405	7					x		9
1406	7	x						-
1407	9						x	7
1408	9						x	7
1409	8	x						3
1410	8					x		3

Nerman fig.	Rim size	Plain	Horizontal lines	Two twisted ribbons	Interlace	Chequer pattern	Animals	Face type
1411	11					?		3
1412	9	x						3
1413	8						(x)	3
1414	9	x						3
1415	10				x			3
1416	10				x			3
1806	8				x			3
1807	10						x	3
1808	10						x	3
1809	9	x						3
1810	10				x			3
1811	12			x				15
1812	11						x	4
1813	15						x	4
1814	12						x	15
1815	12						x	13
1816	9				x			2
1817								13
1818	9				x			2
1819	12				x			13
1820	11						x	13
1821	11						x	13
1822	10						x	13
1824	10	x						13
1825	13				x			13
1826	9				x			2
1827	13			x				13
1828	12						x	13
1829	14						x	13
1830	12						x	13
1831	11						x	13
1832	13						x	15
1833	11						x	15
1834	7	x						-
1835	9				x			1
1836	10	x						12
1837	10						x	12
1838							x	12
1839	10						x	12
1840	10				x			12
1841	11						x	15
1842	13						x	13
1843	12						x	13
1844	12						x	16
2156	13						x	15
2157								14
2158	15			x				14
2159	13			x				3
2160	12			x				14
2161	14						x	14
2162	15						x	14
2163	17						x	14
2164	17						x	14
2165	16						x	14
2166	17						x	14
2167	15						x	18
2168	17						x	18
2169	17						x	18
2170	17						x	18
2171	17						x	17

Fig. 10.12 Catalogue of Gotlandic disc and early box brooches. Based on Nerman 1969.
Katalog der gotländischen Scheibenfibeln und frühen Dosenfibeln.

not unambiguous, but a sorting combining first and second axes is possible. The sorted matrix (Fig. 10.22) thus clearly shows the distinction between the two main groups, a division also clear from the variation along the first axis. These are the necklace-types R3a and R3b (cf. Høilund Nielsen 1987). In the lower part of the sorted matrix also the types R3c and R3d appear. These were distinguished from the second axis.

Now the mainland Swedish material was included, comprising about 5,960 beads in total. The result of the correspondence analyses appears in figure 10.23. Two main groups again appear along the first axis and two further groups along the second axis. The result is thus in accordance with that achieved from the Bornholm material. The matrix was again sorted according to a combination of first and second axes (Fig. 10.24). Closer examination shows a clear resemblance to the Bornholm matrix, and the same groups can be distinguished. Hence, the same necklace-types as used for Bornholm are also applicable to the mainland Swedish necklaces.

Size of rim in mm	Total number	Pattern of rim: Plain	Horizontal lines	Two twisted ribbons	Interlace	Chequer pattern	Animals	Pattern of face: 1	10	5	2	6	8	9	11	3	12	7	13	4	15	16	14	18	17
2	2	2																							
3	7	7						2	1	1															
4	11	6	4	1				4			3														
5	20	8	5	4	3			6		3	6	1	1												
6	7	3	1		3			2	1	3			1												
7	13	3	1		2	5				5			1	1	4										
8	4	1			2	?										4									
9	9	3				6										3	1	2	3						
10	11	2				4	5									5	4		2						
11	7					1	6									1			3	1	2				
12	9			2		1	6										5	2			1	1			
13	6			2		1	3										3		2		1				
14	2						2										1				1				
15	4			1			3												1		2	1			
16	1						1														1				
17	7						7														3	3	1		
Pattern of rim:																									
Plain								6	2	5	2				2	4	1		1						
Horizontal lines								5			5														
Two twisted ribbons								2			2		1												
Interlace								1		4		1	1						1	1		3			
Chequer pattern										2			1	1		6	2	2	5						
Animals																3	3		10	3	4	1	6	4	1

Fig. 10.13 Combinations of size of rim, pattern of rim and faces of Gotlandic disc brooches.
Merkmalskombinationen gotländischer Scheibenfibeln: Randdiche, Randform und Bildseitentyp.

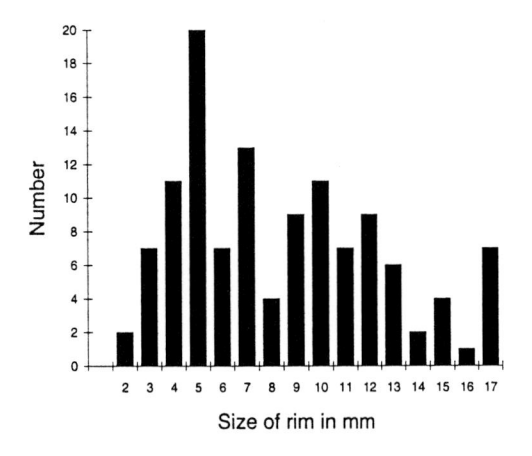

Fig. 10.14 Rim size of all measured disc brooches.
Randdiche aller vermessenen Scheibenfibeln.

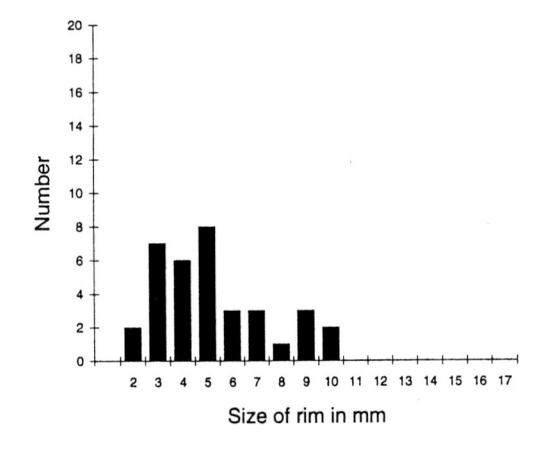

Fig. 10.15 Rim size of all disc brooches with plain rim.
Randdiche der Scheibenfibeln mit glattem Rand.

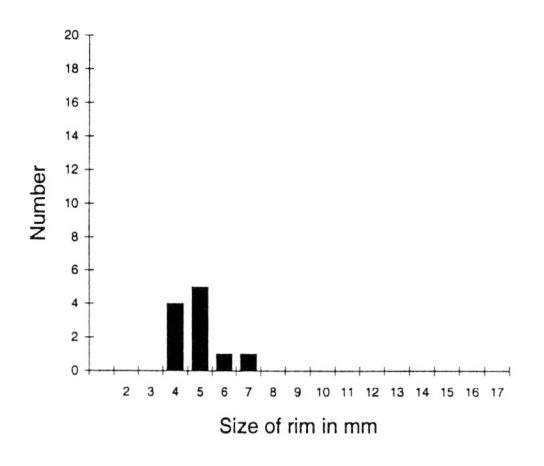

Fig. 10.16 Rim size of all disc brooches with rim with horizontal lines.
Randdiche der Scheibenfibeln mit horizontalen Linien auf dem Rand.

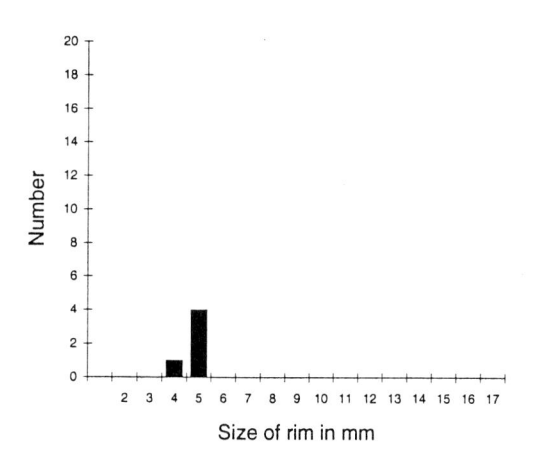

Fig. 10.17 Rim size of all disc brooches with rim with two twisted ribbons.
Randdiche der Scheibenfibeln mit zwei verflochtenen Bändern auf dem Rand.

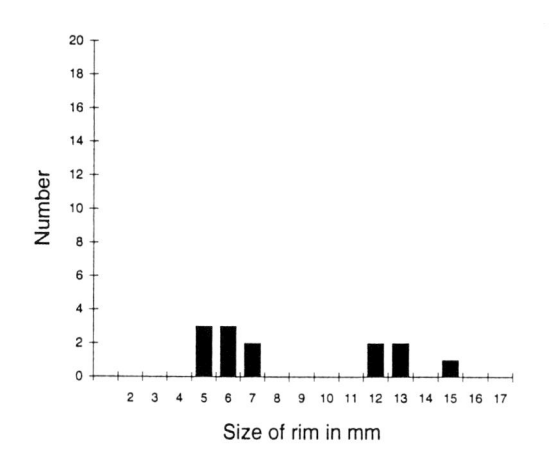

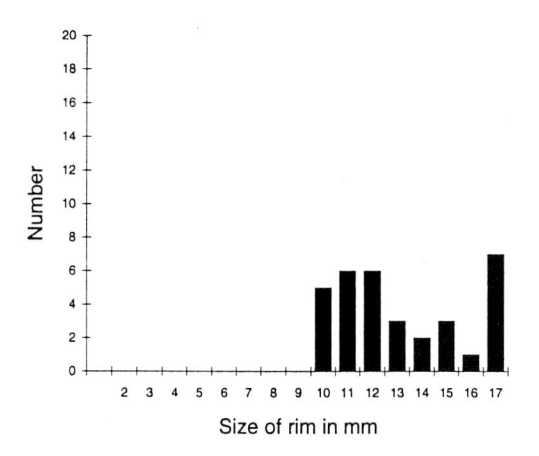

Fig. 10.18 Rim size of all disc brooches with rim with interlace.
Randdiche der Scheibenfibeln mit Flechtband auf dem Rand.

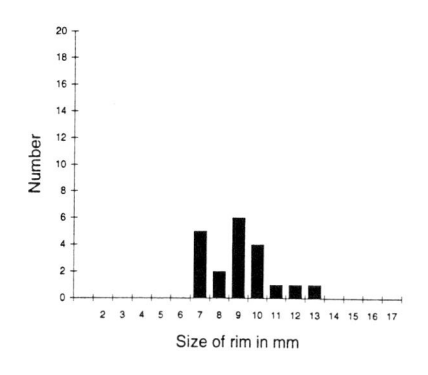

Fig. 10.19 Rim size of all disc brooches with rim with chequer-pattern.
Randdiche der Scheibenfibeln mit Schachbrettmuster auf dem Rand.

Fig. 10.20 Rim size of all disc brooches with rim with animals.
Randdiche der Scheibenfibeln mit Tieren auf dem Rand.

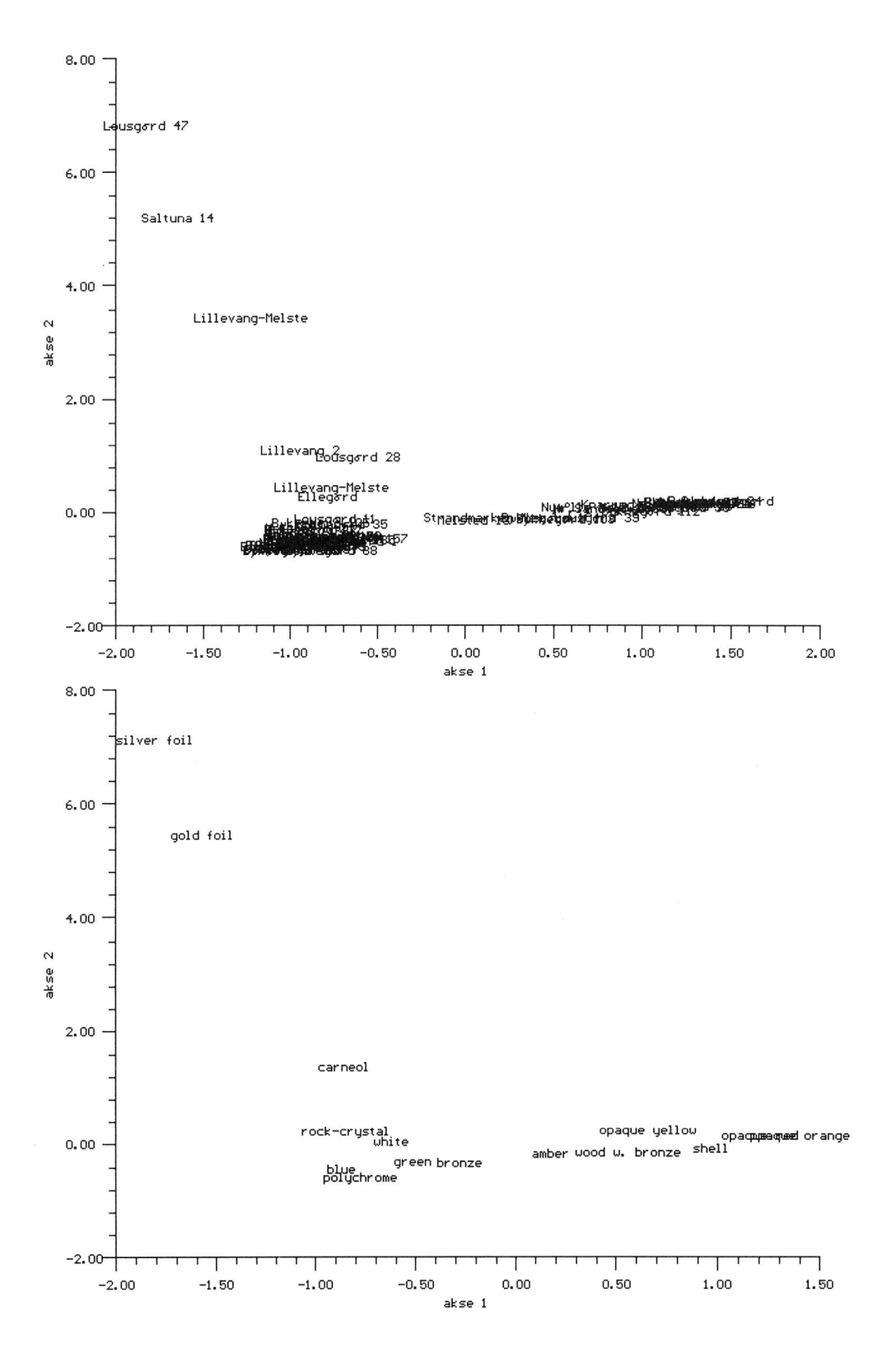

Fig. 10.21 Plot of correspondence analysis of necklaces (upper) and bead-types (lower) from Bornholm.
Korrespondenzanalyse der Halsketten aus Bornholm. Oben werden die Ketten, unten die Typen dargestellt.

	Opaque orange	Opaque red	Shell	Opaque yellow	Wood w. bronze	Amber	Bronze	Green	White	Polychrome	Blue	Rock crystal	Carneol	Gold foil	Silver foil	Group
Gudmingegård	54			2							1					
Bækkegård 24	28	68														
Kobbeå 33	19	12					2									
Bækkegård 90	76	41								4	3					
Lousgård 35	59	65		3			2	3	4							
Kobbeå 2	32	24					6									
Bækkegård 30	24	80						3	2		1					
Bækkegård 23	11	87		1				2	2		1					
Nr.Sandegård 426	37	72	3				1	2	1	4						
Lousgård 36	62	30			2		2		6	2						
Bækkegård 26	79	17				2	2	2		8						
Nr.Sandegård 36	25	7		1						4	1					R3A
Lousgård 40	25	27					2	10								
Bækkegård 112	35	2			6					8						
Knarregård 1	18	9		1		1	2		7							
Nr.Sandegård 397	8	55				11		14	5							
Lousgård 39	18		20				2	5			7					
Nymølle Nexø 2	1	41						8	11							
Bækkegård 103	18			3						10	2					
Bækkegård 104	12	6									12					
Strandmarken Nexø 12	3	51				13		39	26		19					
Melsted 13	11	28				16		14	20		28					
Bækkegård 88	1	1					7			42						
Bækkegård 143										16	17					
Bækkegård 134						2	1	2		24	16					
Bækkegård 66										10	20					
Bækkegård 62										13	24					
Bækkegård 3		1						9		20	5					
Lousgård 10					1	1	1	3		17	22					
Bækkegård 63										7	37					
Bækkegård 59									5	23	36					
Bækkegård 60		3				2			2	22	91					
Bækkegård 85									2	7	19					
Lousgård 6								2	2	9	9					
Bækkegård 5		1				1	1	27		27	34					
Lousgård 12		1					4	4	4	17	8					
Lousgård 9				1			5	2	2	9	16					
Nr.Sandegård 6							1	21	4	21	29	1				
Lousgård 18		1						10	1	8	15					
Lousgård 3								3	3	7	7					
St.Kannikegård 195	1						1	14	5	21	1					R3B
Bækkegård 135	1	1					2	13		10	8					
Bækkegård 76					2				4	5	20					
Lousgård 16		1					16	11	1	3	30					
Bækkegård 50	1								6	17	15	3				
Bækkegård 11						1		39	3	19						
Bækkegård 77		1					1		9	10	25					
Bækkegård 157	1			1			2	6		5	5					
Bækkegård 44								23	8	17						
Nr.Sandegård 4		2					8	34	30	29	61					
Bækkegård 2	1	2				1		2	8	6	32					
Bækkegård 16	1						1	22	2		19					
Bækkegård 153		3					1	7	17	28	9					
Bækkegård 93		1							27	19	32					
Bækkegård 17								7	21	2	24					
Bækkegård 132								11	12		21	3				
Melsted b 2									10		12					
Bækkegård 35		2			1		1		21	6	10					
Bækkegård 105								1	20	5	5					
Lousgård 11						1		2	25	6						
Ellegård		1					8	11	3	6	9			4		
Lillevang-Melsted 4	1							8	12	7				3		R3C
Lousgård 28		2		3			1	5	2	4		5	1	1	2	
Lillevang 2								5	16	3	1		1	5		
Lillevang-Melsted 2								6	19					9	9	
Saltuna 14									1		8			3	11	R3D
Lousgård 47									1		7			18	21	

Fig. 10.22 Sorted matrix of plot in figure 10.21. The matrix is sorted through a combination of the first and the second axes. Necklace-types are indicated.

Geordnete Matrix der Halsketten. Teils nach der ersten Achse der Korrespondenzanalyse Abb. 10.21, teils nach der zweiten Achse sortiert.

The types can briefly be described as follows:

R3a: predominance of red and orange opaque colours

R3b: predominance of blue and green colours

R3c: predominance of blue and green colours combined with gold foil glass beads

R3d: dominated by silver and gold foil glass beads and translucent colourless glass beads.

A further mainland Swedish necklace-type, not included in the analysis, and named R3e, embraces tiny glass beads of various colours. They are much smaller than the usual beads of the period in question, normally just below 10 mm. R3e may probably be connected with Callmer's group F (Callmer 1977, 89).

CHRONOLOGICAL SERIATIONS

A chronology of female grave groups from Bornholm

From 1878 to 1897 Vedel produced a series of chronological analyses of the graves from Bornholm (e.g. Vedel 1886). Ørsnes continued the process (1966). He analysed primarily the material from Bornholm, but included also the remaining Danish material and that from Skåne. The material from Bornholm and the results based on it have, however, always been decisive for Danish chronology, as find-associations are extremely rare in the Danish and Scanian material. The results of Vedel and Ørsnes both agree with my own analyses (both those presented here and those from Høilund Nielsen 1987 – a comparison is presented in Høilund Nielsen 1988).

The difference between the analysis below and that presented in Høilund Nielsen (1987) is the new typology of disc and tortoise brooches. The basis of the analysis is grave groups from Bornholm including primarily material of the Late Germanic Iron Age, plus some of the Early Germanic Iron Age and the Viking Age when this material occurs in grave groups from the same cemeteries. On Bornholm there is a slight possibility of stretching the series back into the Early Germanic Iron Age which is almost impossible in other areas of southern Scandinavia. Önsvala, however, has a grave group from this transitional phase (Larsson 1982), and Lindholm Høje probably has several (Ramskou 1976). To use the latter cemetery for an analysis of the period of transition would, however, imply a thorough analysis of the somewhat poor and plain pottery from that cemetery. Actually the same special type of pot is found with a cruciform brooch, an annular brooch and a beak brooch, which means that it should be possible to bridge the transitional phase here on the basis of the pottery (Høilund Nielsen 1982).

Correspondence analysis was applied to 93 grave groups with 42 types. The result was a good parabola (Fig. 10.25), which indicated that the criterion of continuity is fulfilled. There was furthermore no tendency towards clustering. A phase system is suggested in figure 10.26, showing the sorted matrix. In the phase system special attention has been paid to the necklaces as they are very common. The first phase contains bow brooches of Early Germanic Iron-age types as the characteristic element. Viking-age types are likewise characteristic of the last phase. The system will subsequently be called BOKV (=Bornholm female grave groups). The phases are described below on basis of the most characteristic types (for further information see Fig. 10.26).

BOKV1a: annular brooches (A2e), small equal-armed brooches (F1–3), Ae-pins with simple head (P1), beak brooches (decorated with stamps) (G1–G2).

BOKV1b: Ae-pins with simple head (P1), beak brooches (decorated with stamps) (G1–2), applied disc brooches (I1b), necklace terminals (R1), necklaces of type R3a.

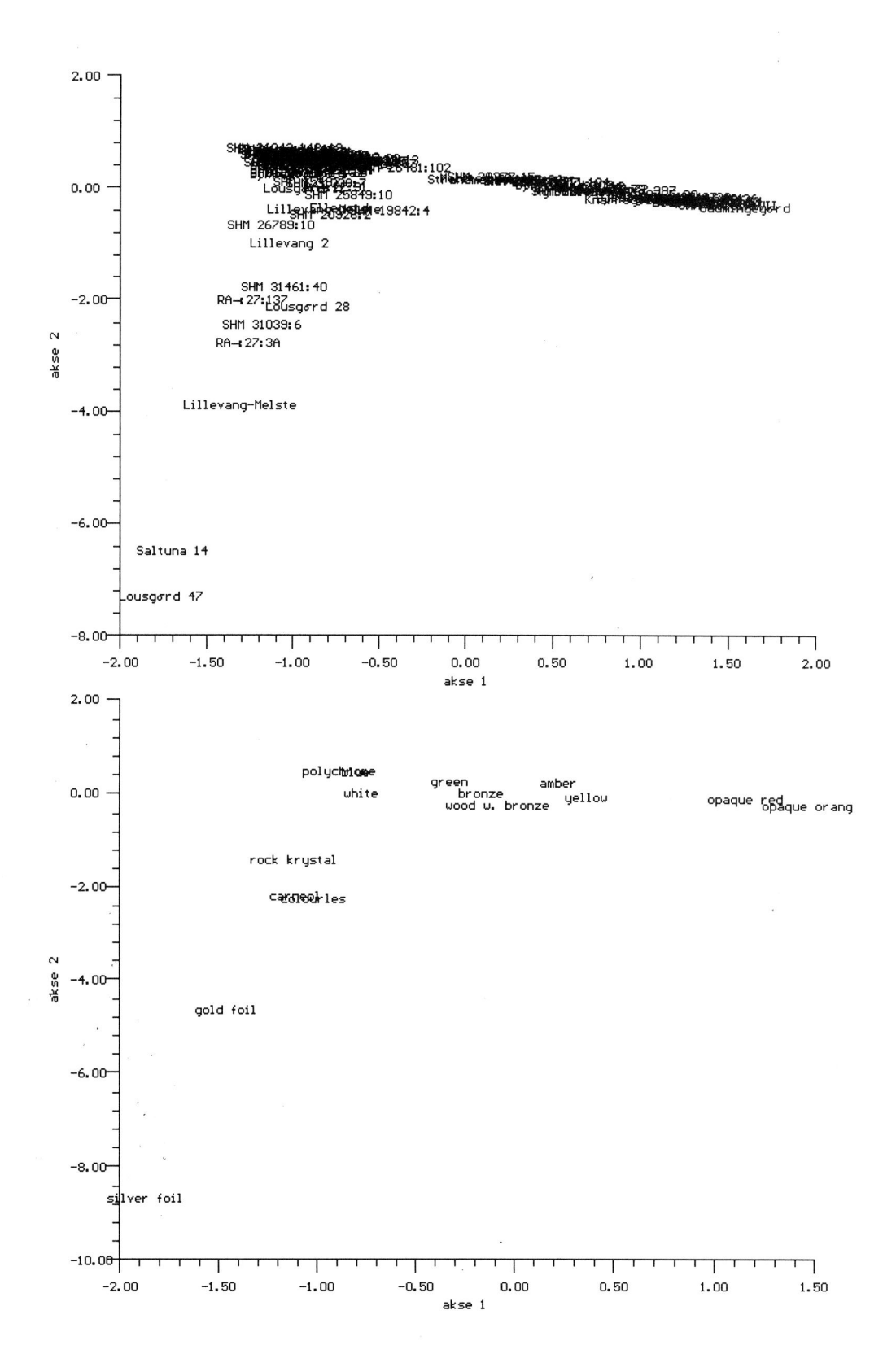

Fig. 10.23 Plot of correspondence analysis of necklaces (upper) and bead-types (lower) from Bornholm and mainland Sweden.
Korrespondenzanalyse der Halsketten aus Bornholm und dem schwedischen Festland. Oben werden die Ketten, unten die Typen dargestellt.

Female grave goods of southern and eastern Scandinavia 175

BOKV1c: beak brooches (G2–3), necklaces of type R3a, plain Fe-pins (P4), spiral bracelets (Q2), chatelaines (S3).

BOKV1d: plain Fe-pins (P4), beak brooches (G3), chatelaines (S3), bird brooches (D), bead string spacer (S1), oval disc brooches (J1d), rectangular brooches (K1B–C), spiral bracelets (Q2), plain bracelets (Q3d), necklace-type R3b.

BOKV2a: necklace-type R3b, disc brooches (I2–3), animal-shaped tortoise brooches (O3–4), ordinary tortoise brooches (N1–2), bracelet (Q3b).

BOKV2b: necklace-type R3c (early grave groups) and R3d (late grave groups), animal-shaped tortoise brooches (O4), ordinary tortoise brooches (N3–5), bracelets (Q3b–c).

BOKV3(=Viking Age): bracelets (Q3c), necklace-type R3d, tortoise brooches (NV), large equal-armed brooches (M3).

As early as his presentation of grave groups from Kyndby, Sjælland, Ørsnes examined finds of the same character from the regions around Denmark and especially from the early part of the Late Germanic Iron Age – the period of the beak brooches – and introduced his tripartite division of the period (Ørsnes-Christensen 1956). The entire Danish material combined with the published Scanian material (Strömberg 1961) was published and discussed by Ørsnes in 1966. Here he expanded his three-phase system. He was further able to split these phases into sub-phases. In comparison with the result produced here the differences are minor. The differences primarily concern the sub-phases and the phase transitions. For comparison, Ørsnes's phases are added to figure 10.26. It appears that Ørsnes's phase 2c disappears, with its finds are divided between the two adjacent phases. The material obviously responds so well to seriation that the results are independent of whoever carries them out.

A chronology of female grave groups from Gotland

Since 1969, the five-phase system based on the Gotlandic grave groups of period VII has been in general use in the Scandinavian Baltic Sea area. *Die Vendelzeit Gotlands* (Nerman 1969), however, is not a verbally described chronology, but a collection of pictures of the entire body of artefacts from Gotland divided into five chronological groups on basis of grave associations. The first text-volume followed in 1975 (Nerman 1975), containing a catalogue of the grave groups organized according to chronology.

Fig. 10.24 (left) Sorted matrix of plot in figure 10.23. The matrix is sorted through a combination of the first and the second axes. Necklace-types are indicated.
Geordnete Matrix der Halsketten. Teils nach der ersten Achse der Korrespondenzanalyse Abb. 10.23, teils nach der zweiten Achse sortiert.

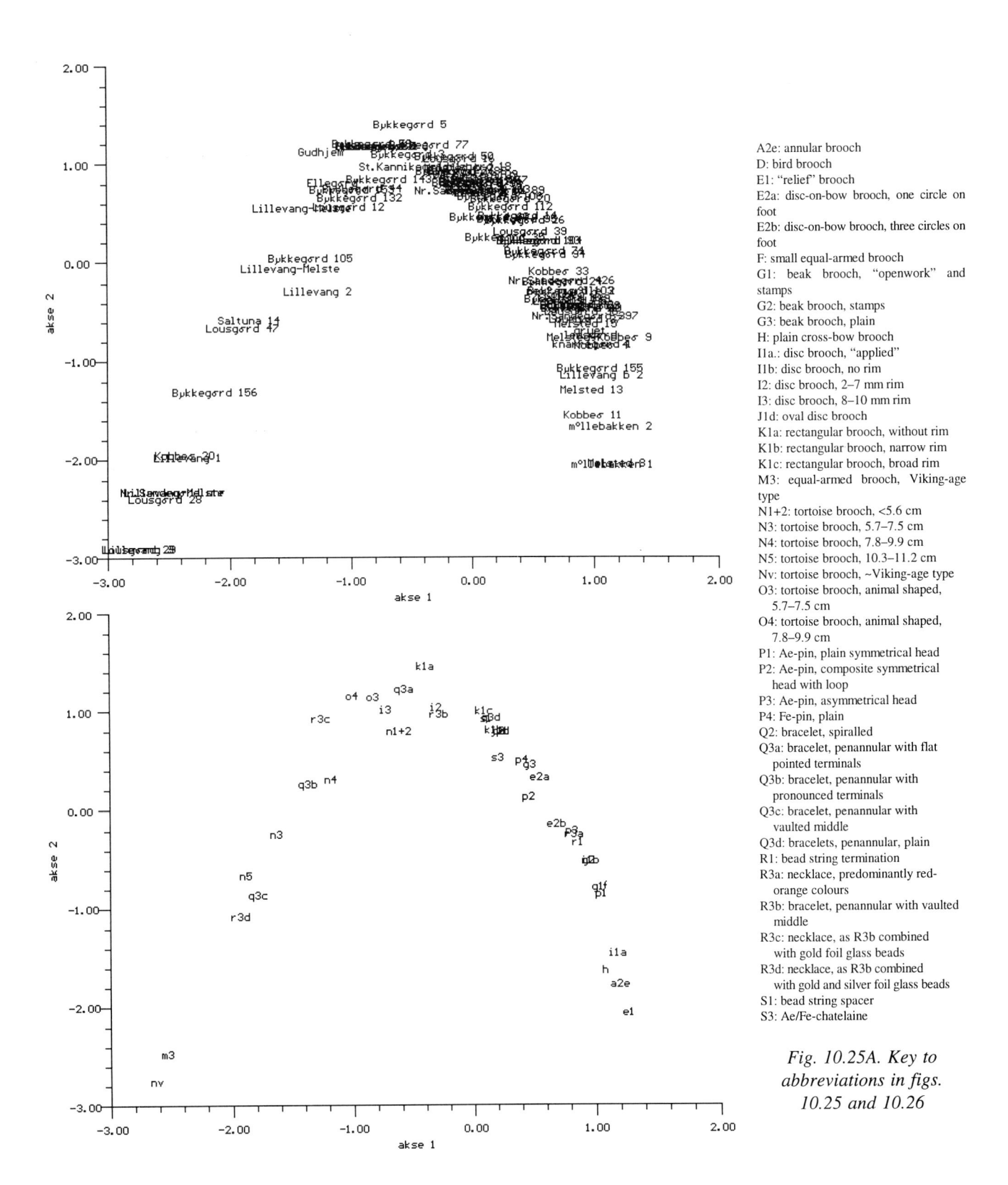

A2e: annular brooch
D: bird brooch
E1: "relief" brooch
E2a: disc-on-bow brooch, one circle on foot
E2b: disc-on-bow brooch, three circles on foot
F: small equal-armed brooch
G1: beak brooch, "openwork" and stamps
G2: beak brooch, stamps
G3: beak brooch, plain
H: plain cross-bow brooch
I1a.: disc brooch, "applied"
I1b: disc brooch, no rim
I2: disc brooch, 2–7 mm rim
I3: disc brooch, 8–10 mm rim
J1d: oval disc brooch
K1a: rectangular brooch, without rim
K1b: rectangular brooch, narrow rim
K1c: rectangular brooch, broad rim
M3: equal-armed brooch, Viking-age type
N1+2: tortoise brooch, <5.6 cm
N3: tortoise brooch, 5.7–7.5 cm
N4: tortoise brooch, 7.8–9.9 cm
N5: tortoise brooch, 10.3–11.2 cm
Nv: tortoise brooch, ~Viking-age type
O3: tortoise brooch, animal shaped, 5.7–7.5 cm
O4: tortoise brooch, animal shaped, 7.8–9.9 cm
P1: Ae-pin, plain symmetrical head
P2: Ae-pin, composite symmetrical head with loop
P3: Ae-pin, asymmetrical head
P4: Fe-pin, plain
Q2: bracelet, spiralled
Q3a: bracelet, penannular with flat pointed terminals
Q3b: bracelet, penannular with pronounced terminals
Q3c: bracelet, penannular with vaulted middle
Q3d: bracelets, penannular, plain
R1: bead string termination
R3a: necklace, predominantly red-orange colours
R3b: bracelet, penannular with vaulted middle
R3c: necklace, as R3b combined with gold foil glass beads
R3d: necklace, as R3b combined with gold and silver foil glass beads
S1: bead string spacer
S3: Ae/Fe-chatelaine

Fig. 10.25A. Key to abbreviations in figs. 10.25 and 10.26

Fig. 10.25 Plot of correspondence analysis of female grave groups (upper) and ornaments (lower) from Bornholm.
Korrespondenzanalyse der Frauengräber Bornholms. Oben sind die Gräber, unten die Typen dargestellt.

	Nv	M3	R3d	N5	Q3c	N3	Q3b	R3c	N4	O4	O3	I3	N1+2	Q3a	K1a	I2	R3b	K1c	S1	Q3d	K1b	S3	Q2	J1d	D	P4	P2	G3	E2a	E2b	P3	R3a	R1	G2	I1b	G1	P1	F	H	I1a	A2e	E1	Ørsnes 1966: 1a	1b	2a	2b	2c	3a	3b	Phases	1st axis	VII
Lillevang 9	2	1																																																	-2,73	
Lousgård 29	2	1																																																	-2,73	
Lousgård 28	2		1																																																-2,51	
Lillevang-Melsted 5	2				1																																													3	-2,47	VIK
Nr.Sandegård unr	2				1																																														-2,47	
Kobbeå 20		1	1																																																-2,35	
Lillevang 1	1				1																																														-2,32	
Bækkegård 156		1			1	1																																													-2,1	
Lousgård 47			1	2	1		1																																										x		-1,87	
Saltuna 14			1	2	2		2																																										x		-1,82	
Lillevang-Melsted 2			1				1		2																																								x		-1,5	
Lillevang-Melsted 4						1	1	1		1																																						x			-1,4	
Bækkegård 105					1	1			1								1																																x	2b	-1,3	D
Lillevang 2			1		1				1													1																											x		-1,25	
Gudhjem								1		1																																						x			-1,21	
Ellegård							2	1			2		1																																			x			-1,12	
Lousgård 12					1		1				1		2				1																															x			-0,99	
Bækkegård 153							1		2			1		1			1																																x		-0,96	
Bækkegård 132							1										1																															x			-0,9	
Bækkegård 44					1						2					1	1																															x			-0,88	
Bækkegård 59										1	2						1																															x			-0,8	
Bækkegård 66										1	1						1																															x			-0,77	
Lousgård 6										1	1						1																															x			-0,77	
Lousgård 11											2						1																															x			-0,71	
Lousgård 3											2						1																															x			-0,71	
Nr.Sandegård 6											2						1																															x		2a	-0,71	C
Bækkegård 143									1							1	1																																x		-0,65	
St.Kannikegård 195													1				1																															x			-0,51	
Bækkegård 3												1				1	1																															x			-0,5	
Bækkegård 5														1	2		1																														x				-0,48	
Bækkegård 77														1	1	2	1								1																					x					-0,33	
Bækkegård 50																1	1			2																												x			-0,12	
Bækkegård 76																1	1									1																				x					-0,11	
Lousgård 16																	1	1	1																												x				-0,08	
Melsted b 2																	1						1																						?	?	?				-0,08	
Nr.Sandegård 4													2			2	1		1			1	1				1						1															x			-0,07	
Bækkegård 88																	1		2			1																								x					-0,01	
Bækkegård 17																	1									1																				x					0,02	
Lousgård 18																	1	1		1	1		1		1																						x				0,06	
Bækkegård 109																	1			1	1	1		1																						x					0,06	
Bækkegård 2																	1				1					1																				x					0,06	
Bækkegård 141																	1							1				1																		x					0,1	
Lousgård 10																	1			1			1					1																		x					0,1	
Bækkegård 93																	1				1	1	1			1																				x					0,11	
Bækkegård 16																	1					1	1		1	1																				x				1d	0,12	
Bækkegård 157																	1		2			1	1		2				1																	x					0,14	
Bækkegård 11																	1									2																				x					0,14	
Lousgård 9																	1	1			1		1				2																				x				0,15	
Bækkegård 163																					1	1																								x					0,16	
Bækkegård 60																	1													1															?	?	?	?	?		0,17	
Bækkegård 106																						1			1	1																				x					0,26	B
Bækkegård 89																				1						2																									0,29	
Bækkegård 35																	1				1	1				1													1							x					0,29	
Bækkegård 20																							1			1		1																		x					0,34	
Bækkegård 112																			1			1	1		2	1	1		1			1													x						0,34	
Bækkegård 95																						1	1			1						1													x						0,39	
Bækkegård 14																						1		1		2		1				1													x						0,4	
Bækkegård 26																				1					2			1				1		1											x						0,46	
Lousgård 39																							1									1														x					0,5	
Bækkegård 104																										1			1			1														x				1c	0,58	
Bækkegård 90																										1			1			1													x						0,58	
Bækkegård 74																										1							1										?	?	?						0,62	
Bækkegård 94																										2						1		1											x						0,63	
Kobbeå 33																												1				1		1										x							0,74	
Nr.Sandegård 426																						1									1	1	2	1											x						0,74	
Bækkegård 24																												1				1				1								x							0,76	
Bækkegård 139																										1										2							x								0,81	
Lousgård 35																						1								1		1	2	3			2							x							0,83	
Saltuna 10																										1										1		1					x								0,83	
Kobbeå 2																													1			1	1		1	1								x							0,84	
Bækkegård 103																																1	1										?	?	?						0,84	
Kobbeå 6																											1					1		2			1						x								0,84	
nexø nymølle 2																																1	1																		0,84	
Lousgård 40																														1		1	2	1	1		1							x						1b	0,88	
Bækkegård 23																																1		2										x							0,91	
gudmingegård																																1		2										x							0,91	
Bækkegård 30																																1	1					1						x							0,92	
Lousgård 36																																1	2	2	1		1							x							0,92	A
Nr.Sandegård 397																																1	2	3			2							x							0,93	
Lousgård 37																																1			1		1						x								0,95	
Melsted 15																																1		1		2							x								0,95	
gryet																																		1		1															0,99	
lensgård																																		1		2							x								1	
Melsted 13																																1							2				?	?							1	
knarregård 1																																1		1		1	1			1				x							1,01	
Kobbeå 11																											1												2		1	1									1,01	
Melsted-Kobbeå 9																																		1			1	1					x							1a	1,03	
Bækkegård 155																																		1						1			x								1,07	
Kobbeå 4																																					1	3					x								1,08	
Lillevang b 2																																				2	1				1		x								1,08	
møllebakken 2																																				1	2		1		1	3									1,15	
møllebakken 1																																							2			2								ÆGJ	1,17	ÆGJ
Melsted 8																																							1	1	1	3									1,21	
1st axis	-2,66	-2,57	-1,96	-1,92	-1,82	-1,66	-1,4	-1,29	-1,21	-1,03	-0,86	-0,76	-0,65	-0,61	-0,44	-0,35	-0,32	0,05	0,07	0,11	0,13	0,17	0,18	0,19	0,21	0,37	0,42	0,43	0,51	0,65	0,77	0,79	0,82	0,91	0,93	0,98	1,01	1,04	1,04	1,14	1,16	1,2										

Fig. 10.26 Sorted matrix of plot in figure 10.25. To the right are listed the datings by Ørsnes, the suggested new phases for Bornholm, the first coordinates of the graves from the plot. Finally the phases in the far right column refer to figure 10.38. The row at the bottom contains the first coordinates of the ornaments.

Geordnete Matrix der Frauengräber Bornholms (nach Abb. 10.25).

Karen Høilund Nielsen

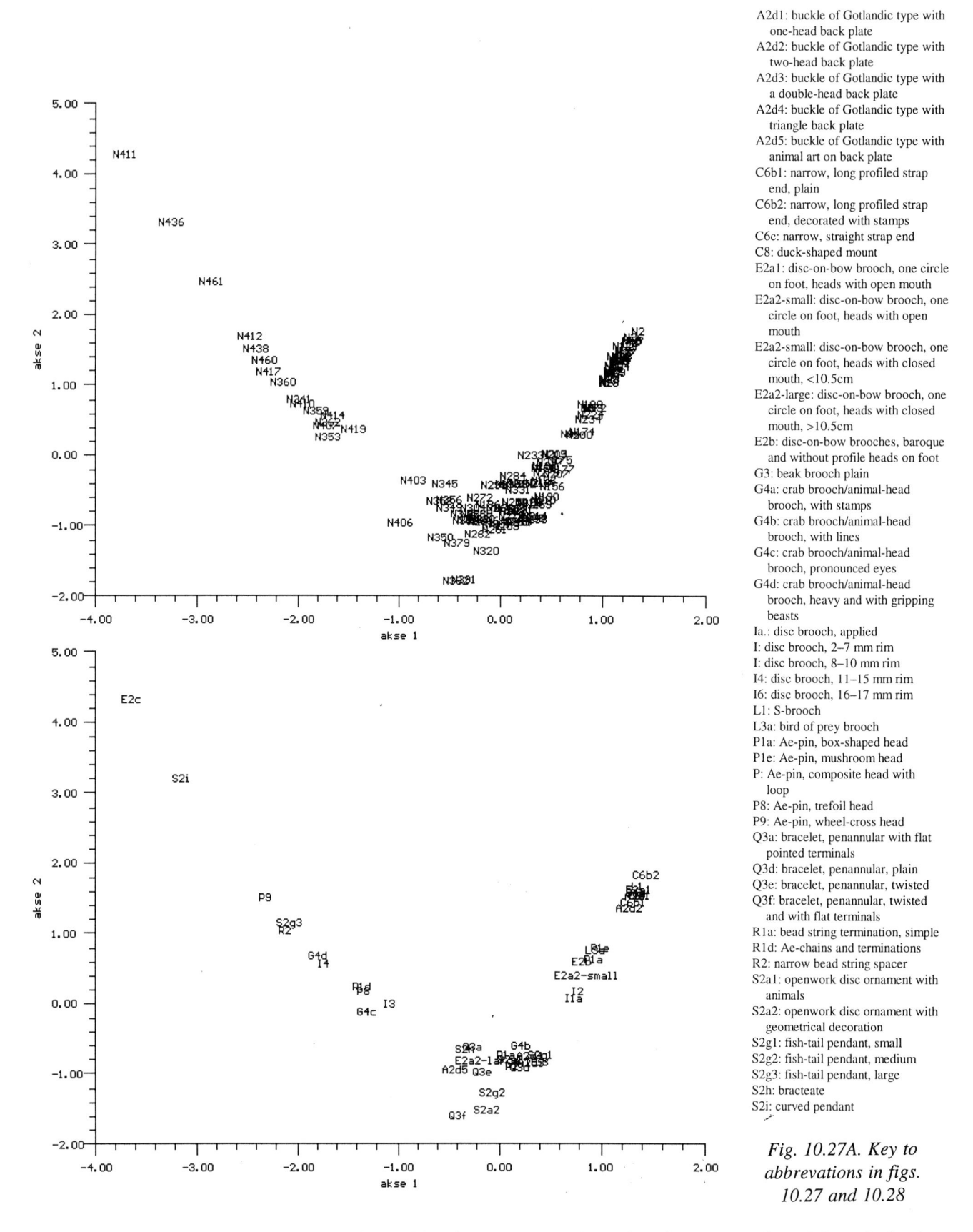

A2d1: buckle of Gotlandic type with one-head back plate
A2d2: buckle of Gotlandic type with two-head back plate
A2d3: buckle of Gotlandic type with a double-head back plate
A2d4: buckle of Gotlandic type with triangle back plate
A2d5: buckle of Gotlandic type with animal art on back plate
C6b1: narrow, long profiled strap end, plain
C6b2: narrow, long profiled strap end, decorated with stamps
C6c: narrow, straight strap end
C8: duck-shaped mount
E2a1: disc-on-bow brooch, one circle on foot, heads with open mouth
E2a2-small: disc-on-bow brooch, one circle on foot, heads with open mouth
E2a2-small: disc-on-bow brooch, one circle on foot, heads with closed mouth, <10.5cm
E2a2-large: disc-on-bow brooch, one circle on foot, heads with closed mouth, >10.5cm
E2b: disc-on-bow brooches, baroque and without profile heads on foot
G3: beak brooch plain
G4a: crab brooch/animal-head brooch, with stamps
G4b: crab brooch/animal-head brooch, with lines
G4c: crab brooch/animal-head brooch, pronounced eyes
G4d: crab brooch/animal-head brooch, heavy and with gripping beasts
Ia.: disc brooch, applied
I: disc brooch, 2–7 mm rim
I: disc brooch, 8–10 mm rim
I4: disc brooch, 11–15 mm rim
I6: disc brooch, 16–17 mm rim
L1: S-brooch
L3a: bird of prey brooch
P1a: Ae-pin, box-shaped head
P1e: Ae-pin, mushroom head
P: Ae-pin, composite head with loop
P8: Ae-pin, trefoil head
P9: Ae-pin, wheel-cross head
Q3a: bracelet, penannular with flat pointed terminals
Q3d: bracelet, penannular, plain
Q3e: bracelet, penannular, twisted
Q3f: bracelet, penannular, twisted and with flat terminals
R1a: bead string termination, simple
R1d: Ae-chains and terminations
R2: narrow bead string spacer
S2a1: openwork disc ornament with animals
S2a2: openwork disc ornament with geometrical decoration
S2g1: fish-tail pendant, small
S2g2: fish-tail pendant, medium
S2g3: fish-tail pendant, large
S2h: bracteate
S2i: curved pendant

Fig. 10.27A. Key to abbrevations in figs. 10.27 and 10.28

Fig. 10.27 Plot of correspondence analysis of female grave groups (upper) and ornaments (lower) from Gotland.
Korrespondenzanalyse der Frauengräber Gotlands. Oben sind die Gräber, unten die Typen dargestellt.

Although the chronology is based only on Gotlandic grave groups and artefacts, many have seen the chronology as valid for much of the Swedish area, without proving that to be the case.

Gotland is the only area apart from Bornholm which is rich in Vendel-period artefacts and grave groups, and thus suitable for detailed chronological analysis. The material has been almost totally published up to the late 1960's (Nerman 1969; 1975). The amount of new finds since then is relatively limited. Nerman's catalogue is therefore so extensive that it is an acceptable basis for a chronological analysis. The only problem is that many types from Gotland are not found in other areas, and most types from mainland Sweden and Denmark are seldom found on Gotland. But there is some exchange of artefacts between the three areas, providing a basis for correlating the various chronological systems.

Nerman divided the grave groups into systematically and unsystematically investigated graves. To begin with only the systematically investigated grave groups were included, but as particular relevant types did not occur in these grave groups some of the unsystematically investigated grave groups were also included — only the most convincing ones, however. This improved the result considerably.

Correspondence analysis was applied to 45 artefact-types and 117 grave groups. The result was a very convincing parabola implying that the criterion of continuity was fulfilled (Fig. 10.27). A phase system is suggested for the sorted matrix (Fig. 10.28). Nerman's chronology also appears in figure 10.28, and can be seen to accord with the seriation produced. The new phase system is given the name GOKV (= Gotland Female Grave Groups), and the phases can be described as follows:

GOKV1: disc-on-bow brooches (E2a1, E2a2<10.5 cm, E2b), stamp-decorated crab/animal-head brooches (G4a), duck-shaped mounts (C8), buckles of Gotlandic type with one head (A2d1), Ae-pins with mushroom head (P1e), Ae-pins with cubic head (P1a).

GOKV2a: Ae-pins with mushroom head (P1e), Ae-pins with cubic head (P1a), disc-on-bow brooches (E2a2<10.5 cm), disc brooches (I1a, I2), small fish-head pendents (S2g1), buckles of Gotlandic type with triangle (A2d4), line-decorated crab/animal-head brooches (G4b).

GOKV2b: small fish-head pendants (S2g1), buckles of Gotlandic type with triangle (A2d4), line-decorated crab/animal-head brooches (G4b), bead string terminations (R1a), plain bracelets (Q3d), Ae-pins with composite head (P2), disc-on-bow brooches (E2a2>10.5 cm), twisted bracelets (Q3e), bracelets with flat pointed terminals (Q3a), Ae-bracteates (S2h).

GOKV2c: buckles of Gotlandic type with triangle (A2d4), line-decorated crab/animal-head brooches (G4b), bead string terminations (R1a), medium fish-head pendants (S2g2), disc-on-bow brooches (E2a2>10.5 cm), twisted bracelets (Q3e), bracelets with flat pointed terminals (Q3a), Ae-bracteates (S2h), twisted bracelet with flat terminals (Q3f).

GOKV3a: buckles of Gotlandic types with Style E (A2d5), animal-head brooches with eyes (G4c), Ae-pins with trefoil head (P8), Ae-chains with terminations (R1d), disc brooches (I4–5), heavy animal-head brooches (G4d), large fish-head pendants (S2g3), simple bead string spacers (R2), Ae-pins with wheel-cross head (P9).

GOKV3b: large fish-head pendants (S2g3), curved pendants (S2i), baroque disc-on-bow brooches (E2c).

The distribution of necklace-types has been added to figure 10.28. It appears that R3a occurs late in GOKV1 and covers all of GOKV2a. R3b appears in GOKV2b and covers all of GOKV2c, whereas R3c is found in GOKV3a. The same typological change in necklace-types as seen for Bornholm and mainland Sweden is thus visible amongst the Gotlandic necklaces.

A chronology of female grave groups from mainland Sweden

For the Swedish mainland, Skåne excepted, it is difficult to establish a chronology, partly because of the large but rather poor body of evidence, consisting of cremation graves with hardly any material left. Several attempts have been made, but have never led to a proper scheme. In 1953 Nils Åberg presented *Den historiska relationen mellan folkvandringstid och vendeltid* ('The Historical Relation between the Migration Period and the Vendel Period'). For the early Vendel Period a handful of richly equipped burials were compiled. This group could be connected to the royal mounds of Old Uppsala. It was interpreted as a first period of the Vendel Period, with gold as a characteristic element. Åberg related another group of graves to the later Vendel Period. They lack the characteristic gold. This division was the basis of a more general analysis of the more ordinary material of the period, the only existing general analysis of the Swedish material. Åberg saw the small equal-armed brooches as a diagnostic type for the early Vendel Period and through these brooches he could tie a series of types to the phase: small equal-armed brooches, cross-bow brooches of Husby type, plain tortoise brooches, annular brooches and pins. Within this phase Åberg saw no further division, but was of the opinion that the plain tortoise brooches developed from the small equal-armed brooches (cf. also Arwidsson 1942) and thus were later. Likewise the Husby brooches were supposed to have developed from Migration-period equal-armed brooches. However he had only two proper Husby brooches. Åberg thought they were from a very early part of the period. The Husby brooches were associated with annular brooches and pins. The small equal-armed brooches were often combined with these types. The crab/animal-head brooches were, in Åberg's opinion, a development of the Husby brooches and thus belonged to the first phase as such. For the small plain

tortoise brooches Åberg thought it possible – on the basis of the strong concentration in Uppland – that they were developed there. Associations were rare, but they were sometimes combined with artefacts such as mounts in late Style II, annular brooches and multi-coloured glass beads. They were dated very cautiously with reference to their occurrence in a cemetery sphere tied to the first phase of the Vendel Period, although Åberg emphasized the dating problems. The ordinary artefacts from the later Vendel Period were not discussed, however. The development in the brooches started with the equal-armed brooches of Migration-period type, from which the Husby brooches had developed. From these and the small equal-armed brooches, plain tortoise brooches developed. The crab brooches should be derivative of the Husby brooches.

Cemetery 150 on Helgö comprises 41 grave groups which Lamm has divided into four phases (Lamm 1970, 217ff.). Husby brooches and annular brooches were included in phase I, buckles of Gotlandic type and beads of glass paste (R3a?) in phase II, small plain tortoise brooches and necklaces of R3b type in phase III and finally tortoise brooches and necklaces of types R3c–d in phase IV. The chronological scheme was based on grave associations from Helgö and its neighbourhood.

In his publication of *Vårby och Vårberg* Ferenius (1971) also discussed the chronological question. His earliest groups belong to the Early Germanic Iron Age, but his third group embraces equal-armed brooches, wheel-cross brooches and necklaces of type R3a which are typical of the Vendel Period.

In the book *Fornfynd och fornlämingar på Lovö* ('Antiquities and Monuments on Lovö') (1973) Lamm analysed some brooch-types, especially Husby brooches, crab brooches and early animal-head brooches. For the Husby brooches he proved that their decoration is very close to that of the small equal-armed brooches. He divided the type into four groups on basis of the bow index. Lamm stated that Husby brooches and variaties of animal-head brooches were produced in the same workshops at Helgö. For a more precise dating of the Husby brooches Lamm referred to the combs and transferred datings from Nerman's chronology for Gotland. He thus achieved a dating for the Husby brooches to VI:2-VII:1. The chronology in Uppland is thus highly dependent on the validity of the Gotlandic chronology. There is a considerable geographical distance between the two areas, which may make the result questionable. Lamm has eight Husby brooches of which five are combined with combs, and seven crab or animal-head brooches of which four are combined with combs. The difference between the combs is that the combs associated with Husby brooches are in one piece whereas the remaining combs are composite. Concentric circles are only found on combs associating Husby brooches. There may be a chronological difference, but if the difference is also significant in a larger region or between regions, it has to appear from the other equipment in the grave groups. From Lamm's list of find associations this seems not to be the case (Lamm 1973, Fig. 2).

As mentioned above, Lamm was of the opinion that Husby brooches were found in the transitional phase between Nerman's phases VI:2 and VII, probably mostly in VI:2, whereas the crab brooches/early animal-head brooches are from phase VII. The small equal-armed brooches (F1–3) should also occur later than the Husby brooches. A comparison with the necklaces, however, shows that the small equal-armed brooches are associated with necklace-type R3a, whereas a single Husby brooch is associated with a necklace-type R3b, according to both the Bornholm chronology and the analysis by Petré (see below) later than R3a. Unfortunately there are no Husby brooches in Petré's own chronological analysis. There are other associated Husby brooches, but not with dated artefacts except for one combined with a bird-of-prey brooch (L3) and a wheel-cross brooch. The bird-of-prey brooch belongs together with necklace-type R3a, and the same applies to the wheel-cross brooch. Compared with the necklaces, the most obvious conclusion would be that the Husby brooches are contemporary with the small equal-armed brooches (F1–3) but outlive them. Altogether, though, it seems difficult to give a certain date on basis of the existing few finds.

Waller too has outlined a chronology for the Mälar Valley (Waller 1977). Her starting point is the pins from Helgö. The finds cover both the Early and the Late Germanic Iron Age. Pins of group III are combined with equal-armed brooches of type F1b; pins of group IV are combined with equal-armed brooches of type F1a, bird-of-prey brooches (L3), animal-head brooches and buckles of Gotlandic type; pins of group V are combined with Husby brooches, animal-head brooches, bird-of-prey brooches (L3), equal-armed brooches (F1b) and buckles of Gotlandic type; and pins of group VI are combined with equal-armed brooches (F1c–d), animal-head brooches and buckles of Gotlandic type. Here too Husby brooches are not so early.

A recent chronological analysis is Petré's analysis of Lovö, especially the cemetery of Lunda (Petré 1984). In this a thorough chronological analysis was made on the basis of the horizontal stratigraphy and an analysis of the beads resulting in a topo-chronology for the cemetery. Petré divided the cemetery into chronological zones, which can be described as follows:

Zone 1: bead-horizon 1
Zone 2: bead-horizon 1+2, a small equal-armed brooch

Fig. 10.28 (left) Sorted matrix of plot in figure 10.27. To the right are listed the necklace-types, the datings by Nerman, the suggested new phases for Gotland, the first coordinates of the graves from the plot. Finally the phases in the far right column refer to figure 10.38. The row at the bottom contains the first coordinates of the ornaments. Geordnete Matrix der Frauengräber Gotlands (nach Abb. 10.27).

Zone 3: bead-horizon 3 (R3A), 5 small equal-armed brooches, 2 bird-of-prey brooches, a snake brooch (single snake), an annular brooch
Zone 4: bead-horizon 3 (R3A), 2 small equal-armed brooches
Zone 5: bead-horizon 4 (R3A), 2 wheel-cross brooches, a snake brooch (double snake)
Zone 6: bead-horizon 5 (R3B), 3 small plain tortoise brooches.

Unfortunately there are no Husby brooches or crab/animal-head brooches in this cemetery. Here again, the small equal-armed brooches are seen in a rather early context, prior to the R3a-necklaces.

The result of the Swedish analyses is that types such as small equal-armed brooches, Husby brooches and early animal-head brooches belong to the early Vendel Period. The early plain tortoise brooches follow later. Petré's analysis reveals an earlier phase with small equal-armed brooches, bird-of-prey brooches, snake brooches (single snake) and annular brooches, and a later phase with snake brooches (double snake) and wheel-cross brooches. Both these phases, but primarily the latter, are combined with necklace-type R3a. Furthermore there is a latest phase with a small plain tortoise brooch associated with necklace-type R3b.

As no proper chronological scheme can be produced including a broad spectrum of types comparable to those from Bornholm and Gotland, I have chosen to make a separate chronology of the mainland Swedish material. For the chronological analysis of this area, as much material as possible was recorded. This consisted primarily of grave groups from the Mälar area. The material suffers from deriving from cremation graves. It was therefore impossible to record the entire corpus of beads and necklaces. There may thus be some doubts about type identifications, especially in relation to the necklaces. The very small number of identifiable types in the individual grave groups constitutes a further problem, as it may be difficult to establish a reliable result.

The material went through correspondence analysis several times before anything sensible turned up. Finally 81 grave groups and 26 associated types were left. The result of the correspondence analysis of this (Fig. 10.29) was not exactly convincing. It looks like something with a beginning and an end, but two ways to get there. There are two possible solutions: the explanation is either parallel sequences, or some grave groups with anomalous associations are included.

A first attempt to improve the result was done by leaving out RAÄ14:4, 13974:7 and 26789:10, and a much better result appeared (Fig. 10.30). There are, however, some problems as the early types do not occur even roughly in the expected order and there is a drop from the parabola containing some of the early types. Although a considerable improvement, the result is still not convincing.

Another possibility was regional variation. To check that, the county (*landskap*) names were added to figure 10.29 (= Fig. 10.31). It transpires that most of the units in the upper row are not from Uppland, whereas most of the units in the lower row are. The grave groups were therefore split into two analyses, one of Uppland and one of the other areas. The analysis of the grave groups from Uppland turned up as a parabola. The further removal of a group of graves characterized by bead string spacers (S1), necklace-type R3b and horse brooches (L4) in fact improved the result (Fig. 10.32). These grave groups will probably fit better into the analysis of the grave groups from outside Uppland. There seems, however, to be no geographical reason. The remaining graves from areas outside Uppland also provided a parabola, and an addition of the above mentioned grave groups from Uppland that did not fit in to the Uppland seriation did not change the result (Fig. 10.34). Also in sorted versions the matrixes were acceptably seriated, although some types were badly represented (Figs. 10.33, 10.35). In figure 10.34 a small group in the upper right corner does not fit too well into the seriation. This group is marked out in grey shading in figure 10.35. It comprises grave groups with combinations of R3a, S2c and L2b: the group that did not fit into the Uppland analysis. It is probably a third group that fits best with the areas outside Uppland, but not perfectly. The differences between the two schemes are a slightly different choice of artefact-types and different datings of some of the types. S1, for example, is very differently dated in the two schemes. The type appears much later in Uppland than in the other areas. The dating outside is in accordance with the dates from Bornholm. There is also a suggestion that annular brooches (A2e) are later in Uppland than in the other areas. In this case too, the areas outside Uppland are more similar to Bornholm.

It is thus clear that the reason why the analysis of all the recorded Swedish grave groups appeared rather odd is regional variation and perhaps concurrent groups (in Uppland). The general direction of the two sub-analyses and that of the entire material are compared alongside the sorted matrix (Fig. 10.36). The dating (into four groups) of the sub-schemes is added to the bottom (artefact-types) and right (grave groups), and shows that the general line of development in the sub-analyses and the analysis of all the recorded grave groups is the same. Thus, even though the parabola is very poor, the chronological scheme for the entire material should be valid as long as it is kept in mind that regional variation and perhaps social groupings are involved too.

Comparison with previous analyses of the Swedish material shows points of both agreement and disagreement. Lamm's analysis of Helgö placed Husby brooches (F4) and annular brooches (A2e) in the same phase. That is not the case in the present analysis. Only in three cases out of sixteen are the two types found together. The combination of small tortoise brooches (N1) and necklaces of type R3b is also seen in the present analysis however. Ferenius relates small equal-armed brooches

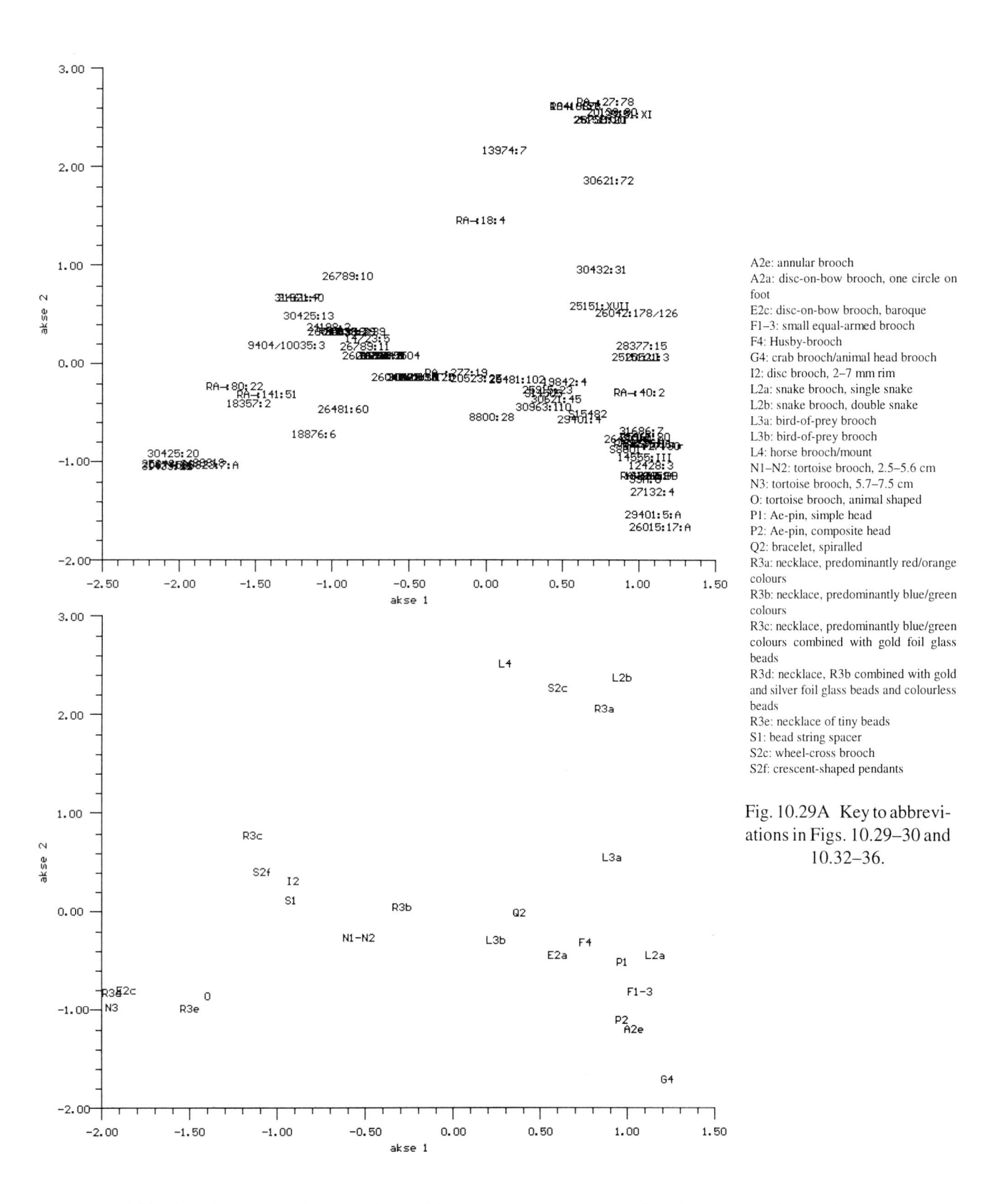

Female grave goods of southern and eastern Scandinavia

A2e: annular brooch
A2a: disc-on-bow brooch, one circle on foot
E2c: disc-on-bow brooch, baroque
F1–3: small equal-armed brooch
F4: Husby-brooch
G4: crab brooch/animal head brooch
I2: disc brooch, 2–7 mm rim
L2a: snake brooch, single snake
L2b: snake brooch, double snake
L3a: bird-of-prey brooch
L3b: bird-of-prey brooch
L4: horse brooch/mount
N1–N2: tortoise brooch, 2.5–5.6 cm
N3: tortoise brooch, 5.7–7.5 cm
O: tortoise brooch, animal shaped
P1: Ae-pin, simple head
P2: Ae-pin, composite head
Q2: bracelet, spiralled
R3a: necklace, predominantly red/orange colours
R3b: necklace, predominantly blue/green colours
R3c: necklace, predominantly blue/green colours combined with gold foil glass beads
R3d: necklace, R3b combined with gold and silver foil glass beads and colourless beads
R3e: necklace of tiny beads
S1: bead string spacer
S2c: wheel-cross brooch
S2f: crescent-shaped pendants

Fig. 10.29A Key to abbreviations in Figs. 10.29–30 and 10.32–36.

Fig. 10.29 Plot of correspondence analysis of female grave groups (upper) and ornaments (lower) from Sweden. Korrespondenzanalyse der Frauengräber aus Schweden. Oben sind die Gräber, unten die Typen dargestellt.

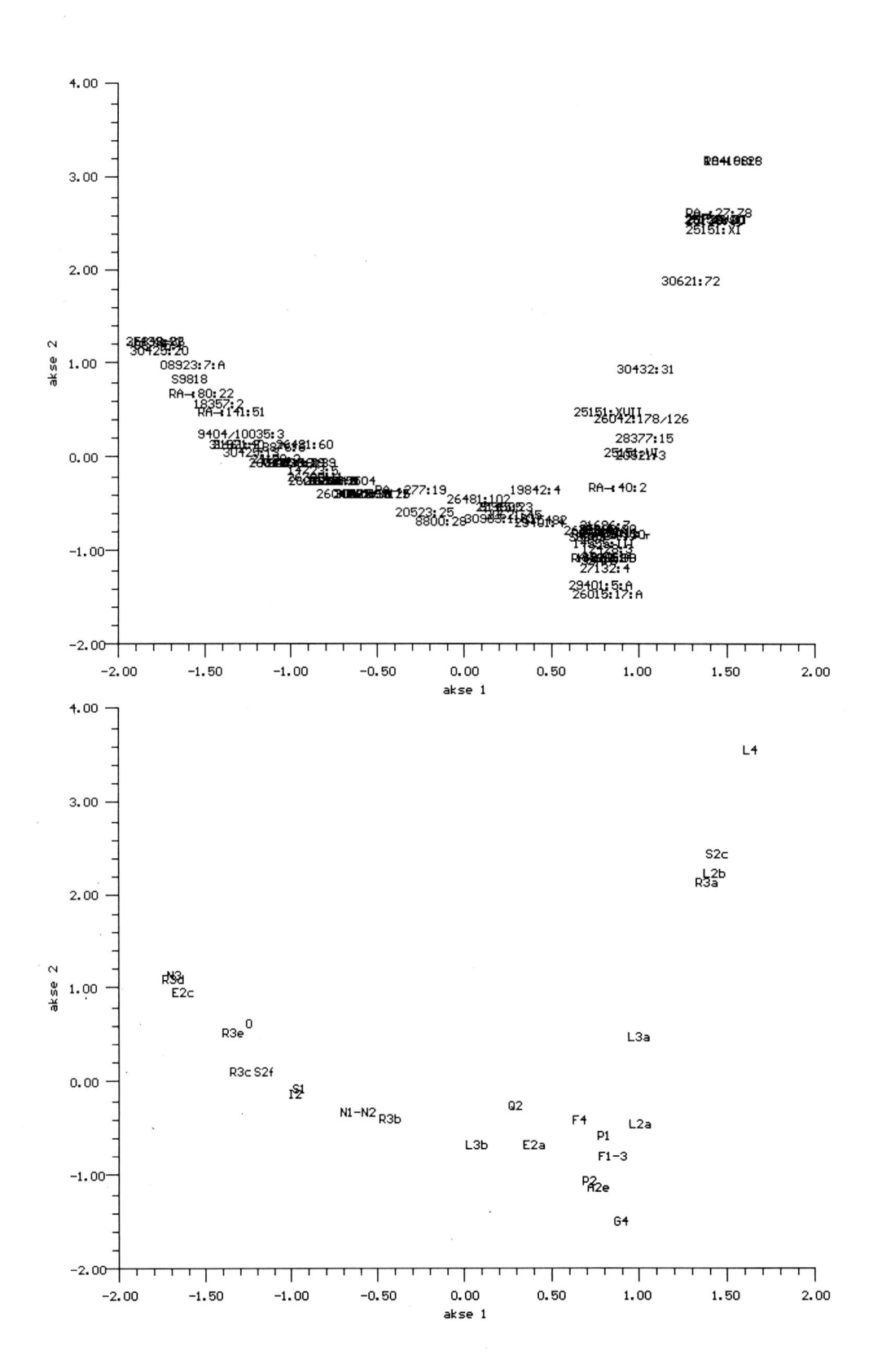

Fig. 10.30 Plot of correspondence analysis of female grave groups (upper) and ornaments (lower) from Sweden. Alternative analysis.
Korrespondenzanalyse der Frauengräber aus Schweden. Alternativer Versuch.

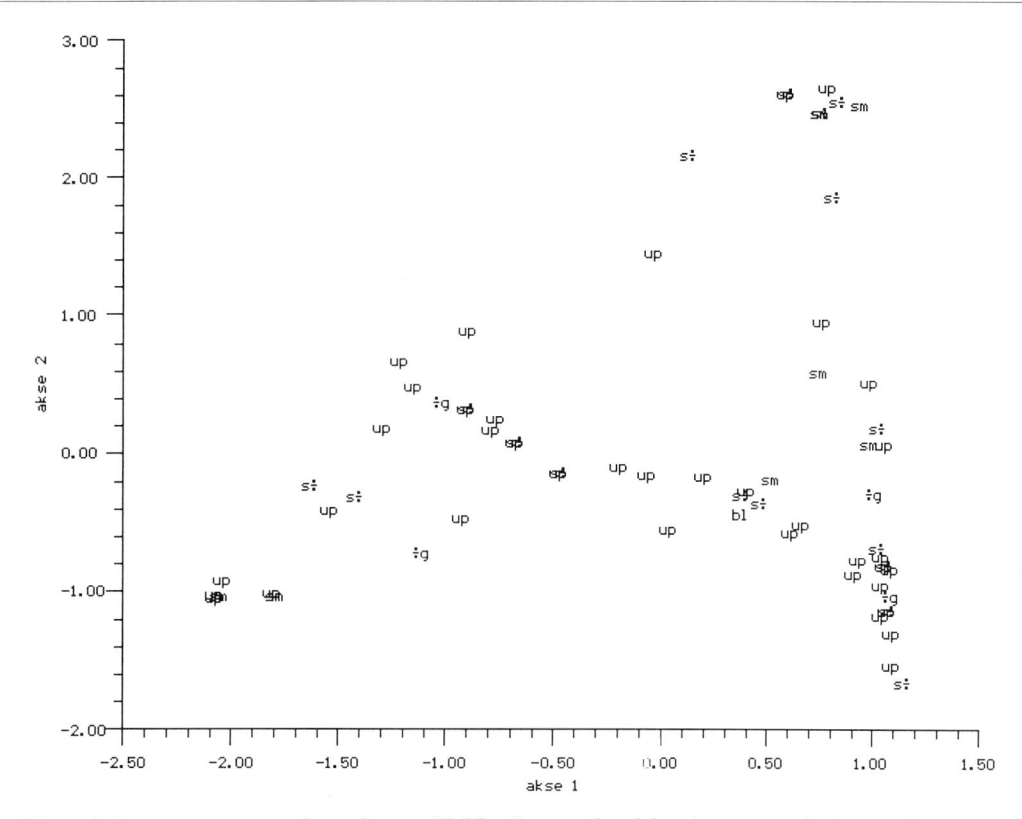

*Fig. 10.31. Plot of the grave groups from figure 10.29. County (*landskap*) names substitutes the names of the units (bl = Blekinge; ÷g = Östergötland; sm = Småland; s÷ = Södermanland; up = Uppland).*
Korrespondenzanalyse der Frauengräber aus Schweden (wie Abb. 10.29). Statt der Gräber werden die jeweiligen Landschaften angezeigt.

(F1–3), wheel-cross brooches (S2c) and necklaces (R3a) to the same phase. R3a is only combined with small equal-armed brooches in two cases out of ten. Wheel-cross brooches are often found together with necklaces of type R3a, but never with small equal-armed brooches. From Lamm's analysis of grave groups from Lovö it was suggested that Husby brooches (F4) are older than small equal-armed brooches (F1–3) and animal-head/crab brooches (G4) which are again older than the small plain tortoise brooches (N1). The only statement proven in the present analysis is that the tortoise brooches are later than the other brooches. Husby brooches are consistently from later contexts, and contemporaneity between the small equal-armed brooches and the crab brooches cannot be proved for certain. Waller's analysis placed the small equal-armed brooches (F1–3) and the Husby brooches later. That fits with the present analysis (see also Høilund Nielsen 1997). It is also correct that some of the bird-of-prey brooches (L3) are contemporary with the small equal-armed brooches. On the other hand the crab brooches are placed too late by Waller. Finally Petré's analysis of Lunda gives an early date to the small equal-armed brooches (F1–3), while bird-of-prey brooches (L3), snake brooches (L2a) and annular brooches (A2e) also occur in the midst of these brooches. That fits with the present analysis. Wheel-cross brooches (S2c) and snake brooches (L2b) are placed later, which is also consistent with the present analysis as well as his dating of the small tortoise brooches (N1). Most cases, especially those with the best and newest evidence, thus agree with the present analysis. I must, however, emphasize that I have not necessarily been able to record all finds that would be qualified for the analysis: this was the largest sample I could get hold of in 1986.

The chronological scheme achieved thus seems to be acceptable for further analysis. The chronological scheme is named SVKV (= Swedish Female Grave Groups). The phases can be described as follows:

SVKV1a: early animal-head brooches/crab brooches (G4a), snake brooches, single animal (L2a), small equal-armed brooches (F1–3), annular brooches (A2e), Ae-pins with simple head (P1), Ae-pins with composite head (P2), bird-of-prey brooches (L3a).

SVKV1b: Ae-pins with simple head (P1), bird-of-prey brooches (L3a), snake brooch, double animal (L2b), necklaces (R3a), Husby brooches (F4), wheel-cross brooches (S2c).

SVKV1c: spiral bracelets (Q2), bird-of-prey brooches with animal-style decoration (L3b), horse brooches (L4), necklaces (R3b).

SVKV2a: necklaces (R3b), small tortoise brooches (N1–2 – though, all are of the type N1a).

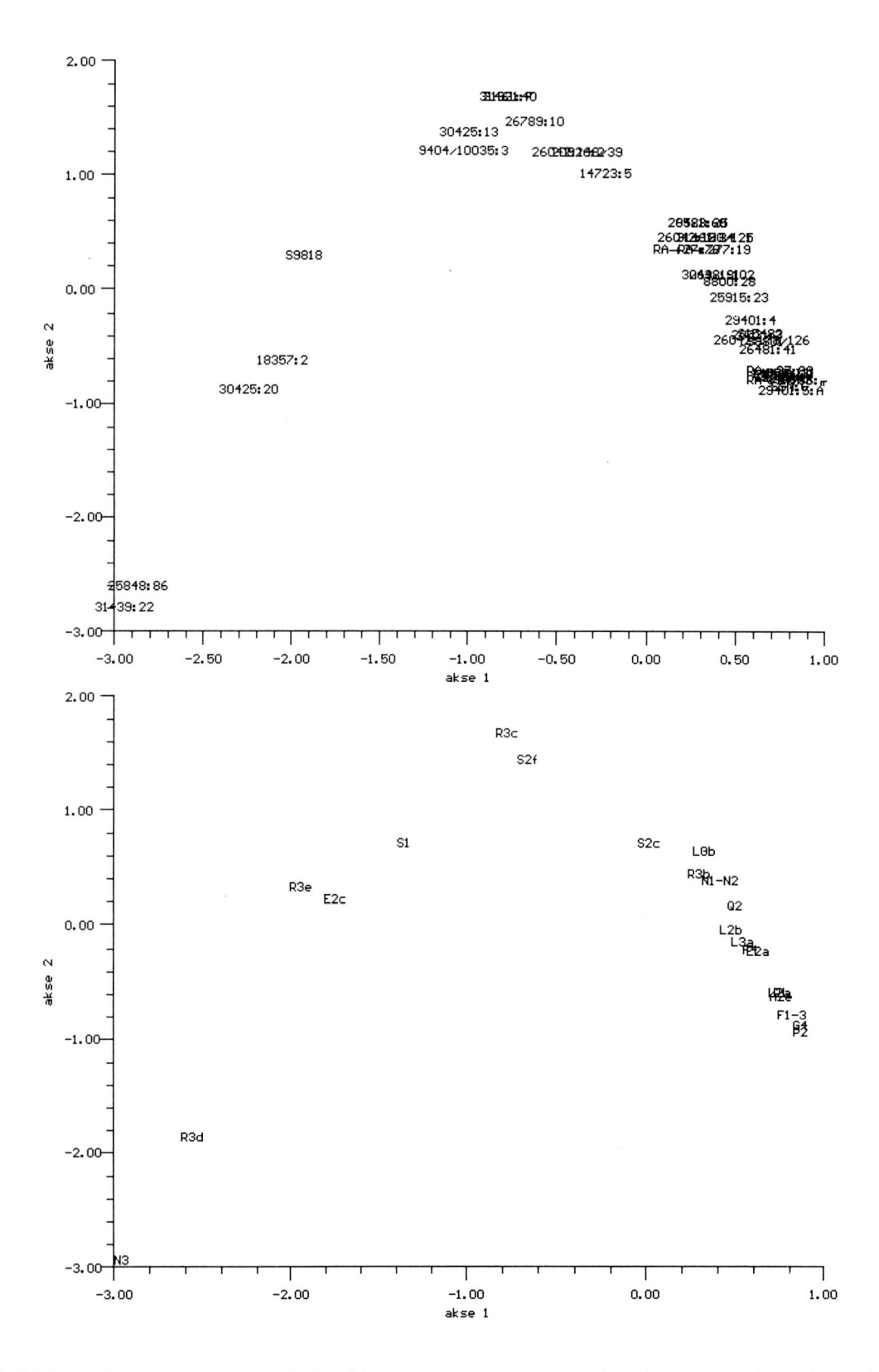

Fig. 10.32 Plot of correspondence analysis of female grave groups (upper) and ornaments (lower) from Uppland.
Korrespondenzanalyse der Frauengräber aus Uppland. Oben sind die Gräber, unten die Typen dargestellt.

	N3	R3d	R3e	E2c	S1	R3c	S2t	S2c	R3b	O	N1-N2	L2b	Q2	L3a	F4	E2a	L2a	P1	A2e	F1-3	G4	P2	
31439:22	2	1																					-2,92
25848:86	1	1																					-2,85
30425:20		1	1																				-2,23
18357:2		1		1																			-2,03
S9818			1	1																			-1,92
9404/10035:3			1	1	4																		-1,01
30425:13			1	1	1																		-0,98
31331:7				1	1																		-0,76
31461:40				1	1																		-0,76
26789:10				1	4	1																	-0,61
26042:148/39					2		1																-0,37
20926:2					2		1																-0,37
14723:5					1	1																	-0,21
RAÄ:27:78					1				1														0,24
20523:25						1	1																0,3
26481:60						1		1															0,3
26042:173/125						1			1														0,34
26042:180/121						1			1														0,34
31461:34						1			1														0,34
30432:31						1					1	1											0,38
RAÄ:277:19						1			1	1													0,39
26481:102						1				1													0,44
8800:28						1	1												1				0,48
25915:23						1			1				1	1									0,54
29401:4						1							1		1	1							0,6
20521:3													1		1								0,64
S15482								1						1	2		1						0,65
26042:178/126										1					2								0,66
S8801											1			1									0,68
26481:41										1			1	1									0,7
RAÄ:27:30													1	1	1	1							0,77
14555:III														1	1	1							0,78
27132:4													1		1		1						0,79
14723:14														1	1								0,79
25915:80													1			1							0,79
31461:20														1		1							0,79
RAÄ:27:36:Ö														1		2							0,8
SSM:H														1		2							0,8
SSM:G																		1	1		1		0,82
29401:5:A																			1		1	1	0,83
	-2,98	-2,58	-1,97	-1,77	-1,38	-0,8	-0,68	0,28	0,31	0,31	0,39	0,46	0,48	0,52	0,57	0,61	0,73	0,73	0,74	0,8	0,83	0,85	

Fig. 10.33 Sorted matrix of plot in figure 10.32. To the right are listed the first coordinates of the graves from the plot. The row at the bottom contains the first coordinates of the ornaments.
Geordnete Matrix der Frauengräber Upplands (nach Abb. 10.32).

SVKV2b: necklaces (R3b), disc brooches (I2), bead string spacer (S1), crescent-shaped pendants (S2f).

SVKV2c: bead string spacer (S1), crescent-shaped pendants (S2f), necklaces (R3c).

SVKV2d: tortoise brooches (N3), disc-on-bow brooches, baroque (E2c), necklaces (R3d).

Synchronisation of the three chronological schemes

For further analysis of a cross-regional nature, it is a problem to have three seperate chronological schemes with hardly any common types as a shared means of dating. some synchronisation of the three chronological schemes presented above is needed, if at all possible. A supra-regional system could be coarser, but still better than to try to work with three systems. The more detailed schemes behind it would still exist.

The basis for such calibration resides in three types of data: the data from the three chronological schemes; associated artefact-types not included in the seriations but nonetheless present in the grave groups analysed; and secondarily associated artefact-types not included in the seriations and not present in the analysed grave groups but combined with artefact-types loosely datable within one of the three schemes. The two latter groups are listed below:

Buckle of Gotlandic type A2d1: Viken (cem. 57) 5, combined with animal-head brooch G4a – SVKV1a; Lundby, combined with G4a – SVKV1a

Buckle of Gotlandic type A2d4: Linta 17, combined with small equal-armed brooch F1b – SVKV1b; Lousgård 39, combined with oval disc brooch J3 – BOKV1c

Bird brooch D: Ålsta gård 63, combined with necklace R3b – SVKV1c–2b; Hjortsberga, combined with small equal-armed brooch F1b and beaked brooch G3 – SVKV1c

Beak brooch G3: Hjortsberga, combined with small equal-armed brooch F1B and bird brooch D2b2 – SVKV1c

Rectangular brooch K1: Karby, combined with tortoise brooch N1a – SVKV2a

S-brooch L1a: Melsted 4, combined with beaked brooch G1c – BOKV1a

Snake brooch L2a: Nerman 59 – GOKV2a

Bead string spacer S1: Nerman 345 – GOKV2c

Openwork disc ornament S2a (geometric): Nørre Sandegård 4 – BOKV2a

Wheel-cross brooch S2c: Nerman 299 – GOKV2c

Crescent-shaped pendant S2f: Nerman 460 – GOKV3a

Bracelet Q3b: Nerman 350 – GOKV2c

Bracelet Q3c: Nerman 360 – GOKV3a

Nerman 414 – GOKV3a

The original idea was for a systematic comparison of grave groups and artefact-types of common appearance and their relationship to the chronological schemes of the relevant other areas. The results with correspondence analysis, however, encouraged me to try this for such an analysis as well.

The phases of the three chronological schemes are used as units, the variables being the types occurring in each phase with the number of occurrences of each separate type is used as its value. Only types occurring in at least two of the schemes are included. The analysed types are thus to some extent shared across the area. The associated artefact-types listed above are also included. If the idea works, the phases, after correspondence analysis, should appear parallel, in the original order. It is thus a precondition that the single phases of the original schemes appear in the right order. Otherwise the method does not work in this case.

A correpondence analysis was applied to the 20 phases and 33 artefact-types and a convincing result appeared (Fig. 10.37). The phases within each scheme appear in the

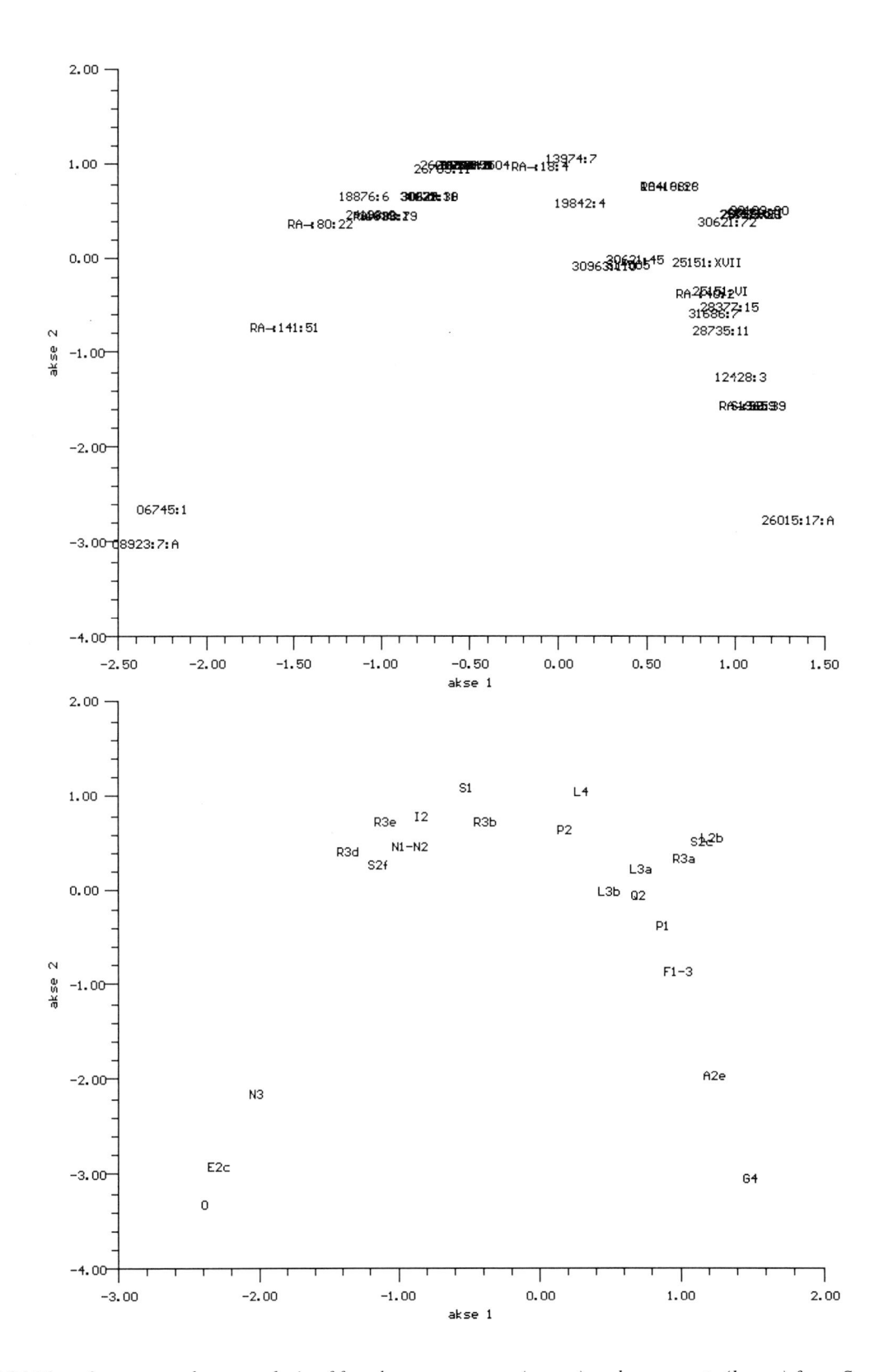

Fig. 10.34 Plot of correspondence analysis of female grave groups (upper) and ornaments (lower) from Sweden except Uppland.
Korrespondenzanalyse der Frauengräber aus Schweden mit Ausnahme Upplands. Oben sind die Gräber, unten die Typen dargestellt.

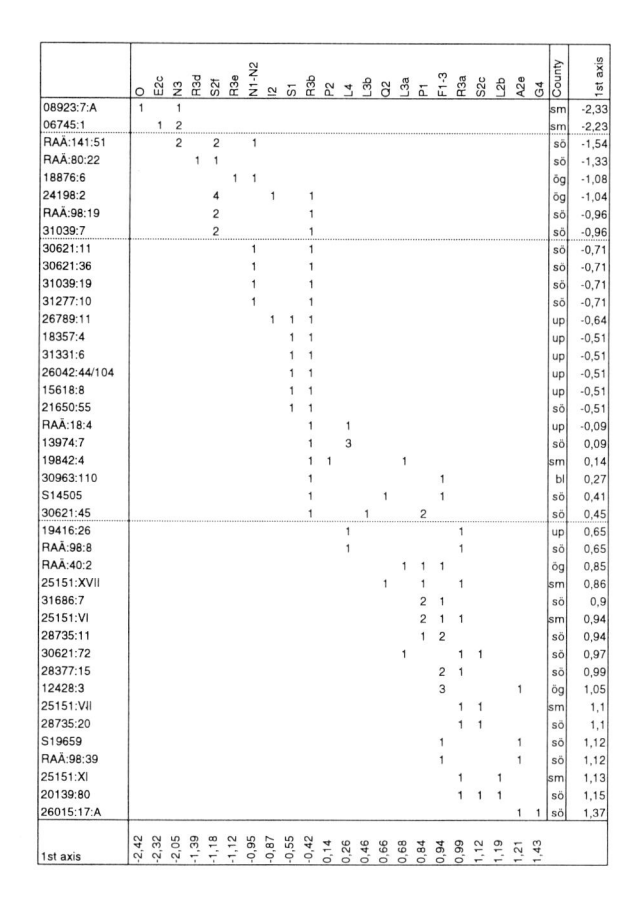

Fig. 10.35 Sorted matrix of plot in figure 10.34. To the right are listed the county (landskap) codes and the first coordinates of the graves from the plot. The row at the bottom contains the first coordinates of the ornaments.
Geordnete Matrix der Frauengräber aus Schweden mit Ausnahme Upplands (nach Abb. 10.34).

right order. The plot and the sorted matrix (Fig. 10.38) can be used for the establishment of supra-regional phases. A primary requirement is that at least one phase from each scheme must occur in each phase. With that starting point and on basis of the plot, the sorted matrix and the other types associable with the phase-units analysed a four-phase system was established. It is named VII in accordance with Nerman. The phases are thus named VIIA–D. Below, the most characteristic artefact-types of each phase are listed (see also Fig. 10.39) – the absolute dates are only preliminary:

Phase VIIA – middle and later 6th century. Covers the periods GOKV1–2a, SVKV1a–b and BOKV1a–b. Small equal-armed brooches (F1–3), beak brooches with stamps (G1–2), crab brooches/early animal-head brooches (G4a), Husby brooches (F4), snake brooches (L2), S-brooches (L1), wheel-cross brooches (S2c), disc brooches, rim 2–7 mm (I2), bird-of-prey brooches (L3a), disc-on-bow broches <10.5 cm long and with profile heads on foot (E2a1, E2a2-small, E2b).

Phase VIIB – 7th century. Covers the period GOKV2b–c, SVKV1c–2a and BOKV1c–d. Beak brooches, plain (G3), rectangular brooches (K1), oval disc brooches (J1), bird brooches (D), animal-head brooches with lines (G4b), disc brooches, rim 8–10 mm (I3), tortoise brooches <4.7 cm, plain (N1a), horse brooches (L4), openwork disc ornaments with animals (S2a), disc-on-bow brooches >10.5 cm long (E2a2-large), bird-of-prey brooches with animal style (L3b).

Phase VIIC – first half 8th century. Covers the periods GOKV3a, SVKV2b–c and BOKV2a. Tortoise brooches <5.6 cm (N1b–N2), animal-shaped tortoise brooches (O), animal-head brooches with eyes or heavy cast (G4c–d), disc brooches, rim 11–15 mm (I4), bead string spacer (S1), crescent-shaped pendants (S2f).

Phase VIID – second half 8th century. Covers the periods GOKV3b, SVKV2d and BOKV2b(-3). Tortoise brooches <11.2 cm (N3–5), disc brooches, rim 16–17 mm (I5), disc-on-bow brooch >10.5 cm, baroque and without profile heads on foot.

CONCLUSIONS

The result of the analyses of the southern and eastern Scandinavian material is that it has been possible to establish a sort of supra-regional chronology for the area on the basis of the seriation principle and by means of correspondence analysis, in the way that local chronologies were initially made in the three sub-areas. For mainland Sweden the situation was complicated, as local variation and perhaps contemporary social divisions blur the picture. The three chronological schemes have only some artefact-types in common. Secondly, therefore, on basis of these and of 'foreign contamination', a basis for a further seriation of the entire area was established, this time by analysing the phases of the three chronological schemes. The result was successful and yielded a sort of supra-regional chronology for the area.

The example shows how it was possible to make a supra-regional chronology on basis of three rather different chronological schemes. The supra-regional chronology is not so detailed as the local chronologies, but for cross-regional analysis there has to be some elasticity to counter-balance the earlier discussed problems.

A future inclusion of the recent analysis of Lars Jørgensen (in Jørgensen and Nørgård Jørgensen 1997) may refine a few details. Further analyses ought also to go in the direction of including the Norwegian material (the Finnish material is probably too different) and a better correlation of male and female graves. The first steps have already been taken by Jørgensen and Nørgård Jørgensen (1997, 38, Fig. 26) for Bornholm. It ought to be done for the entire region, and all of Scandinavia as well, which with almost the entire Scandinavian material as a basis ought to be possible. Furthermore, in the future, the absolute

	N3	R3d	E2c	R3e	O	R3c	S2f	S1	I2	N1-N2	R3b	L3b	L4	Q2	E2a	S2c	F4	R3a	L3a	P1	L2b	P2	A2e	F1-3	L2a	G4	Phases	1st axis	County	Uppland	Sörmland etc	VII
31439:22	2	1																										-2,06	up	4		
25848:86	1	1																										-2,06	up	4		
06745:1	2		1																									-2,04	sm		4	
30425:20		1	1																								2d	-2,02	up	4		D
S9818			1	1																								-1,79	up	4		
08923:7:A	1				1																							-1,78	sm		4	
RAÄ:80:22		1					1																					-1,62	sö		3	
18357:2		1						1																				-1,52	up	4		
RAÄ:141:51	2						2			1																		-1,41	sö		3	
9404/10035:3			1				4	1																				-1,28	up	3		
31331:7						1	1																				2c	-1,2	up	3		
31461:40						1	1																					-1,2	up	3		
30425:13						1	1	1																				-1,13	up	3		
18876:6				1						1																		-1,1	ög		3	
24198:2							4		1		1																	-1	ög		3	
26481:60					1						1																	-0,91	up	2		
26042:148/39							2				1																	-0,89	up	3		
RAÄ:98:19							2				1																	-0,89	sö		3	C
31039:7							2				1																	-0,89	sö		3	
20926:2							2				1																2b	-0,89	up	3		
26789:10						1	4									1												-0,88	up	3		
26789:11								1	1		1																	-0,77	up		2	
14723:5							1				1																	-0,75	up	3		
21650:55								1			1																	-0,66	sö		2	
18357:4								1			1																	-0,66	up		2	
31331:6								1			1																	-0,66	up		2	
26042:44/104								1			1																	-0,66	up		2	
15618:8								1			1																	-0,66	up		2	
30621:11										1	1																	-0,46	sö		2	
30621:36										1	1																	-0,46	sö		2	
31039:19										1	1																	-0,46	sö		2	
26042:173/125										1	1																2a	-0,46	up	2		
26042:180/121										1	1																	-0,46	up	2		
31461:34										1	1																	-0,46	up	2		
31277:10										1	1																	-0,46	sö		2	
RAÄ:277:19										1	1			1														-0,18	up	2		
20523:25											1	1																-0,05	up	2		
RAÄ:18:4											1		1															-0,02	up		2	
8800:28										1	1												1					0,05	up	2		B
13974:7											1		3															0,14	sö		2	
26481:102											1						1											0,22	up	2		
S14505											1			1										1				0,39	sö		2	
30963:110											1													1			1c	0,39	bl			
25915:23											1			1	1					1								0,42	up	2		
30621:45											1	1								2								0,48	sö		2	
19842:4											1								1			1						0,54	sm		2	
RAÄ:98:8													1					1										0,6	sö		2	
19416:26													1					1										0,6	up		1	
29401:4											1						1			1			1					0,63	up	2		
S15482											1				1					2				1				0,68	up	1		
25151:XVII														1				1		1								0,76	sm		1	
28735:20																1		1										0,76	sö		1	
25151:VII																1		1										0,76	sm		1	
30432:31																1	1		1									0,78	up	2		
RAÄ:27:78																1					1						1b	0,8	up	2		
30621:72																1		1	1									0,82	sö		1	
20139:80																1		1			1							0,84	sö		1	
S8801																	1						1					0,93	up	1		
26481:41																	1			1			1					0,95	up	1		
25151:XI																		1			1							0,95	sm		1	
26042:178/126																				2	1							1	up	1		
25151:VI																		1		2				1				1	sm		1	
RAÄ:40:2																			1	1				1				1,02	ög		1	
31686:7																				2				1				1,04	sö		1	
28377:15																		1						2				1,04	sö		1	
14555:III																				1			1	1				1,06	up	1		A
25915:80																				1				1				1,06	up	1		
31461:20																				1				1				1,06	up	1		
SSM:G																						1	1	1				1,06	up	1		
28735:11																				1				2				1,07	sö		1	
RAÄ:27:36:Ö																				1				2			1a	1,07	up	1		
20521:3																			1						1			1,07	up	1		
SSM:H																				1				2				1,07	up	1		
S19659																							1	1				1,09	sö		1	
RAÄ:98:39																							1	1				1,09	sö		1	
14723:14																							1	1				1,09	up	1		
12428:3																							1	3				1,1	ög		1	
RAÄ:27:30																				1			1	1	1			1,1	up	1		
27132:4																				1			1			1		1,11	up	1		
29401:5:A																						1	1			1		1,11	up	1		
26015:17:A																							1			1		1,16	sö		1	
1st axis	-1,96	-1,96	-1,88	-1,52	-1,42	-1,16	-1,11	-0,94	-0,93	-0,56	-0,31	0,23	0,28	0,36	0,58	0,58	0,74	0,85	0,89	0,95	0,95	0,95	1,02	1,05	1,14	1,19						
Uppland	4	4	4	4	2	3	3	34		2	23	2		2	12	23	12		12	1	12	1	1	1	1	1						
Södermanland etc	34	3	4	3	4		3	2	23	23	23	2	2	12		1		1	1	1	1	2	1	1		1						

*Fig. 10.36 Sorted matrix of plot in figure 10.29. To the right are listed the the suggested new phases for mainland Sweden, the first coordinates of the graves from the plot, the county (*landskap*) codes, the phases of the two sub series. Finally the phases in the far right column refer to figure 10.38. The two rows at the bottom contain the first coordinates of the ornaments and the phases of the two sub series.*

Geordnete Matrix der Frauengräber aus Schweden (nach Abb. 10.29).

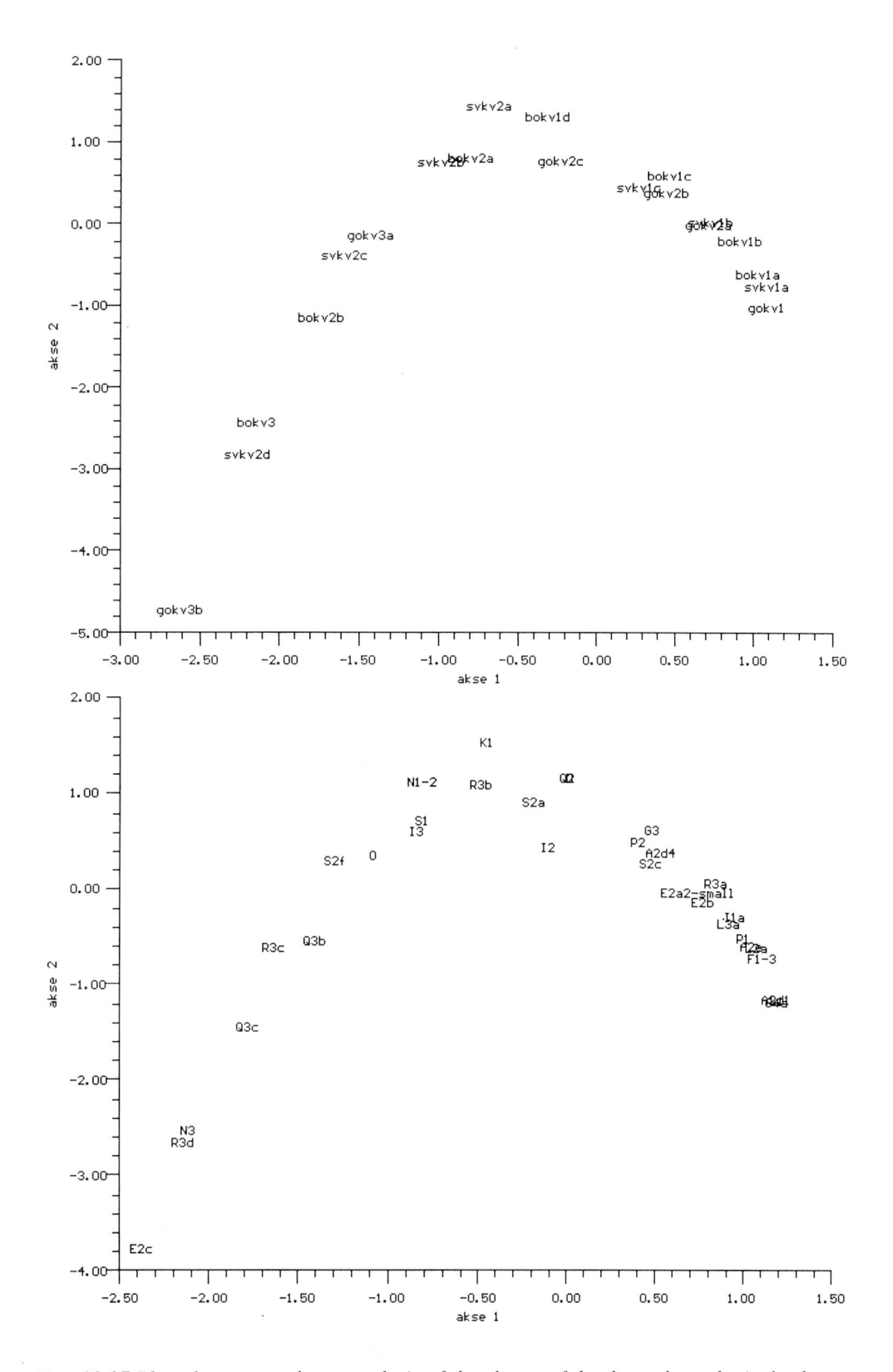

Fig. 10.37 Plot of correspondence analysis of the phases of the three chronological schemes.
Korrespondenzanalyse aller Phasen der drei Regionalchronologien.

Fig. 10.38 Sorted matrix of plot in figure 10.37. To the right are listed the first coordinates of the phases and the new phases in the supraregional chronology. The row at the bottom contains the first coordinates of the ornaments.
Geordnete Matrix der drei Regionalchronologien (nach Abb. 10.37).

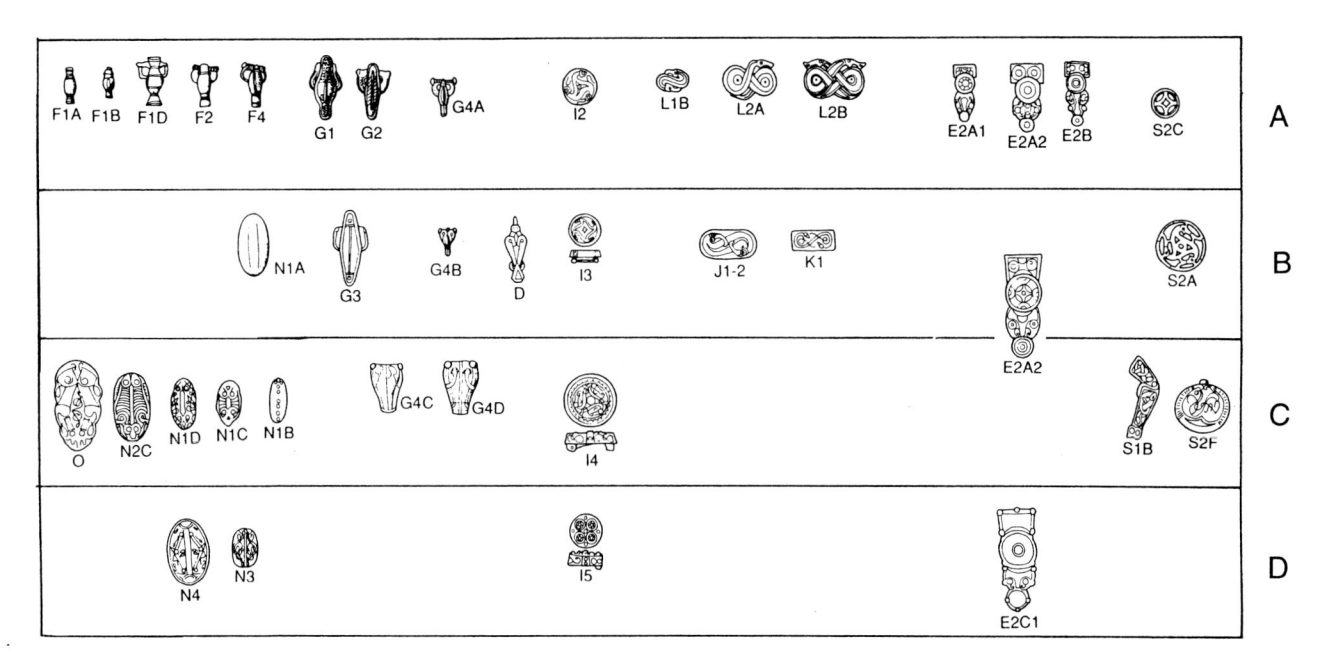

Fig. 10.39 Scheme of supra-regional chronology covering Bornholm, mainland Sweden and Gotland.
Übersicht der überregionalen Chronologie für Bornholm, Gotland und dem schwedischen Festland.

chronology ought to be analysed much more deeply than it has previously been (Høilund Nielsen 1987; Ørsnes 1966); and not just for Bornholm as already done by Jørgensen and Nørgård Jørgensen (1997), but for the whole of Scandinavia. The recent investigations of Continental material – the Alamannic, and most of that from the Rhineland – should provide us with a better basis for comparisons.

Acknowledgements

The project of which this is one of the results was funded by the Carlsberg Foundation to whom I am very grateful. My thanks also to the University of Aarhus for financial support to further visits to Sweden, and to the following Museums for giving me access to their archaeological material: Nationalmuseet, Copenhagen; Bornholms

Museum, Rønne; Fyns Stiftsmuseum, Odense; Den Antikvariske Samling, Ribe; Aalborg Historisk Museum; Horsens Museum; Statens Historiska Museum, Stockholm; Upplands Museet för Nordiska Fornsaker, Uppsala; Kalmar Läns Museum; Lunds Universitets Historiska Museum.

DEUTSCHE ZUSAMMENFASSUNG

Ziel dieses Beitrages ist es, einen methodischen Ansatz für das Problem aufzuzeigen, aus sehr unterschiedlichen Regionalchronologien eine überregionale Chronologie zu entwickeln. Dazu wird zunächst ein Abriß der bisherigen dänischen und schwedischen Forschung zur Chronologie der Späten Germanischen Eisenzeit (bzw. Vendelzeit) gegeben. Es schließt eine methodische Diskussion an, die insbesondere auf das Verfahren der Korrespondenzanalyse eingeht.

Für die angestrebte Korrespondenzanalyse wird zunächst eine neue Typologie erarbeitet, vor allem für die Schildkrötenfibeln, Scheiben- und Dosenfibeln sowie die Halsketten. Sie führt zu einer detaillierteren und auch weitergreifenden Gliederung des Fundstoffes. Anschließend werden -jeweils für sich- die Frauengräber von Bornholm, von Gotland und vom schwedischen Festland seriiert. Die resultierenden Chronologien für Bornholm und Gotland weichen von den älteren Studien von Ørsnes und Nerman nur in geringen Details ab. Das Ergebnis für Festlandschweden ist etwas unklar; dies liegt vermutlich an einer größeren regionalen Variabilität verbunden mit einer größeren sozialen Differenzierung innerhalb Upplands. Immerhin zeigt der Vergleich mit den beiden anderen Regionen und mit älteren Forschungen, daß das Resultat in seinem Kern durchaus zuverlässig ist.

Diese drei Detailanalysen werden anschließend miteinander verknüpft, erneut mit Hilfe einer Korrespondenzanalyse. So kann das Gesamtmaterial in vier Hauptphasen A – D gegliedert werden, von denen jede eine oder mehrere Phasen der drei Regionalanalysen einschließt. Als Ergebnis kann eine brauchbare überregionale Chronologie vorgestellt werden.

NOTE

1   The analyses reported on in this paper were made in 1990 on material recorded between 1985 and 1988. Finds later than that have not been included.

BIBLIOGRAPHY

Åberg, N. 1922. 'Stil II'. *Fornvännen* 1922, 1–26.
Åberg, N. 1953. *Den historiska relationen mellan folkvandringstid och vendeltid*. Kungl. Vitterhets Historie och Antikvitets Akademiens Handlingar 83, Stockholm.
Andresen, J. and Ilkjær, J. 1989. 'Ordnung muss sein'. *KARK-nyhedsbrev* 1989 nr. 3, 34–9.
Arne, T. J. 1919. 'Gravar från "Vendeltid" vid Lagerlunda i Kärna socken, Östergötland'. *Fornvännen* 1919, 1–20.
Arwidsson, G. 1942. *Vendelstile. Email und Glas im 7.–8. Jahrhundert*. Valsgärdestudien I, Uppsala.
Baxter, M. J. 1994. *Explorative Multivariate Analysis in Archeology*. Edinburgh.
Brøndsted, J. 1934. Inedita aus dem dänischen Nationalmuseum. *Acta Archaeol* 5, 145–82.
Callmer, J. 1977. *Trade Beads and Bead Trade in Scandinavia ca 800–1000 A.D.* Lund.
Engelhardt, C. 1867. *Kragehul Mose 1751–1865*. Copenhagen.
Ethelberg, P. 1986. *Hjemsted — en gravplads fra 4. og 5. årh. e.Kr.* Skrifter fra Museumrådet for Sønderjyllands Amt 2. Haderslev.
Ferenius, Jonas 1971. *Vårby och Vårberg. En studie i järnålderns bebyggelsehistoria*. Stockholm.
Gräslund, B. 1974. 'Relativ datering. Om kronologisk metod i nordisk arkeologi'. *TOR* 16.
Hildebrand, Hans 1878. 'Jernåldern på Gotland'. *Vitterhetsakademiens Månadsblad* 1878, 733–57.
Høilund Nielsen, K. 1982. *Ældre germansk jernalder i Jylland*. Aarhus Universitet cand. phil. thesis. Unpublished.
Høilund Nielsen, K. 1987. 'Zur Chronologie der jüngeren germanischen Eisenzeit auf Bornholm. Untersuchungen zu Schmuckgarnituren'. *Acta Archaeol* 57 (1986), 47–86.
Høilund Nielsen, K. 1988. 'Correspondence analysis applied to hoards and graves of the Germanic Iron Age'. In T. Madsen (ed), *Multivariate Archaeology. Numerical Approaches in Scandinavian Archaeology*, 37–54. Højbjerg.
Høilund Nielsen, K. 1991. 'Centrum og periferi i 6.-8. årh. Territoriale studier af dyrestil og kvindesmykker i yngre germansk jernalder'. In P. Mortensen and B. M. Rasmussen (eds), *Fra Stamme til Stat i Danmark 2. Høvdingesamfund og kongemagt*, 127–54. Højbjerg.
Høilund Nielsen, K. 1992a. *Dyrestil og Samfund i Syd- og Østskandinavien i 6.-8. årh. e.Kr.f.* Unpublished.
Høilund Nielsen, K. 1992b. Præsentation af KARK-nyhedsbrev 1989–1990. *Fornvännen* 87, 181–5.
Høilund Nielsen, K. 1996. 'The burial ground'. In *Lindholm Høje Burial ground and village*, 27–38. Aalborg.
Høilund Nielsen, K. 1997. 'Die frühmittelalterliche Perlen Skandinaviens. Chronologische Untersuchungen'. In U. von Freeden and A. Wieczorek (eds), *Perlen. Archäologie, Techniken, Analysen. Akten des Internationalen Perlensymposiums in Mannheim vom 11. bis 14. November 1994*, 187–96. Bonn.
Jensen, C. K. and Høilund Nielsen, K. 1997. 'Burial Data and correspondence analysis'. In C. K. Jensen and K. Høilund Nielsen (eds), *Burial and Society. The Chronological and Social Analysis of Archaeological Burial Data*, 29–61. Aarhus.
Jensen, J. 1988. 'Christian Jürgensen Thomsen og tre periodesystemet'. *Aarb nordisk Oldkyndighed og Historie* (1988), 11–18.
Jørgensen, L., and Nørgård Jørgensen A. 1997. *Nørre Sandegård Vest. A Cemetery from the 6th-8th Centuries on Bornholm*. Nordiske Fortidsminder 14. Copenhagen.
Lamm, J. P. 1973. *Fornfynd och fornlämningar på Lovö. Arkeologiska studier kring en uppländsk järnåldersbygd*. Stockholm.
Lamm, K. 1970. 'Summary Concerning Cemetery 150. Archaeological Analyses'. In W. Holmqvist (ed), *Excavations at Helgö III*, 217–27. Stockholm.
Larsson, Lars 1982. 'Gräber und Siedlungsreste der jüngeren Eisenzeit bei Önsvala im südwestlichen Schonen, Schweden'. *Acta Archaeol* 52 (1981), 129–268.
Lindqvist, S. 1926. *Vendelkulturens ålder och ursprung*. Stockholm.
Malmer, M. P. 1963. *Metodproblem inom järnålderns konsthistoria*. Lund.

Montelius, O. 1872. *Svenska Fornsaker*. Stockholm.
Montelius, O. 1892. 'Öfversikt öfver den nordiska forntidens perioder intill kristendomens införande. *Svenska fornminnesföreningens tidskrift* 8, 127–63.
Montelius, O. 1893. 'De förhistoriska perioderna i Skandinavien'. *Vitterhetsakademiens Månadsblad* 22. Bihang. Stockholm.
Müller, S. 1897. *Vor Oldtid. Danmarks forhistoriske Archæologi almenfatteligt fremstillet*. Copenhagen.
Nerman, B. 1919. 'Gravfynden på Gotland under tiden 550–800 e.Kr.'. *Antik Tidskrift för Sverige* 22:4, 27–101.
Nerman, B. 1969. *Die Vendelzeit Gotlands II*. Stockholm.
Nerman, B. 1975. *Die Vendelzeit Gotlands I:1*. Stockholm.
Nielsen, J. N. 1980. 'En jernalderboplads og -gravplads ved Sejlflod i Østhimmerland. En orientering efter et års undersøgelser'. *Antikvariske stud* 4, 83–102.
Nielsen, J. N. et al. 1985. 'En rig germanertidsgrav fra Sejlflod, Nordjylland'. *Aarb nordisk Oldkyndighed og Historie* (1983), 66–122.
Nørgård Jørgensen, A. 1992a. 'Kobbeågravpladsen, en yngre jernaldergravplads på Bornholm'. *Aarb nordisk Oldkyndighed og Historie* (1991), 123–83.
Nørgård Jørgensen, A. 1992b. 'Weapon sets in Gotlandic grave finds from 530–800 A.D.: A Chronological Analysis'. In L. Jørgensen (ed), *Chronological Studies of Anglo-Saxon England, Lombard Italy and Vendel Period Sweden*, 5–34. Copenhagen.
Ørsnes, M. 1966. *Form og stil i sydskandinaviens yngre germanske jernalder*. Copenhagen.
Ørsnes, M. 1970. 'Südskandinavische Ornamentik in der jüngeren germanischen Eisenzeit'. *Acta Archaeol* 40, 1–131.
Ørsnes-Christensen, M. 1956. 'Kyndby. Ein seeländischer Grabplatz aus dem 7.-8. Jahrhundert nach Christus'. *Acta Archaeol* 26, 70–162.

Olsén, P. 1981. 'Om "vendeltidens" ålder och ursprung'. *Fornvännen* 76, 8–15.
Petré, B. 1984. *Arkeologiska undersökningar på Lovö*. Stockholm.
Ramskou, T. 1976. *Lindholm Høje Gravpladsen*. Nordiske Fortidsminder 2. Copenhagen.
Salin, B. 1904. *Die Altgermanische Thierornamentik. Typologische Studie über germanische Metallgegenstände aus dem IV. bis IX. Jahrhundert, nebst einer Studie über irische Ornamentik*. Stockholm.
Steuer, H. 1982. *Frühgeschichtliche Sozialstrukturen in Mitteleuropa*. Göttingen.
Stjerna, K. 1905. 'Bidrag till Bornholms befolkningshistoria under järnåldern'. *Antik Tidskrift för Sverige* 18:1, 1–296.
Stolpe, H. 1884. 'Vendelfyndet'. *Antik Tidskrift för Sverige* 8:1, 1–34.
Street-Jensen, J. 1988. 'Thomsen og tredelingen — endnu engang'. *Aarb nordisk Oldkyndighed og Historie* (1988), 19–28.
Strömberg, M. 1961. *Untersuchungen zur jüngeren Eisenzeit in Schonen* I–II. Lund.
Vedel, E. 1878. 'Nyere undersøgelser angaaende Jernalderen paa Bornholm'. *Aarb nordisk Oldkyndighed og Historie* 1878, 73–258.
Vedel, E. 1886. *Bornholms Oldtidsminder og Oldsager*. Copenhagen.
Vedel, E. 1890. 'Bornholmske Undersøgelser med særligt hensyn til den senere Jernalder'. *Aarb nordisk Oldkyndighed og Historie* 1890, 1–104.
Waller, J. 1977. 'Der Übergang von der Völkerwanderungszeit zur Vendelzeit im östlichen Mälartal'. In G. Kossack and J. Reichstein (eds), *Archäologische Beiträge zur Chronologie der Völkerwanderungszeit*, 65–8. Bonn.
Worsaae, J. J. A. 1865. *Slesvigs eller Sønderjyllands Oldtidsminder. En sammenlignende Undersøgelse*. Copenhagen.

Synposis of discussion

The discussion of the papers on Scandinavian chronology focussed primarily on the problems of defining and dating the sequence of change in the 6th century, not only in terms of the transition between the most distinct periods (the Migration Period, *alias* Early Germanic Iron Age, versus the Vendel Period, Merovingian Period or Late Germanic Iron Age), but also in terms of the sharpness and utility of stadia or phases within these. A wide range of categories of evidence that would repay further study and/or which ought to be taken account of in investigating this controversial subject were proposed. These included particular brooch-types and techniques of decoration (especially garnet jewellery); still open questions concerning the stylistic development of bracteate designs, particularly D-bracteates; and the largely neglected possibility of establishing pottery sequences which could remedy problems within the metalwork sequences. In the latter respect it was noted that, in a way partly comparable with Anglo-Saxon England (see above, p. 89), central Swedish cemeteries could provide an insight into the sequence from the 6th century into the 7th if a means of including cremation burial contexts could be found.

Male graves, and weaponry in particular, also proved to be a special focus for discussion, starting from the issues raised by comparison with the Continental material and the sequences worked out for that. The question of whether the same phenomena ought to have the same date was a particularly keen one in this context. The general impression held was that while there were some significant correspondences in developments between Scandinavia and the Continent, there were also some striking discrepancies, for instance in the range covered by ring swords and the provision of axes. As noted in Anne Nørgård Jørgensen's paper (this volume), a discrepancy between the two areas in the dating of buckles similar to her type GU3 was felt to be particularly awkward.

The possible effect of social and economic factors on weapon deposition was noted as potentially relevant to these problems. It was argued, however, that not only are we in a position to predict with confidence which artefact-types – such as prestige weapons – are most likely to be affected by the 'heirloom factor' but also that it was usually easy enough to recognize instances of it when they occur.

One further issue raised was that of the impact of the large quantity of artefactual evidence produced by metal-detecting, particularly in Denmark, in recent years. It was commented that more could yet be done to take full account of this material in the general archaeological perspective, although some individual and specially focussed initiatives had already produced important results.

JH